AF904956

SINTERING APPLICATIONS

Edited by **Burcu Ertuğ**

Sintering Applications
http://dx.doi.org/10.5772/56064
Edited by Burcu Ertuğ

Contributors

Faming Zhang, Eberhard Burkel, Yujie Quan, Christoph Schweigel, Olaf Kessler, Narla Sharma, Hongfang Zhang, Chee-Leung Mak, Helen Lai Wa Chan, Xi Yao, Rosidah Alias, José Holanda, Chunping Zhang, Baojun Zhao, Nilgun Kuskonmaz, Guido Falk, Adolfo Fernández, Ramón Torrecillas, J.L Menendez, Marta Suárez, Waldemar Monteiro, Amin Salem, Burcu Ertug

Published by InTech

Janeza Trdine 9, 51000 Rijeka, Croatia

Copyright © 2013 InTech

All chapters are distributed under the Creative Commons Attribution 3.0 license, which allows users to download, copy and build upon published articles even for commercial purposes, as long as the author and publisher are properly credited, which ensures maximum dissemination and a wider impact of our publications. However, users who aim to disseminate and distribute copies of this book as a whole must not seek monetary compensation for such service (excluded InTech representatives and agreed collaborations). After this work has been published by InTech, authors have the right to republish it, in whole or part, in any publication of which they are the author, and to make other personal use of the work. Any republication, referencing or personal use of the work must explicitly identify the original source.

Notice

Statements and opinions expressed in the chapters are these of the individual contributors and not necessarily those of the editors or publisher. No responsibility is accepted for the accuracy of information contained in the published chapters. The publisher assumes no responsibility for any damage or injury to persons or property arising out of the use of any materials, instructions, methods or ideas contained in the book.

Publishing Process Manager Danijela Duric
Technical Editor InTech DTP team
Cover InTech Design team

Additional hard copies can be obtained from orders@intechopen.com

Sintering Applications, Edited by Burcu Ertuğ
p. cm.
ISBN 978-953-51-0974-7

We are IntechOpen, the world's leading publisher of Open Access books

Built by scientists, for scientists

4,200+
Open access books available

116,000+
International authors and editors

125M+
Downloads

151
Countries delivered to

Our authors are among the
Top 1%
most cited scientists

12.2%
Contributors from top 500 universities

CLARIVATE ANALYTICS BOOK CITATION INDEX INDEXED

WEB OF SCIENCE™

Selection of our books indexed in the Book Citation Index
in Web of Science™ Core Collection (BKCI)

Interested in publishing with us?
Contact book.department@intechopen.com

Numbers displayed above are based on latest data collected.
For more information visit www.intechopen.com

For EU product safety concerns:
IN TECH d.o.o., Prolaz Marije Krucifikse Kozulić 3, 51000 Rijeka, Croatia,
info@intechopen.com or visit our website at intechopen.com.

Meet the editor

Burcu Ertuğ received her M.S. in Metallurgical and Materials Engineering from Istanbul Technical University (ITU), Maslak, Istanbul, Turkey in 1999.Then she worked as teaching and research assistant at Istanbul Technical University, Istanbul Turkey until 2001. She received PhD. in Ceramics Programme from Istanbul Technical University (ITU) in 2008. Her thesis research dealt with porous barium titanate humidity sensors. Although she is interested in research area such as barium titanate humidity sensors and silicate glasses and glass ceramics, she had also worked on boron nitride ceramics. She is currently interested in dental glasses and glass ceramics, silicate glasses and barium hexaferrite production.

Contents

Preface

Materials science and engineering depends on three classes of materials namely metals, ceramics and polymers. The aim of this book is to help the graduate and doctoral students in materials science and related disciplines on the sintering of several materials.

During the sintering process, changes occur in the microstructure because of decomposition or phase transformations. Three major changes commonly ocur during sintering. The grain size increases, pore shape and pore size changes. The result is a decrease in the porosity after sintering. The fabrication process for ceramics involves several different steps. The first step in ceramics processing is compaction. In the compaction stage, ceramic powders are pressed in a die to shape the powder into the desired form. Sintering is one of the final stages of ceramics fabrication and is used to increase the strength of the compacted material. In the Sintering of Ceramics section, the fabrication of electronic ceramics and glass-ceramics were presented. Especially dielectric properties were focused on. In other chapters, sintering behaviour of ceramic tiles and nano-alumina were investigated. Apart from oxides, the sintering of non-oxide ceramics was examined.

Powder metallurgy manufacturing technology consists of three steps; mixing elemental or alloy powders, compacting those powders in a die at room temperature and then sintering or heating the shape in a controlled atmosphere furnace to bond the particles together metallurgically. In the Sintering of Metals section, two sections dealt with copper containing structures. The sintering of titanium alloys is another topic focused in this section. The chapter on lead and zinc covers the sintering in the field of extractive metallurgy. Finally two more chapter focus on the basics of sintering,i.e viscous flow and spark plasma sintering.

Dr.Burcu Ertug
Istanbul Technical University,
Department of Metallurgical and Materials Engineering,
Istanbul, Turkey

Section 1

Sintering Of Ceramics

Chapter 1

Oxide Ceramic Functional Thin Layer Processing by Thermal and Laser Sintering of Green Layers

Guido Falk, Katrin Klein and Christoph Rivinius

Additional information is available at the end of the chapter

http://dx.doi.org/10.5772/53257

1. Introduction

The growing scientific and technological significance of ceramic thin layer processing methodologies is not only due to steady growth of protection coatings against corrosion, friction and wear. By entering the micro-scale, the subject of tailoring functional interfaces raises many new questions, which are connected with an improvement of topological and microstructural characteristics of thin ceramic single layers, double layers as well as multilayers for advanced applications such as electrochemical energy storage and conversion. In this way the global importance of efficient and functional ceramic thin layers has already been recognized.

Ceramic thin film processing methodologies were also asked to be sustainable and environmentally compatible. The focus of this book chapter is therefore on laser and thermal processing of ceramic thin films from aqueous suspensions. The ceramic thin film systems assigned to the application areas of anti-wear and electrochemical functional coatings are displayed in relationship to the desired microstructural, morphological as well as compositional characteristics. In particular, the study covers comparative research results on laser sintering of dip coated substrates as well as thermal processing of electrophoretically deposited ceramic thin films. In both cases colloidal dispersed particles suspended in aqueous solutions serve as starting levels.

Particle-based processing of ceramic thin films has already been proposed for a wide range of applications. The singular advantage of this processing methodology involves the abandonment of any vacuum or inert gas conditions. Furthermore, the relatively low temperatures used result in a simple processing scheme to create complex oxide and non-oxide structural and functional ceramic coatings.

From the point of view of colloidal process engineering of structural and functional ceramics this chapter gives on overview of the technological feasibility and the potential of applications of laser and thermal ceramic thin films made from aqueous suspensions.

INTECH
open science | open minds

© 2013 The Author(s). Licensee InTech. This chapter is distributed under the terms of the Creative Commons Attribution License http://creativecommons.org/licenses/by/3.0), which permits unrestricted use, distribution, and reproduction in any medium, provided the original work is properly cited.

We consider the two applications segments of anti-wear and electrochemical functional coatings with the strongest potential to be the unique characteristics of the as-processed layered systems. Interested readers may find, both, further information about the state-of-the-art of the topic considered as well as information on future development trends.

As efficient and sustainable production technologies of ceramic thin films involve new problem solving strategies, it would be a great success for the co-authors when the reader could be motivated and activated to maximize all efforts in pursuing new avenues in the challenging research field of ceramic thin processing.

In this connection we would like to thank Rolf Clasen for his ever-lasting encouraging support and his pioneering visions that are closely connected with this displayed study. The co-authors also express their gratitude to the numerous colleagues of the institute for their effort, in particular to Christian Oswald and Doris Ranker, because, without their support the project would not have succeeded.

1.1. Structural ceramic thin films for anti-wear and anti-corrosion applications

Tasks of ceramic coatings are very diversified. It is expected that they protect metals from oxidation or corrosion, erosion wear, as well as thermal and electric insulation. According to the application one or more functions of the coatings stand in the forefront. The oldest process is enameling by which glass melts are deposited on a metal substrate [1, 2]. The enameling mixture, also referred to as frit, contains a high portion of oxides, specifically SiO_2. The production is made mostly on powder metallurgical processing. Metal substrates are coated by dipping or spraying, and then the enamels are melted within the temperature range between 700 °C and 900 °C.

High temperature enamels for special applications include additives of Cr_2O_3, Al, Si, Si_3N_4, SiAlON or AlN, which allow application temperatures up to 1000 °C [3]. When conventional porcelain enamels are used, the basic enamel composition contains CoO or NiO, which ensure improved adhesion onto steel surfaces as a result of interface roughening by selective corrosion. In practice cracks and spallings (fish scales defect) can occur resulting from a recombination of hydrogen diffusing from the steel interface. High temperature coatings can be produced by thermal spray process [4-7]. In this case powdered coating materials (usually granulated ceramics, agglomerate size about 30 µm – 100 µm), partly of wholly melted, are deposited with high speed onto adequate substrates.

The bond to the substrate is usually determined by the mechanical interlocking of the sprayed layer. In individual cases chemical reactions can contribute to the adhesion. We distinguish between flame spraying (use of fuel gases) and plasma spraying (use of partly ionized gases within an arc). In the latter case high melt-point materials can be melted due to much higher temperature in the range of about up to 30.000 K. Complex ceramic composites, such as TiC-TiN-$MoSi_2$-Cr_3C_2 or Al_2O_3 or ZrO_2 can be applied as substrate materials.

Meanwhile, even monolithic ceramic components can be processed by thermal spray process and higher dimensions (< 1 m) have already been realized due to the near-net shaping character of this processing method [8]. In this case the deposition of material is applied layer by layer onto a cooled preform.

Furthermore, ceramic coatings can be processed by physical (PVD) or chemical (CVD) vapor deposition methods. Here, thermal loading of substrates is reduced compared to other coating processes [9, 10]. PVD is used to be the collective name for reactive evaporation (ion plating), sputtering, and arc discharge atomization with subsequent deposition by glow discharge. Here, also several target materials can serve as material suppliers simultaneously.

Dense hard coatings can be processed by PVD from temperatures of 300 °C. For example, columnar structured PVD-coatings are mainly processed for use in thermal insulation of highly loaded turbines.

For the CVD process, one or more reactants are chemically converted in gas atmospheres and are deposited onto a substrate in the temperature range between 600 °C and 1000 °C. The term chemical vapor infiltration (CVI) refers to the internal coating of porous preforms. Well-known examples are CVD/CVI synthesis of SiC from $SiCl_4/CH_4$ or CH_3SiCl_3, being used for example for the processing of large-scale high-temperature SiC matrix fiber composites or for CVD-coatings of diamond-like carbon (DLC) [11].

In a few special cases sol-gel process can be used for coating of metal substrates. In this case the coatings are characterized by complex chemical composition, such as ceramic high temperature superconductors.

1.2. Functional ceramic thin film processing in energy materials applications

The processing of thin ceramic layers for advanced energy storage and conversion succeeds for example with a vacuum slip casting process (VSC) [12, 13]. Disadvantages of the VSC process (such as geometrical confinement, coating quality depending on substrate characteristics) involve the development of alternative procedures, suitable for processing of advanced ceramic electrolyte layers, such as for solid oxide fuel cell (SOFC) applications. Special consideration is given to the cost factor of the processing methodologies used, as soon as large-scale production is implemented.

For the case of the tubular SOFC concept of Siemens-Westinghouse, the processing of the functional SOFC layers is therefore realized by electrochemical vapor deposition (EVD). Here, in a first step, salts are evaporated, which then react onto the substrate to the desired product under adequate boundary conditions (reactants such as oxygen or water). The pores of the substrate are closed. The further growth of the layer is determined by the (electrochemical) oxygen transport of the first layer [14-16]. This process is technically complex and not economical for a desired mass-production of SOFC systems.

The processing of SOFC electrolyte layers by laser-assisted PVD (physical vapor deposition) is subject of current research and scientific discussions. Here, a target material chemically composed of the electrolyte stoichiometry is evaporated by laser energy. The vapor is deposited onto the SOFC substrate as electrolyte layer [17].

The achieved coated surface area is currently fairly low (e.g. 10 x 10 mm^2). The equipment requirement with its associated costs is relatively high due to the necessary heating of the substrate and the required multiple coating processes when gas-tight coatings should be achieved.

Increased equipment investment are required for plasma spray process too [18, 19]. Furthermore, the as-processed coatings are relatively thick and porous.

Processing of SOFC coatings from liquid phases succeeds by spray pyrolysis. Here, a precursor is sprayed onto a heated substrate. Evaporation of solvent and reaction of remaining salt result in desired phase formation. Layer thicknesses of up to 30 μm can be achieved by a multi-layer spray pyrolysis processing [20, 21].

Wet powder spraying is a simple, easy to automate process for production of porous SOFC-cathode layers [22-24]. Layer thicknesses of up to 50 μm and more have already been realized on substrates of almost any size. The liquid precursor is sprayed by a spray gun. The processing of the cathode layer onto a substrate sized 300 x 300 mm^2 is usually executed within 5 minutes. Largely independent of the substrate geometry the motion unit of the spray gun (x-y-positioning system) is controlled by a central computer.

Another process for ceramic thin layer production often reported and discussed in literature is electrophoretic deposition (EPD) [25-28]. In this case the liquid suspension used is exposed to an externally applied electric field. Colloidal stabilized and charged particles move in direction to the oppositely charged electrode or a membrane, where they are collected and form a layer. Coatings of thicknesses up to several millimeters have already been realized depending on the deposition time. This fast deposition process is specifically adequate for deposition of nano-sized particles since the deposition kinetic is not dependent on the particle size.

Advanced EPD processes are based on aqueous suspensions exclusively, other EPD process methodologies use organic solvents such as acetic acid, acetone, acetyl-acetone and ethanol [29, 30].

1.3. Sintering of ceramic thin films

A fundamental understanding of ceramic thin film sintering on rigid substrates including a comprehensive theoretical model for the phenomenological assessment of shear and densification characteristics is given in the literature [31, 32]. A comparative study reports the resulted sintering stress calculations on the condition that the assumptions of linear viscous material behaviour and a rigid substrate material are fully respected [33]. In the event that the thin layer should not be deformable during the sintering process and any dimensional change should be completely restricted (constrained sintering) the sintering stresses as well as the strain rate can be calculated with this model too.

Other theoretical calculations comprise the sliding of the thin layer on the substrate when an equivalent friction coefficient for the interface is assumed [34-36]. This coefficient depends on the film thickness and the characteristics of the substrate.

Provided that constrained sintering conditions are fulfilled, viscous Poisson´s ratio for LTCC tapes can be determined. The change of particle geometry subjected to constrained sintering conditions has already been modeled at the micro-scale. Parameters such as dihedral angle, grain size of the particles, thin film densification rate as well as surface and grain boundary energy are taken account in the model [37, 38]. The described reproduces the equilibrium shape of the initially spherical particles during the thin film densification process. The distance between the centers of gravity of the particles is assumed to be constant.

Another approach is based on a Monte Carlo study in order to solve the viscous flow problem during densification process of thin layers [39]. Based on this work numerical simulations have been elaborated [40, 41]. Close to free edges, sintering density is inconsistent after sintering process and deformation of the sintered geometry is observed. In the immediate vicinity of the free edge the density is uniform, however. This free edge mechanism is particularly critical as soon as the dimension of the sintered structure is reduced and the thickness is not negligible in comparison to the lateral dimension. However, the accuracy and applicability of the simulated densification behaviour of constrained sintered layers should always be put into relation with experimental data and observed findings.

If conditions of constrained sintering are met significant sintering stresses will appear within thin layers that result in reduced densification rates[42, 43]. Additionally, in the specific case of sintering thick layers the constrained sintering condition is only applied alongside surfaces of the sintered body. By that way, the free surfaces such pores within the sintered layer are not perfectly constrained resulting in a curvature and distortion of surficial areas.

Tensile stresses that reduces driving sintering forces have already been modeled by Finite Element Analysis (FEA) [42]. In the exemplary case of Al2O3/glass-interfaces it was shown that these tensile stresses decrease in through-thickness direction towards the middle of the interface as well as in close vicinity to free edges. These simulations have been verified by experimental data and showed good correlation between modeled and experimental data.

Owing to the option of applying significant quantities of targeted heat into surficial layers of a material by laser energy, laser processing methodologies are well suited to sintering of particulate thin films coatings. Thus one can coat ceramic substrates with metal layers as well as ceramic coatings on metal substrates. From a technological point of view the latter case is easier to implement. For example, ferrite powder is easily sintered onto PZT (lead zirconate titanate) substrates by laser energy [44]. The advanced densification of ceramic coatings and thin films on metal substrates by laser energy is more sophisticated. Here, care must be taken to ensure that the interfacial temperature between the layer and the substrate does not exceed the melting temperature of the supporting material. When precise process engineering is guaranteed then sintering of a ceramic protection layer, for example yttrium stabilized zirconia (YSZ) and alumina (Al_2O_3) on high-temperature steel substrates is possible.

However it must be noted that the laser sintering process must be subjected to the prior powder accumulation and pre-sintering in order to keep the exposed overall surficial laser energy as low as possible [45]. However, the use of sinter active powders will facilitate the laser sintering process and might lead to the bypassing of complex powder accumulation and pre-sintering processing steps.

For example, electrophoretically deposited alumina green layers have already been laser sintered on standard chromium and chromium-nickel steels, though under inert gas atmosphere conditions [46]. Moreover, laser sintering of powdered YSZ green layers on TiAl6V4-alloy substrates has already been reported [47].

2. Advanced deposition of oxide ceramic thin films from aqueous solutions

2.1. Electrophoretic deposition of tubular ceramic thin films

Electrophoretic deposition of nano and sub-micron sized particles represents a technologically attractive and relatively simple method. The shaping of homogeneous 2D and 3D green bodies is achieved during the deposition process (electrophoretic transport and electroosmotic interfacial kinetics) without the need of additional operating steps [48]. Further advantages compared with other procedures are shortest possible process times and the simple design of the deposition equipment. The EPD process consists of two sub-processes. The charged particles are attracted first to the electrode or interface of the opposite polarity (electrophoretic transport) under the influence of an external electric field. In a second step, particles are deposited onto the polar substrate under the influence of interfacial electrohydrodynamic kinetics (electroosmosis of the second kind). Since the deposition rate is independent of the particle size, the EPD methodology is specifically suitable for shaping of complex green bodies made of nano-sized particles. Especially in this case, green densities achieved are significantly higher compared to other shaping methods [49]. Furthermore, processed green bodies exhibit excellent homogeneities of particle packing [50]. Within the following investigations the EPD processed green thin film layers are exclusively produced from aqueous suspensions. The substrates are micro-tubular pre-sintered NiO-YSZ composites processed by powder injection molding (PIM). The experimental EPD-setup is represented in the following Figure 1. The polarizable NiO-YSZ substrate is fixed to the centered support by means of a clamp mechanism and is stabilized by a central Pt-wire (ξ 0.5 $\mu\mu$) working as anode. At the same time the NiO-YSZ substrate subdivides the setup into two chambers. Inside the micro-tubular substrate there is the compensation chamber housing the electrolyte solution. The suspension chamber is arranged between the outer tubular shaped cathode made of graphite and the micro-tubular NiO-YSZ support.

The electrical conductivity of the electrolyte solution is 10-fold higher than the electrical conductivity of the aqueous suspension. The inside diameter of the EPD cell amounts 15 mm, with a height of 70 mm.

As the parallel processing of green thin film layers enables higher throughput, another experimental EPD setup has been realized, as it is shown in Figure 2. Thereby, seven micro-tubular substrates can now be coated simultaneously.

http://dx.doi.org/10.5772/53257

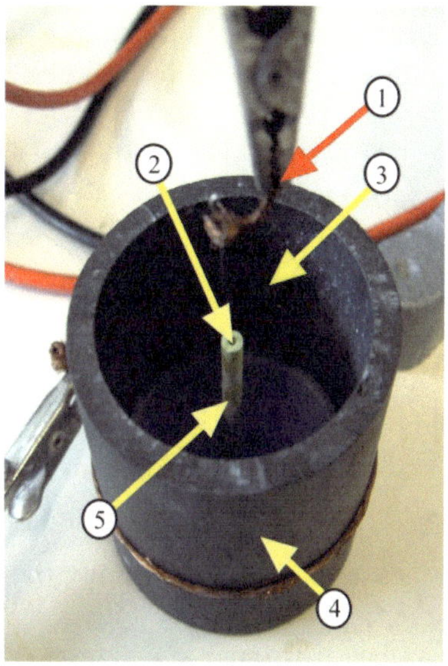

Figure 1. Experimental setup of the membrane type EPD cell arrangement. The negatively charged YSZ particles of the suspension are deposited onto the outer surface of the porous Ni-YSZ micro-tube (1: anode; 2: compensation chamber; 3: suspension chamber; 4: cathode; 5Ni-YSZ micro-tube).

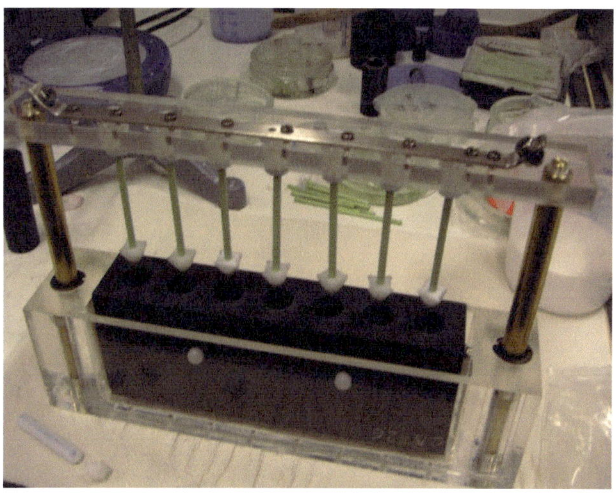

Figure 2. Experimental multi-cell EPD setup for the simultaneous EPD-coating of seven micro-tubular cells.

Even in this case, the Pt-wire anode is characterized by a diameter of 0.5 mm. The cathode also consists of a graphite block CB26 (Carbon Industrie-Produkte GmbH, Germany) (dimensions: 185.5 mm wide, 67 mm deep, and 36 mm thick). Seven center-holes in the block of 15 mm in diameter are arranged. The experimental setup enables a reduction of the dead volume in the deposition cell, so contributing to reducing waste suspension. In order to avoid that graphite particles become detached, the graphite block was flushed with cleaning solutions and cleaned in an ultrasonic unit. Holders and rails are made of POM (polyoxymethylene) with sealing made of duplicating silicone. Various centering gauges made of POM, Teflon and POM/duplicating silicone combinations have been tested. In Figure 3 the brackets for the micro-tubular substrates is displayed. Compared to the single-cell EPD setup this multi-cell setup the compensation chamber is not perfectly sealed off from the suspension chamber. By that way, the suspension is taking over the function of the electrically conductive electrolyte too.

The micro-tubular substrates to be coated are soaked in an electrolyte solution of 20-fold increased electrical conductivity in relation to the suspension. Subsequently, the micro-tubular substrates are positioned on the Pt-electrode wire and the centering gauges are put in place at the pipe endings.

The Pt-wire threaded micro-tubes are lowered down into the suspension and the voltage is applied. After switching off the electric field, the rails are removed from the suspension chamber und the coated micro-tubes are mounted in a frame and stored for drying.

The following Figure 3 displays the EPD coated micro-tubular substrates immediately after the deposition process.

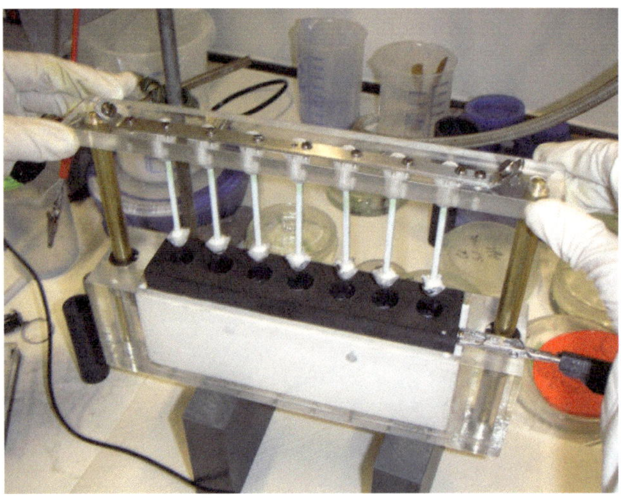

Figure 3. YSZ-coated NiO-YSZ micro-tubular substrates immediately after processing by the multiple-cell EPD deposition setup.

2.2. Dip coating

According to the dip coating procedure described herein Al_2O_3 thin films are processed onto borosilicate glass flat specimen. Suspensions used have been prepared by aqueous colloidal processing route. Hanging on the rails the substrates become immersed in the suspension and are retracted after a short deposition period. The parameters such as feed speed, plunge speed, holding time, and drawing speed are controlled automatically by a remote control system.

Start and end points of the immersion process are freely selectable and are only limited by the dimensions of the experimental setup.

By that way it is possible to coat various-sized substrates with the same settings. Pilot trial experiments designed to test rheological characteristics of the suspensions have been made on soda lime glass substrates. Laser sintering experiments were conducted with borosilicate substrates later on with the same proven parameters again. The following Figure 4 depicts the experimental dip coating setup used.

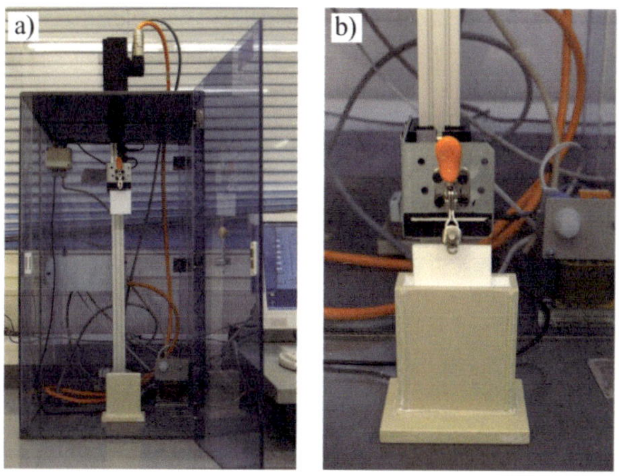

Figure 4. Experimental setup of the remote controlled dip coating setup (a) and removal of a borosilicate substrate during the dip-coating process (b).

In dip-coating technology, the properties of gelatine forming additives in terms of film formation, thermoreversibility and adhesion are particularly important. Therefore, the use of preheated substrates within the temperature range between 80 °C and 100 °C have been investigated. This processing option is hereinafter referred to as "thermogelation".

The coated substrates are hung on a specific sample holder and dried for 24 hours under ambient laboratory conditions after being pulled out of the dipping chamber. The dip-coating drawing speed was varied within a certain experimental matrix in order to determine the optimum thin layer characteristics suitable for laser sintering. The remote controlled drawing speeds used were within the designated interval of 8.33 mm/s to 33.3 mm/s.

In order to establish a correlation between dip-coating drawing speed and rheological characteristics of the suspension, it is necessary to analyze viscosity with a conventional rheometer at shear-rate conditions that are relevant for the dip-coating process. For this, the trajectory velocity on the outside of the cylindrical body has to be calculated out of the rotation speed of the cylindrical body. This calculated speed can be compared with the dip-coating drawing speed realized during the dip-coating process. The quantitative derivation of these equations is given in the literature [51] and consequently there is a direct connection between the shear-rate and the dip-coating drawing speed.

The following Table 1 provides the dip-coating drawing speeds as well as the calculated shear-rates of the experimentally realized dip-coating conditions.

drawing speed (mm/s)	shear rate (s^{-1})
33.3	24.1
16.6	12.1
8.3	6.0

Table 1. Dip-coating drawing speeds and corresponding shear-rates realized from the dip-coating processes

3. Laser assisted surface engineering processes

3.1. Laser sintering system and components

The laser processing of the dip coated specimens is made by a prototypical CO_2-laser setup constructed and designed by Fa. Auratech (Austria). The applied CO_2 slab laser (Rofin Sinar Ltd.) delivers an output power of 100 watts at pulse frequency of 20 kHz and an output stability of ± 7%. The wave length of the laser is 10.6 μm. The scanner unit, which is installed downstream the laser emission enables to precisely control the physical travel of the laser beam within the x- and y-direction. By that way, the user can home in on any location of an area of 100 mm x 100 mm. The laser travel speed can be controlled up to 7500 mm/s. The following Figure 5 depicts the laser equipment used for the laser sintering and densification of green ceramic thin films.

Figure 5. Prototypical CO_2-laser sintering facility used for sintering and densification of green thin film ceramic layers.

http://dx.doi.org/10.5772/53257

Through the use of scanner unit and most modern machines and hardware equipment the good focussing of the laser beam allows power densities that can barely reached by conventional weld processes. The height adjustable and heated sample changer allows the variation of the focal spot diameters in the range between 350 μm and 7.5 mm.

The heated sample holder enables adjustable substrate temperatures up 540 °C in order to reduce the risk of thermal stress formation during laser operation. Therefore the installed electric heating plate (Thermolyne Cimarec, Thermo Fisher Scientific Inc., USA) is characterized by thermostatic control and a protecting plate made of alumina (see Figure 6).

Since laser processing of green ceramic thin films is a function of a number of substance and device specific parameters, an adequate parameter matrix has been tested. Key parameters are focal spot diameter of the laser, and laser power. These two parameters are applied for the calculation of the power density. The travel speed of the laser, v_{scan}, is another important parameter in achieving optimum heat transfer rates. The pulse frequency was kept at a constant level of 20 kHz for all displayed results. Furthermore, the retaining speed of the scanning beam was not changed during laser operation and was set to the factory default level of 7477 mm/s.

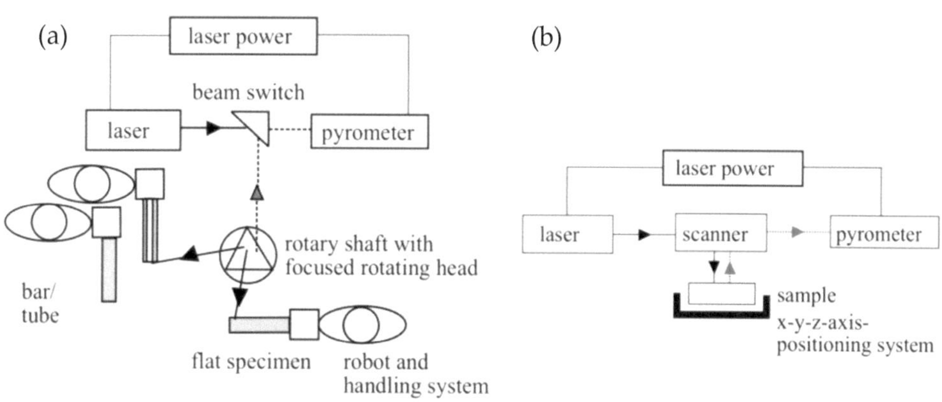

Figure 6. Schematic of the experimental laser sintering and densification setup with the option of defocused operation (a) and focused operation mode (b).

3.2. Laser sintering and laser consolidation

Processes of laser consolidation and laser sintering comprise the scanning the sample surface by a focused laser beam. The motion system is usually numerically controlled. Here, small line spacing resulting in a high overlap and energy input result in balling. As soon as the scan spacing exceeds the focal diameter the as-sintered and as-consolidated thin films become fragile.

It should also be underlined that the measured distance between two lines on the sample surface can deviate significantly from the line spacing fixed with a software programmed input mask. This is because the scanning unit shows significant aberrations such as astigmatism for angular incident beams caused by the specific dimensions of the rotating mirrors used. Depending on the present distance between sample and focal plane of the scanning unit the real line spacing increases or decreases. These aberrations may be offset with correction files controlled by the software settings. Each correction files only correct the error at a single working plane. In order to guarantee a high flexibility during laser processing the implementation of correcting files is omitted in this study. Later in this study it distinguishes between "line spacing" conventionally set via the software and "real line spacing" at the sample surface. The following Figure 7 depicts a schematic of the varied adjustment and activation options of the scanner unit. The schematic shown in Figure 7b represents the only possibility to guarantee a homogeneous temperature distribution along the entire sample surface.

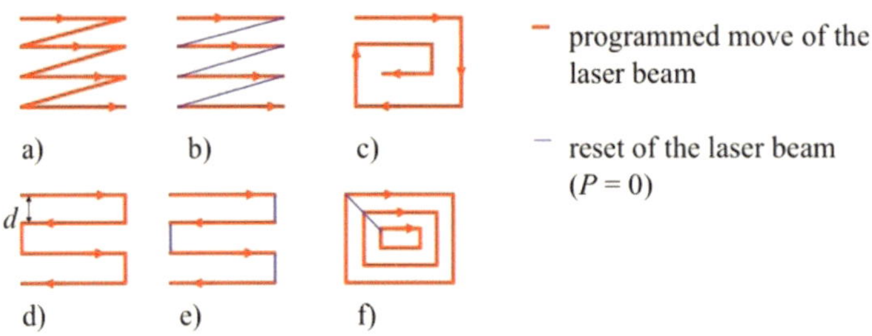

Figure 7. Schematic of varied adjustment and activation options of the scanner unit used for laser sintering and laser consolidation.

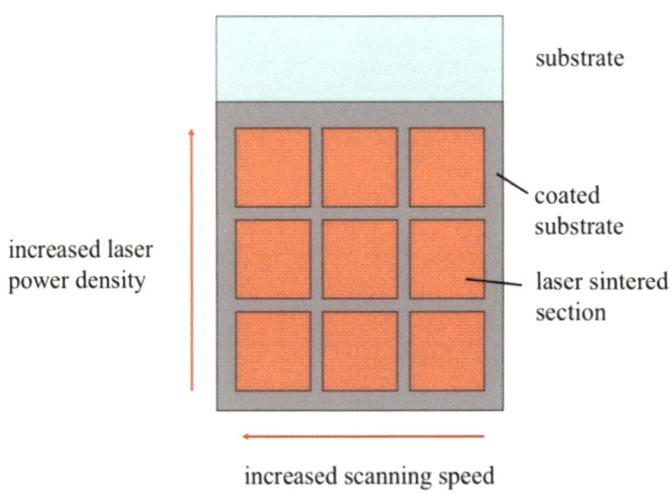

Figure 8. Schematic of the procedure for the determination of energy densities realized by laser irradiation of ceramic thin films made of Al_2O_3 deposited onto borosilicate substrates.

The laser scanning shown in Figure 7a, 7d and 7e results in an irregular temperature distribution on the edge of the laser scanned sample surface due to the heat flow transmitted from the irradiated surface towards the surrounding, cooled zones of the material. The scanning given in Figure 7c and 7f lead to a rise in temperature in the center of the laser irradiated sample area. The following Figure 8 depicts the procedure for the determination of the energy densities that are at least required in order to realize complete melting of the deposited thin film layer, and in this way achieving the necessary transparency.

With the laser scanning technology used the substrate is laser irradiated on nine sample segments characterized by different laser parameter combinations. In particular, the laser treated lines are irradiated at constant laser energy density and the laser treated columns are irradiated at constant laser travel speed. Not only we adapt suitable combinations of laser parameters to improve the surface characteristics of the irradiated sample surfaces, but we also use multi-laser irradiation procedures. In the latter case, the irradiated area is scanned multiple times with the same parameters. Only scanning directions are varied in this case.

3.3. Laser ablation

Simple line scanning experiments are performed in order to find essential basics and to understand the opportunities and limits of micro-structuring of surfaces by laser ablation. For example, enabling complex 2D and 3D geometries the scanner software used offers two options: First, specially created and designed vector graphic files can be used as template. On the other hand, any Graphic files can be imported and converted to scanner bitmaps. The conversion is necessary, because the scanner unit is not able to process gray scale pictures. Therefore, the optimal solution is to use graphic templates in duplex format, in other words, graphics that are composed exclusively from white and black pixels. JPEG and BMP-formatted templates usually have smoothed edges by using different gray scales (see Figure 9a). Visible artifacts appear by using such templates for converting into a scannerbitmap as it is shown in Figure 9b. The development of artifact formation also depends on the sized of the given pixels. The smaller the pixel size is, the less important is the artifact formation. The pixel size, however, is limited downward by the focal-spot: If the pixel size is smaller then the focal spot of the laser beam, there will be multiple irradiation of the surface area and in consequence to an increased material removal.

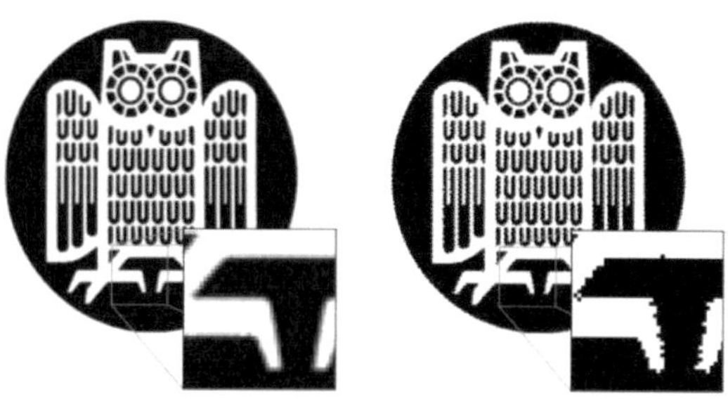

a) b)

Figure 9. Formation of artifacts caused by the conversion of a graphic template into scanner bitmaps: template (a) and converted bitmap (b)

Laser travel speed as well as laser power will be adopted after the conversion of the template into the scanner bitmap format. However, line spacing is influenced by the pixel size. The scanning unit controls line by line the individual pixels. This process refers to the scanning with a laser travel speed v_{scan} analogous to the laser sintering process. Depending on the pre-configuration the black or white pixels are irradiated.

3.4. Laser micro-structuring

In addition to laser sintering, laser consolidation and laser ablation, the laser equipment depicted and described above is also suitable to process micro-structured decals used in the processing step of ceramic surface finishing. In general, three different processing routes have already been verified. First, the decals are processed by selective laser sinter (SLS) onto flat glass samples. Two alternatives processes are depicted in the following Figure 10.

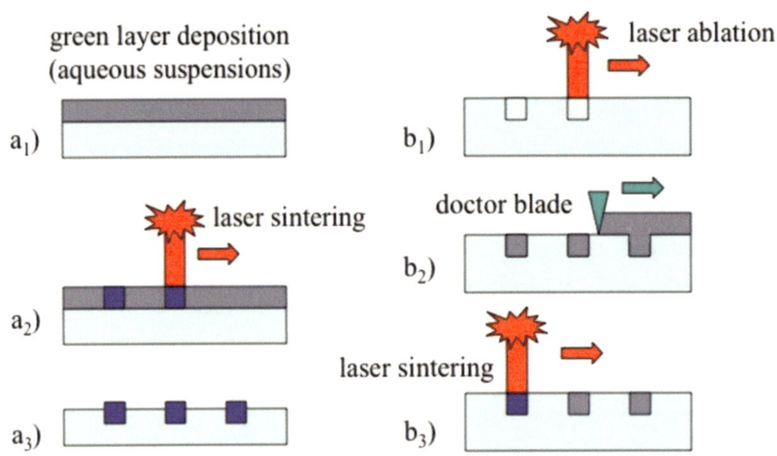

Figure 10. Alternative micro-structuring of glass surfaces by laser energy.

For the decal laser processing, the process depicted on left side of Figure 10, comprise the deposition of a green layer made onto the glass substrate (a_1). Such a system can be established by means of dip-coating, spray processing or painting. Selective areas are irradiated by laser energy after drying (a_2). By that way, the particles are intended to sink into the glass surface. After laser processing the selective irradiated surface sites remain structured by the coated powder (a_3). The process depicted on the left side of Figure 10 represents a combination of laser ablation and laser sintering. First, pits, recesses, and anomalies are processed into the surface by means of laser ablation (b_1). The injuries are then filled up with the dry or liquid coating material. During wiping of the excess coating material with a squeegee, it should be guaranteed that the deposition of coating material is exclusively restricted to the laser ablated surface sites. The non-treated surface remains almost free of deposited material (b_2). The surface is irradiated by laser energy after drying and consequently, a proper bonding between the coating material and the substrate is achieved (b_3).

http://dx.doi.org/10.5772/53257

4. Thermal processing of thin YSZ-electrolytes for SOFC applications

4.1. Sintering of YSZ thin films

We investigated the influence of sintering parameters on micro-structure of electrolyte and anode substrate respectively for given EPD boundary conditions: the electrical field strength was set to 30 V/cm, and the deposition was fixed to 30 seconds. PIM processed and pre-sintered NiO-YSZ substrates were characterized by mean pore diameter in the range between of 150 nm and 200 nm. The following figure 11 below depicts an example of the macrostructure of the EPD-coated micro-tubular composite after sintering at 1400 °C (heating rate 10 K/min and holding time 10 min).

Figure 11. YSZ coated and sintered NiO-YSZ micro-tubular substrate.

As with all thermal sintering investigation considered in this study the sintered electrolyte layers were transparent. The following Figure 12 shows an EDX-line-scan of a sample sintered at 1380 °C (heating rate 5 K/min, holding time 30 min).

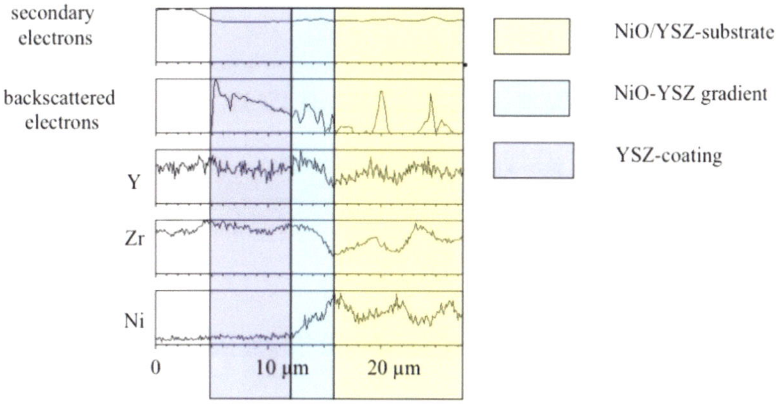

Figure 12. EDX-line-scan of the cross section of a YSZ coated NiO-YSZ micro-tubular substrate.

In this specific case the EPD deposition parameters were fixed to electric field strength of 20 V/cm and deposition time of 60 s. The multilayer structure of the coating composed of the anode substrate, a gradient interlayer and the electrolyte is clearly visible.

4.2. YSZ thin film morphology and green density correlation

In the following Figure 13 SEM topographies of green layers are compared with sintered layers thermally treated at 1400 °C (heating rate of 10 K/min, holding time 10 min). In the light of these data and analysis results provided by the study, it may therefore by concluded that high green layer densities lead to higher grain size.

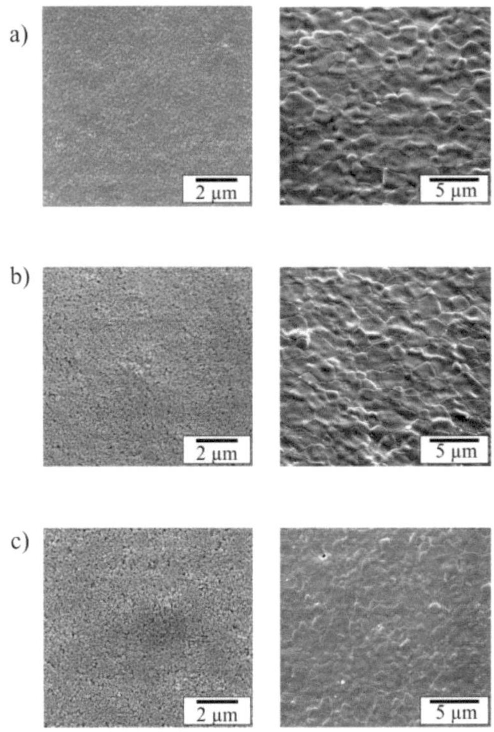

Figure 13. SEM topography of green electrolyte layer (left) and sintered electrolyte layer (right). Suspension and EPD characteristics are set to (a) 5 wt.-% YSZ, 3 wt.-% polyelectrolyte additive, E = 30 V/cm, t = 30 s; (b) 15 wt.-% YSZ, 1 wt.-% polyelectrolyte additive, E = 30 V/cm, t = 10 s; (c) 10 wt.-% YSZ, 0.3 wt.-% polyelectrolyte additive, E = 35 V/cm, t = 5 s.

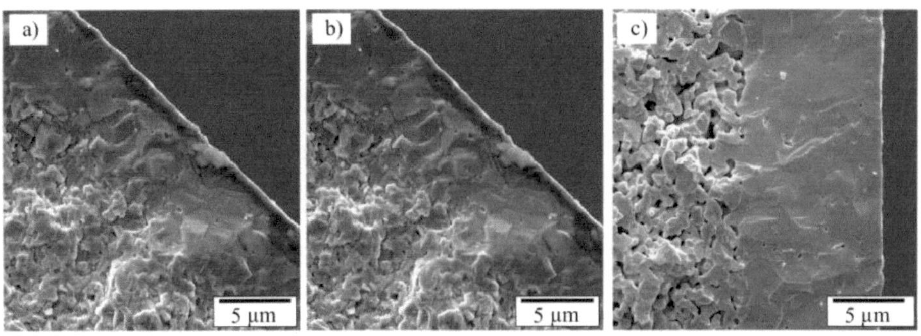

Figure 14. SEM topographies of breaking edges of three thermally sintered YSZ layers. The processing conditions correspond to those given in Figure 13.

SEM topographies of breaking edges of the samples shown in Figure 13 depicts that different green densities do not influence the sintering densities and the residual porosity of the electrolyte layer (see Figure 14).

4.3. Two-step thermal processing of YSZ thin films

Initial steps toward consolidating the layered microstructure were implemented successfully by two-step-sintering approach in order to achieve high densification rates of the sintered electrolyte layers at reasonable mean porosity values of the anode substrate. In addition to this objective, the two-step-sintering approach intends to achieve high densification rates at relatively low grain growth rates. Thermal analytical methods, such as thermogravimetry, differential thermogravimetry and dilatometry up to 1500 °C was used, in order to compare sintering shrinkage of anode substrates between one-step sintering and two-step-sintering approach. The following Figure 15 shows the porosities of the anode substrates as well as the YSZ grain sizes achieved as a function of the thermal sintering methodology used.

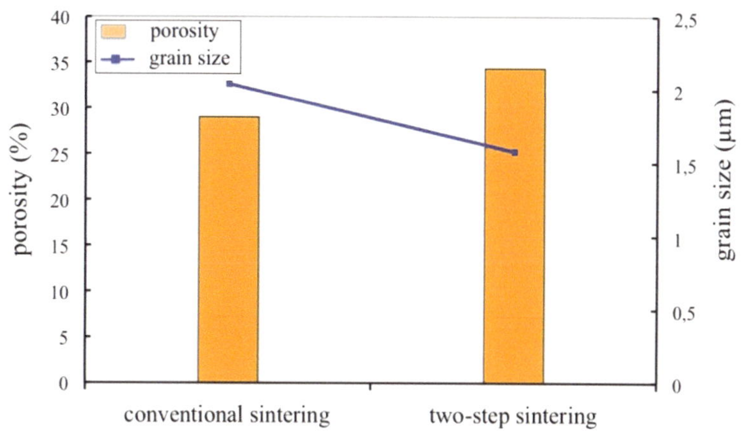

Figure 15. Porosities of the anode substrate and YSZ grain sizes of micro-tubular NiO-YSZ anode substrates sintered conventionally at 1400 °C (holding time 10 min) and sintered by two-step-sintering at 1300 °C (holding time 30 min) / 1200 °C (holding time 10 hrs.)

It is clearly shown that the shrinkage by two-step-sintering has been significantly reduced. According to these investigations, the conventionally sintered electrolyte layers are characterized by a mean grain size in the range of 2 µm, whereas the mean grain size of the two-step-sintered electrolyte layers is 1.6 µm. On the basis of the experimental sintering data and the analysis of the breaking edges shown in the following Figure 16, it is evident that electrolytes sintered by two-step-sintering show significantly increased densification. The better suitability of two-step-sintered electrolyte layers can easily be seen in Figure 16 b. Here, a full densification of the electrolyte thin film was only achieved during the second sintering step.

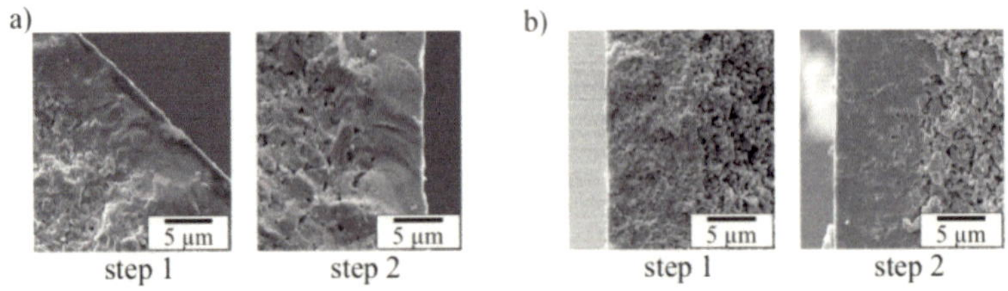

Figure 16. SEM topographies of breaking edges of two-step-sintered YSZ electrolyte thin films processed by EPD. (a) two-step-sintering at 1400 °C (holding time 10 min) / 1250 °C (holding time 10 hrs.) and (b) two step sintering at 1250 °C (holding time 30 min) /1150 °C (holding time 10 hrs.).

The effects of different two-step-sintering conditions on the porosity and shrinkage behavior of the NiO-YSZ anode substrates are depicted in the following Figure 17.

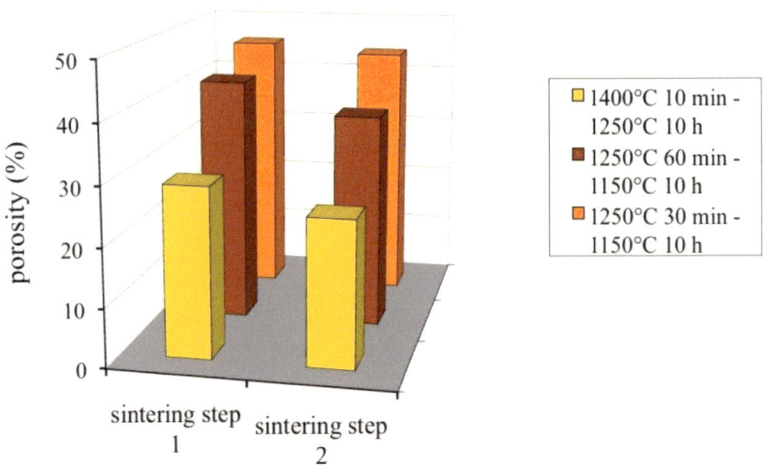

Figure 17. Densification behavior of NiO-YSZ substrates, two-step-sintered at different sintering conditions.

Investigations regarding the film thicknesses of the as-sintered layers revealed mean values in the range between 3.5 µm and 15.5 µm at EPD coating conditions of 30 V/cm (applied electric field) and deposition time of 30 s. Different thin film thicknesses were almost impossible to be correlated with the sintering programs used, because over the entire sample cross section the thickness variation was in the range of about 8µm. Specific coating condition (35 V/cm at deposition time of 5 s) lead to a reduced thickness variation in the range of 3 µm.

4.4. SEM/EDX characterization of YSZ thin layers

Initial tests of SOFC-systems, made of micro-tubular anode-electrolyte composites presented in this study, resulted in a performance of 1.7 W at operating temperature of 850 °C, which corresponds to a power density of 400 mW/cm^2.

A prototype of the cell is shown in the following Figure 18a. Figure 18b depicts the SEM topography of the cross-sectioned structure of NiO-YSZ substrate, YSZ electrolyte layer and LSM cathode processes by wet powder spraying.

Figure 18. Prototype of a micro-tubular SOFC cell (a) and SEM topography of the cross-sectioned structure of NiO-YSZ-substrate, YSZ electrolyte and LSM cathode.

The following Figure 19 shows the EDX mapping of the three layer systems of NiO-YSZ anode support, YSZ electrolyte and LSM cathode. It is shown that faulty anode edges are automatically smoothed by the EPD processed YSZ layer.

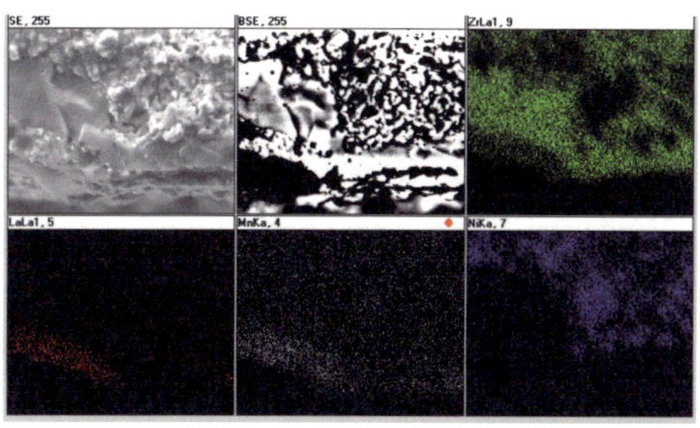

Figure 19. EDX mapping of the three-layer systems of NiO-YSZ anode support, YSZ electrolyte and LSM cathode. Above, from left to right: secondary electron image, back scattered electron image, Zr element distribution. Bottom, from left ro right: La, Mn and Ni element distribution.

5. Laser processing of anti-wear resistant alumina thin layers

The laser processing investigations should take all reasonable steps to ensure that wet deposition and subsequent laser irradiation of different Al_2O_3 coating specifications would result in superior wear resistance and anti-scratch behavior of coated borosilicate glass surfaces.

On these aspects Al_2O_3 suspensions have been prepared according to precautionary specifications, whilst at guaranteeing optimum quality of processing.

5.1. Powder, substrates and substrate pretreatment

Tailor-made submicron Al_2O_3-particle suspensions doped with 10 wt.-% methylcellulose (MC) binder additives have been produced. Furthermore all suspensions have been dispersed with 2 wt.-% anti-foaming agents. The following Figure 20 shows the viscosity of three different Al_2O_3 suspension qualities as a function of the shear rate.

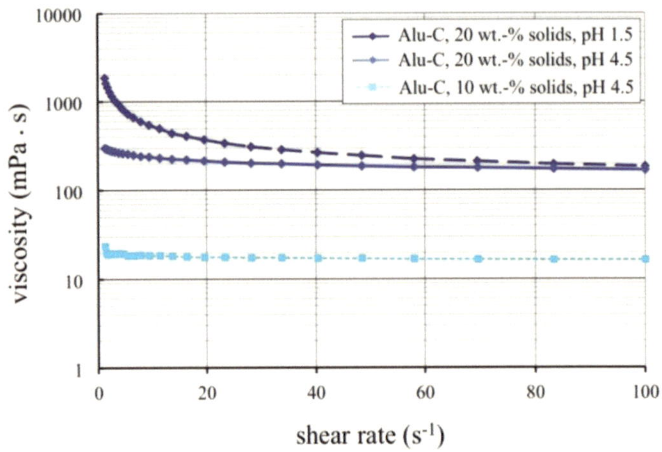

Figure 20. Viscosity as a function of the shear of three different Al_2O_3-supension qualities. The variations pertain pH value and particle concentration.

Suspensions with a particle concentration of 10 wt.-% are characterized by almost Newtonian flow behavior. Due to increase of particle concentration to 20 wt.-%, a tenfold and upward increase in viscosity is even possible. Therefore, it is not possible to further increase the particle concentration of Al_2O_3-suspensions, because the resulting viscosity increase makes it impossible to use these suspensions for dip coating. Second, the lowering of the pH-value to below 1.8 to 1.5 causes significant increase of viscosity, especially at small shear rates. This impact is almost compensated for shear rates exceeding values of 80 s^{-1}. Since the relevant shear rates realized during dip-coating deposition process attain values in the range of about 24 s-1, tailoring of the suspensions was performed in consideration of mean shear rate levels below of 24 s^{-1}.

5.2. Surface characterization

The mean diameter of agglomerates formed in the suspension must be smaller than the thickness of the processed layer, in order to realize homogeneous dip-coated layers. Figure 21 provides an overview with the best possible combination of coatings processed with different Al_2O_3 powder qualities characterized by different mean particle diameters d_{50} as well as specific BET surface areas A_{spec}.

The following Figure 22 depicts the curves of the as-processed layer thicknesses as a function of the distance from the sample lower border and the different powder qualities, in order to quantify the different impact of the various processing conditions.

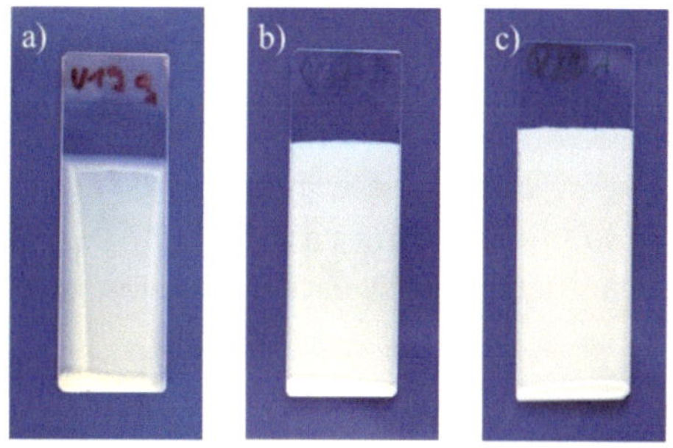

Figure 21. Green thin film layers processed by dip-coating onto silica glass substrates. The Al_2O_3 qualities used are (a) Alu-C, Evonik Degussa, Germany, d_{50} = 13 nm, A_{spec} = 50 m²/g; (b) AKP-50, Sumitomo Chemical Ltd., Japan, d_{50} = 200 nm, A_{spec} = 10.6 m²/g; (c) Al_2O_3 Alfa Aeasar, Germany, d_{50} = 2.07 µm, A_{spec} = 11.3 m²/g. The dip-coating was performed at a particle concentration of 30 wt.-%, a drawing speed of 33.3 mm/s and pH 4.5.

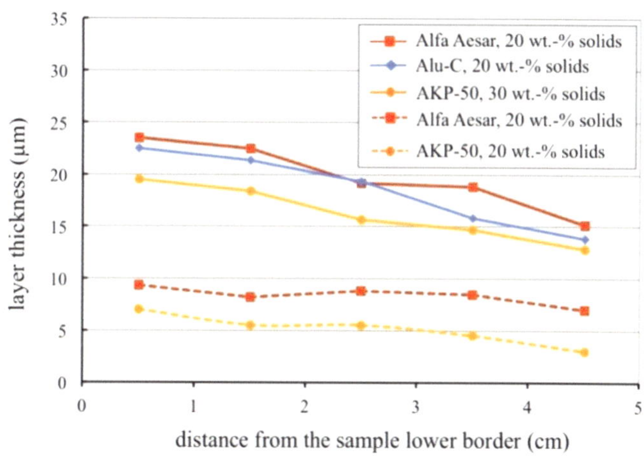

Figure 22. Comparison of the as-processed layer thicknesses of different dip-coated alumina layers.

This Figure shows that the largest layer thickness can be achieved using a combination of Alfa Aesar powder quality as well as a particle concentration of 30 wt.-%. Values at similar levels were determined for Alu-C suspensions with a particle concentration of 20 wt.-%. The layer thicknesses having achieved with AKP-50 suspensions with a particle concentration of 30 wt.-% were below these corresponding values mentioned above. Layers made of AKP-50 and Alfa-Aesar suspensions with particle concentration of 20 wt.-% were not included in the decision to determine the most suitable suspension characteristics, because the laser processing of scratch resistant Al_2O_3 layers requires the greatest possible layer thicknesses.

The green Alu-C layers, when viewed with the naked eye, appear as homogenous coatings. Light-microscopy images, however, reveal a tendency to form drying cracks. According to these investigations, drying cracks appear only occasionally at reduced layer thicknesses. For thicker coatings, produced of high solid loaded suspensions, drying cracks extend throughout the full depth of the layer. The characteristics of the substrate chosen have no effect on how many cracks are generated.

By contrast, the layers produced of AKP-50 and Alfa Aesar powders are crack-free and homogeneous. The surface topographies of these layers produced of 20 wt.-% AKP-50 and Alfa Aesar powders are depicted in the following Figure 23. Suspensions with increased proportions of solids do not significantly impede surface morphology of coatings.

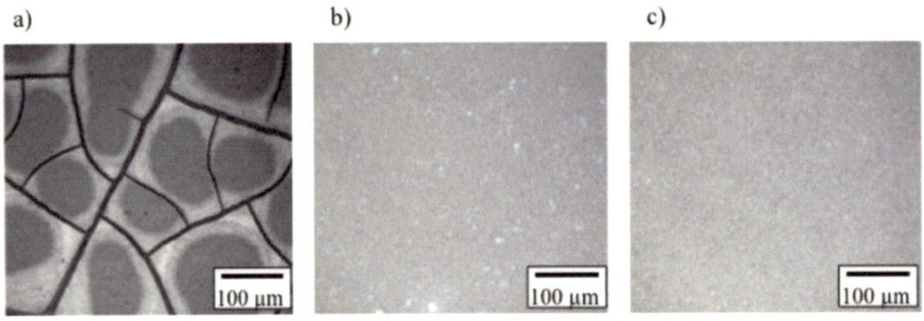

Figure 23. Al_2O_3 green layers: Alu-C green layers produced of suspensions showing a solid concentration of 20 wt.-% (a), AKP-50 green layers (b) and Alfa Aesar green layers (c) produced of suspensions showing a solid concentration of 20 wt.-%. All green layers shown have been processed onto soda lime glass substrates.

5.3. Laser processing of alumina thin films and characterization

The following Figure 24 depicts light optical microscope images of laser processed Alu-C layers. The parameters were not changed with the exception of the scanning speed.

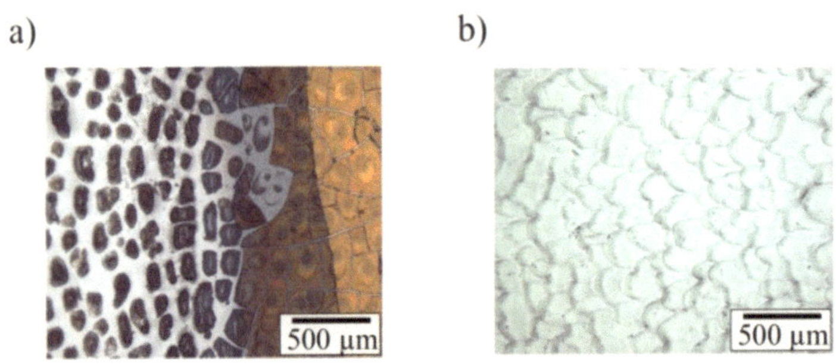

Figure 24. Light optical microscope images of Alu-C layers processed by dip-coating and subsequent CO_2-laser technique. For the structures shown laser line spacing was 0.1 mm, focused laser beam diameter was 1.0 mm and laser power density was set to 4000 W/cm^2. The scanning speed was varied as follows: 400 mm/s (a) and 200 mm/s (b).

Unlike other topographies, the Figure 24a shows an incomplete laser processed layer. Building bridges between laser sintered and green, not sintered areas are clearly visible. This effect is caused by the considerably high scanning speed of 400 mm/s in which the cracks already contained in the green layer were propagated. Simultaneously, the green layered,

http://dx.doi.org/10.5772/53257

floe oriented microstructures resulting in brownish to black discoloration and complete intransparency. The discoloration of the coating results from incomplete thermal degradation of polymer additives used.

The microstructure of laser processed green layers remains almost unchanged by reducing the scanning speed. The coloration of the floe-oriented microstructure only is varied from dark brown to light brown. A further decrease of the scanning speed down to 300 mm/s finally caused the floe-oriented microstructure to be increased and completely transparent. Singular floe-oriented microstructures are characterized by remaining pin holes and bubbles. From a scanning speed of 250 mm/s and less these heterogeneities disappeared, even though, inbetween singular floe-oriented micro zones not coated areas are observed. This surficial micro-structure is thus termed in the following as "island structure". Finally, a further increase of the specific laser power density by a further reduction of the scanning speed down to 200 mm/s causes completely densified and transparent coatings. These coatings are characterized by a shed-like external micro-structure. The orientation of the sheds takes place perpendicular to the scanning direction of the laser beam. Consequently the laser scanning lines of the microstructure shown in Figure 24b are run from the top to the bottom.

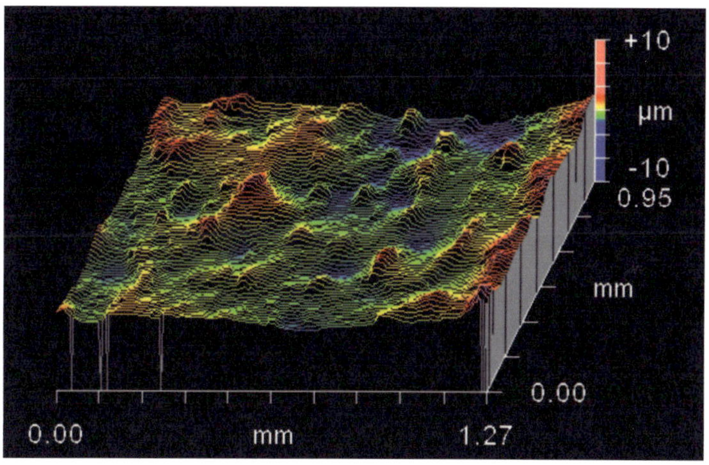

Figure 25. Topography of laser scanned alumina coating onto borosilicate substrates analysed by white light interferometry (WLI). Dual laser processing at corresponding scanning direction results in a decreased surface roughness of the irradiated sample

Multiple scanning was performed in order to avoid the formation of the shed-like microstructures and, by that way, to improve the transparency of the as-laser processed layers. The following Figure 25 depicts the sample topography after dual laser scanning moves. The laser powder density used has to be reduced to 3000 W/cm^2, in order to restrict the laser energy quantity absorbed by the irradiated surface. The scanning speed was fixed at 200 mm/s, as mentioned before. The smoothed surface of Figure 25 is twofold laser irradiated. In this case first and second laser run are characterized by consistent scanning direction. Here, the mean roughness is in the range of 0.7 μm.

In another specific case, the scanning direction of the second laser run has been rotated 90° with respect to the first run. Here, the surface roughness is significantly reduced to 0.3 μm.

The WLI and SEM analysis show that the thickness of the anti scratch alumina coatings is difficult to determine. Therefore, EDX mappings of breaking edges of the as-laser processed and alumina coated borosilicate samples were captured.

The EDX mappings show that the thickness of the scratch-resistant layer processed by laser irradiation is in the range between 20 μm and 40 μm. Moreover, it was found by the analysis results that the coating does not consist of the pure alumina phase, but also a certain Si-concentration in the layer has been assessed.

Consequently, pure alumina green bodies were processed, laser sintered and analysed, in order to investigate the interaction between the laser beam and the coated substrate surface. The following SEM topography shown in Figure 26 depicts a breaking edge of a laser sintered alumina green body with detail magnification. The following Figure 26 shows the edge of a layer processed by high scanning rate of 350 mm/s causing surface-near isolated floe-oriented microstructures of alumina. Alumina rich zones, however, are largely free from Si intercalations.

The laser irradiation parameters given in Figure 27 lead to the processing of completely densified alumina coatings. First, the SEM topography shows that the thickness of the coatings obtained varies between 150 μm and 200 μm.

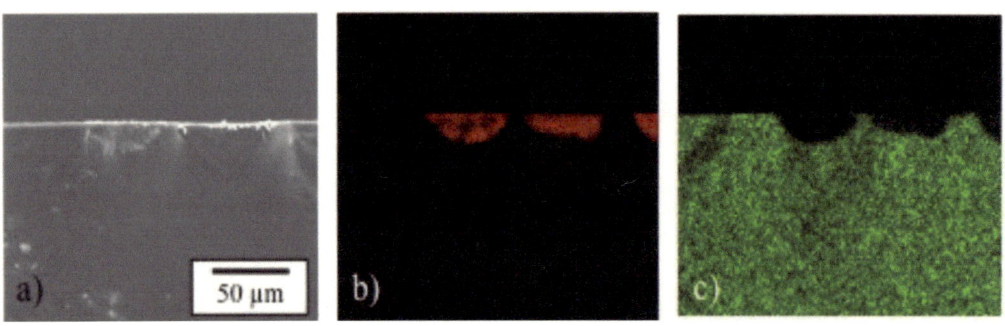

Figure 26. SEM topography of front edge Alu-C layer showing near-surface floe-oriented alumina microstructures (a), EDX mapping showing red coloured dots of Al-k-α peaks (b) and EDX mapping showing green coloured dots of Si-k-α peaks (c).

Detail magnification shown in Figure 27a represents a completely densified layer. The transition between completely densified layer and porous residual structure is depicted in detail magnification of Figure 27b.

Even at a distance of 100 μm below the densified layer the thermal impact of the laser irradiation and absorption of laser energy is observed as it is shown in detail magnification of Figure 26c. Compared to the untreated green structure (see detail magnification of Figure 26d) the grain structure of Figure 26c reveals significant grain growth effect caused by laser energy absorption.

http://dx.doi.org/10.5772/53257

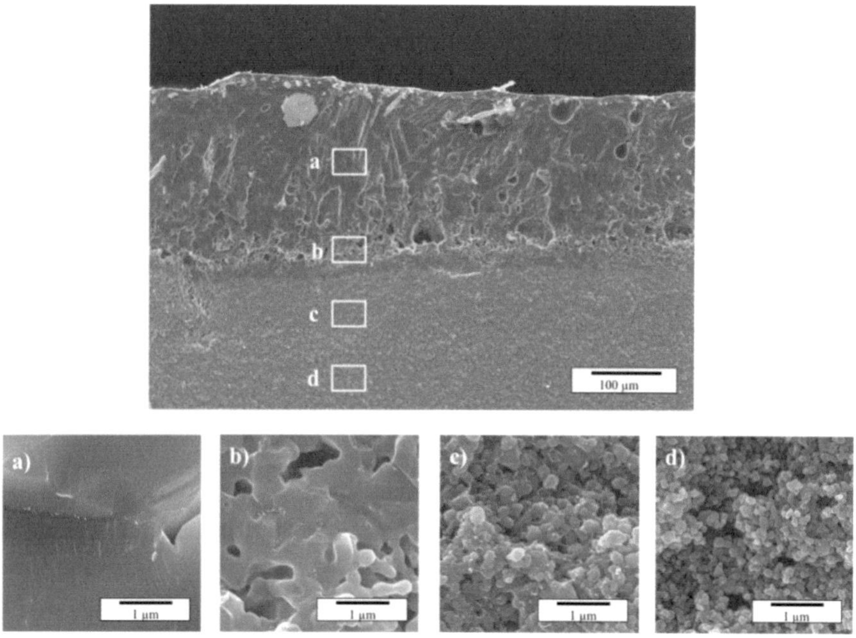

Figure 27. SEM topography of a breaking edge of casted alumina green body made of AKP-50 powder after laser irradiation (v_{scan} = 200 mm/s; laser beam energy density 4000 W/cm^2).

X-ray diffraction (XRD) patterns of a laser irradiated alumina green body and a laser irradiated alumina coating onto borosilicate glass substrate indicate the pure α-alumina phase. The crystalline state of the green layer is converted to an amorphous structure after laser irradiation as soon as Si atoms diffuse into the alumina phase and significantly high cooling rates are achieved. However, laser sintering of glass substrates coated with alumina layers enables the effective and sustainable processing of scratch resistant protecting layers. Within these application areas, laser processing is expected to develop as a sustainable production methodology. Furthermore, the study showed that laser sintering model systems are difficult to evaluate, because material characteristics and specific absorption behavior of laser energy controls the as-processed coatings substantially.

6. Conclusion

Comparative investigations include the processing of dense YSZ electrolyte layers in thicknesses between 5 µm and 20 µm by means of electrophoretic deposition of aqueous YSZ suspensions and subsequent thermal co-sintering. The resulting micro-tubular NiO-YSZ/YSZ-composite has shown promising potential to be used as SOFC key components.

In this study, it was shown that electrophoretic deposition enables the processing of green, crack-free YSZ thin layers, based on sustainable and simple processing methodologies. The layer thickness, however, was not homogeneous over the entire cross section. Thermal sintering programs used resulted in crack-free co-sintered composites. During sintering investigations, a specific impact of the green in the sintering microstructure was observed: Increased green densities generally result in increased grain growth effects. The two-step-sintering approach enables the complete densification of the sintered YSZ layers at a maximum sintering temperature of 1250 °C. The mean grain size achieved at these conditions is below 1 μm. However, a sintering temperature of 1350 °C is necessary in order to guarantee the electrical conductivity of the NiO-YSZ anode support.

Reducing sintering atmospheres would result in an increase of the anode porosity, independent of the sintering programs used previously. The total porosity after anode reduction to the Ni-YSZ-system consists of the porosity achieved after two-step-sintering and the porosity formed according to the reduction process. Three layered systems composed of micro-tubular Ni-YSZ anode supports, YSZ electrolyte layers and LSM cathodes have already been tested successfully as prototype cells for SOFC applications.

The framework of the efforts to deal with thermal and laser sintering problems of green layer densification include the processing of transparent, scratch-resistant coatings made of alumina and the production of thin electrolyte layers made of YSZ for energy applications.

The processing of transparent scratch-resistant alumina coatings started with the evaluation of suitable suspension conditioning methods. It was shown that in the light of the experiences gained, three different alumina powders were tailored in terms of their suitable aqueous suspension characteristics. These starting materials were used to develop alumina dip coatings on borosilicate glass substrates.

Because of the different powder characteristics, the process development succeeds provided that a careful tailoring of specific suspension characteristics (particle concentration, pH, additive type and concentration, viscosity) is considered.

The different layers derived from the developed suspensions are optimized so that homogeneous and thick green layers are processed.

The results of the laser irradiation experiments showed first that is was not necessary to debind the methylcellulose additive within a subsequent debinding step. Temperatures formed under the influence of laser absorption cause complete degradation and evaporation of polymer additives without any structural change of the coatings. The subsequent laser irradiation experiments showed that reduced laser powder densities result in an incomplete densification of the processed coatings. Once a critical laser power density is achieved (4000 W/cm^2 in this specific case) a complete densification of the coated surface is observed. Optical characteristics of laser-irradiated coatings, above all transparency, have been affected by single laser irradiation processing caused by significant roughness of the surficial coating layers. The subsequent dual laser sintering methodology lead to significantly reduced surface roughness and increased transparency. Based on comparative sintering investigations with casted alumina green bodies, EDX mappings and XRD analysis showed that Si atoms diffuse into the alumina layer during laser densification of the coating. Caused by significantly high cooling rates as-processed and densified oxide Si-Al-layers are generally referred to as amorphous. However, caused by the extended Al atoms within the layer, a increase of strength, toughness and scratch resistance is observed.

Author details

Guido Falk*, Katrin Klein and Christoph Rivinius

*Address all correspondence to: g.falk@nanotech.uni-saarland.de

Saarland University, Chair Powder Technology of Glass and Ceramics, Saarbruecken, Germany

References

[1] Petzold A, Pöschmann H. Email und Emailliertechnik. Stuttgart: Deutscher Verlag f. Grundstoffindustrie; 1992.

[2] Dietzel AH. Emaillierung. Berlin: Springer-Verlag; 1981.

[3] Hennicke HW, Padel A. Emaillierungen mit kristallinen Einlagerungen für Einsätze bei hohen Temperaturen (Hochtemperaturemails). Mitteilungen des Vereins Deutscher Emailfachleute e.V. 1991; 39 25-36.

[4] Sölter H-J, Müller U, Lugscheider E. High-Speed Temperature Measurement for On-Line Proess Control and Quality Assurance during Plasma Spraying. Therm. Spraying 1992; 24 169-175.

[5] Lugscheider E, Jungklaus H, Wielage B, Henker A. Plasmaspritzen von Titanhartstoffen-Neue Möglichkeiten zum Verschleißschutz. Schweissen & Schneiden 1995; 47 822-831.

[6] Fehringer G, Janes S, Wildersohn M, Clasen R. Proton-conducting ceramics as electrode/electrolyte-materials for SOFCs: Preparation, mechanical and thermal-mechanical properties of thermal sprayed coatings, material combination and stacks. J. Eur. Ceram. Soc. 2004; 24 705-715.

[7] Dubourg L, Lima RS, Moreau C. Properties of alumina-titania coatings prepared by laser-assisted air plasma spraying. Surf. Coat. Technol. 2007; 201 6278-6284.

[8] Lutz EH. Microstructure and Properties of Plasma Ceramics. J. Am. Ceram. Soc. 1994; 77 1274-1280.

[9] Bennett MJ, Knights CF, Ayres CF, Tuson AT, Desport JA, Rickerby DS, Saunders SRJ, Coley KS. Corrosion-resistant silica coatings obtained by plasma-assisted chemical vapour deposition. Mater. Sci. Eng. 1991; A139 91-102.

[10] He JL, Chu CH, Wang HL, Hon MH. Corrosion protection by PECVD-SiOx as a top coating on TiN-coated steel. Surf. & Coatings Technol. 1994; 63 15-23.

[11] Lee J-S, Liu K-S, Lin I-N. Deposition of diamond films on SiO_2 surfaces using a high power microwave enhanced chemical vapor deposition process. J. Appl. Phys. 1997; 81 486-491.

[12] Menzler NH, Tietz F, Uhlenbruck S, Buchkremer HP, Stöver D. Materials and manufacturing technologies for solid oxide fuel cells. J. Mater. Sci. 2010; 45 3109-3135.

[13] Tietz F, Buchkremer H-P, Stöver D. 10 Years of Materials Research for Solid Oxide Fuel Cells at Forschungszentrum Julich. J. Electroceram. 2006; 17 701-707.

[14] Pal UB, Singhal SC. Electrochemical Vapor Deposition of Yttria-stabilized Zirconia Films. J. Electrochem. Soc. 1990; 137 2937-2941.

[15] Haart LGJd. 26. IFF-Ferienkurs in Jülich: Elektrokeram. Materialien, Grundlagen und Anwendungen.: conference proceedings, 1995.

[16] Steinberger-Wilckens R, Blum L, Buchkremer H-P, Gross S, Haart LBd, Hilpert K, Nabielek H, Quadakkers WJ, Reisgen U, Steinbrech RW, Tietz F. Overview of the development of solid oxide fuel cells at Forschungszentrum Juelich. Int. J. Appl. Ceram. Technol. 2006; 3 470-476.

[17] Honegger K, Engler J. PVD and Thermal Spray Deposition of electrolyte/electrode system. Jahresbericht 1999. Sulzuer HEXIS AG; 1999

[18] Henne R, Fendler E, Lang M. 1. European SOFC Forum: conference proceedings, Luzern/Ch. 1994.

[19] Friis M, Persson C, Wigren J. Influence of particle in-flight characteristics on the microstructure of atmospheric plasma sprayed yttria stabilized ZrO_2. Surf. Coat. Technol. 2001; 141 115-127.

[20] Zhao B, Zhang R, Lu L, Xie H. Preparation of $La_xSr_{1-x}MnO_3$ ultrafine powder by spray pyrolysis. Mater. Sci. & Eng. 1997; B49 36-41.

[21] Maric R, Ohara S, Fukui T, Inagaki T, Fujita J. High-performance Ni-SDC cermet anode for solid oxide fuel cells at medium operating temperature. Electrochem. Solid State Lett. 1998; 1 201-203.

[22] Jansen H, Buchkremer HP, Stöver D, Wippermann K. Third international Symposium of Solid Oxide Fuel Cells: conference proceedings, Honolulu. Electrochemical Society; 1993.

[23] Wilkenhöner R, Hauber T, Malléner W, Stöver D. Herstellung von elektroaktiven Schichten für die Hochtemperatur-Brennstoffzelle mittels Wet Powder Spraying. Fortschrittsber. Dtsch. Keram. Ges. 1995; 10 263-269.

[24] Schüller E, Vaßen R, Stöver D. Thin Electrolyte Layers for SOFC via Wet Powder Spraying (WPS). Adv. Eng. Mat. 2002; 4 659-662.

[25] Mayen-Mondragon R, Falk G, Clasen R. Electrophoretic Impregnation/Deposition Complemented with Polymeric Templating for the Fabrication of Functionalized-Porosity Layered- Ceramics: A Solid-Oxide-Fuel-Cells Approach. J. Am Ceram. Soc. 2011; 95 593-599.

[26] Cherng JS, Sau JR, Chung CC. Aqueous electrophoretic deposition of YSZ electrolyte layers for solid oxide fuel cells. J. Solid State Electrochem. 2008; 12 925-933.

[27] Hosomi T, Matsuda M, Miyake M. Electrophoretic deposition for fabrication of YSZ electrolyte film on non-conducting porous NiO-YSZ composite substrate for intermediate temperature SOFC. J. Eur. Ceram. Soc. 2007; 27 173-178.

[28] Falk G, Böhm N, Delaporte P-G, Clasen R. 31st International Conference on Advanced Ceramics and Composites: conference proceedings, Daytona Beach American Ceramic Society; 2007.

[29] Boccaccini AR, Roether JA, Thomas BJC, Shaffer MS, Chavez E, Stoll E, Minay EJ. The electrophoretic deposition of inorganic nanoscaled materials - a review. J. Ceram. Soc. Japan 2006; 114 1-14.

[30] Neirinck B, Fransaer J, Biest OVd, Vleugels J. Aqueous electrophoretic deposition in asymmetric AC electric fields (AC-EPD). Electrochem. Commun. 2009; 11 57-60.

[31] Bordia RK, Raj R. Sintering behavior of ceramic films constrained by a rigid substrate. J. Am. Ceram. Soc. 1985; 68 287-292.

[32] Bordia RK, Scherer GW. On constrained sintering - III. rigid inclusions. Acta metall. 1988; 36 2411-2416.

[33] Scherer GW, Garino T. Viscous sintering on a rigid substrate. J. Am. Ceram. Soc. 1985; 68 216-220.

[34] Jagota A, Mikeska KR, Bordia RK. Isotropic constitutive model for sintering particle packings. J. Am. Ceram. Soc. 1990; 73 2266-2273.

[35] Jagota A. Simulation of the Viscous Sintering of Coated Particles. J. Am. Ceram. Soc. 1994; 77 2237-2239.

[36] Jagota A, Scherer GW. Viscosities and Sintering Rates of Composite Packings of Spheres. J. Am. Ceram. Soc. 1995; 78 521-528.

[37] Wakai F, Aldinger F. Equilibrium configuration of particles in sintering under constraint. Acta Mater. 2003; 51 641-652.

[38] Wakai F, Brakke KA. Mechanics of sintering for coupled grain boundary and surface diffusion. Acta Mater. 2011; 59 5379-5387.

[39] Bordère S, Gendron D, Bernard D. Improvement in the accuracy of calculated interface morphologies within Monte Carlo simulations of sintering processes. Scripta Mater. 2006; 55 267-270.

[40] Zhao Y, Dharani LR. Theoretical model for the analysis of a ceramic thin film sintering on a non-sintering substrate. Thin Solid Films 1994; 245 109-114.

[41] Olevsky EA, Tikare V, Garrino T. Multi-scale study of sintering: A review. J. Am. Ceram. Soc. 2006; 89 1914-1922.

[42] Tzeng S-Y, Jean J-H. Stress development during constrained sintering of alumina/glass/alumina sandwich structure. J. Am. Ceram. Soc. 2002; 85 335-340.

[43] Mohanram A, Lee S-H, Messing GL, Green DJ. Constrained sintering of low temperature co-fired ceramics. J. Am. Ceram. Soc. 2006; 89 1923-1929.

[44] Amin Z, Dalgarno KW, Comyn TP, Tavernor AW. Metal-electroceramic bonding in PZT through the selective application of laser energy. J. Mat. Sci 2006; 41 2831-2838.

[45] X. Wang, P. Xiao, M. Schmidt, Li L. Laser processing of yttria stabilized zirconia/alumina coatings on Fecralloy substrates. Surface & Coatings Technol. 2004; 187 370-376.

[46] Exner H, Reinecke AM, Nieher M. Laser beam sintering of thin alumina coatings on metals. J. Ceram. Proc. Res. 2002; 3 66-69.

[47] Kurella A, Dahorte NB. Laser induced multi-scale textured zirconia coating on Ti-6Al-4V. J. Mater. Sci.: Mater. Med. 2006; 17 565-572.

[48] Sarkar P, De D, Rho H. Synthesis and microstructural manipulation of ceramics by electrophoretic deposition. J. Mater. Sci. 2004; 39 819-823.

[49] Harbach F, Nienburg H. Homogeneous Functional Ceramic Components through Electrophoretic Deposition from Stable Colloidal Suspensions - I Basic Concepts and Application to Zirconia. J. Eur. Ceram. Soc. 1998; 18 675-683.

[50] Wittwer H, Krüger HG. Möglichkeiten und Grenzen der Elektrophorese. cfi/Ber. DKG 1995; 72 556-560.

[51] Fehringer G. Herstellung von Schichten aus Nanopulvern über das Dip-Coating Verfahren mit wässrigen Suspensionen, über atmosphärisches Plasmaspritzen und Elektroschmelzsprühen. Dissertation. Universität des Saarlandes Saarbrücken; 2008.

Chapter 2

Powder Preparation, Properties and Industrial Applications of Hexagonal Boron Nitride

Burcu Ertuğ

Additional information is available at the end of the chapter

http://dx.doi.org/10.5772/53325

1. Introduction

Boron nitride was first prepared in 1840's by an English chemist, W.H.Balmain, by using molten boric acid and potassium cyanide, but unfortunately this new compound was unstable and required many attempts to obtain a stable boron nitride. For nearly a hundred years studies on boron nitride remained in laboratory scale because of the technical difficulties of different production techniques and high cost of the material which is obtained with these synthetic methods but in 1950's Carborundum and Union Carbide companies managed to prepare boron nitride powder on an industrial scale and fabricated shaped parts of boron nitride for commercial applications with sophisticated hot pressing techniques [1].

2. Synthesis of boron nitride

Several methods have been studied for the synthesis of boron nitride. A chart prepared by D.Fister showing the possible routes to produce boron nitride is given in Figure 1.

A partial list showing generalized methods to produce hexagonal boron nitride is also given here :

1. Direct nitridation of boron according to

$2B + N_2 = 2BN$

reaction. This method is limited since pure boron is an expensive element and nitridation of the element is not homogeneous even at high temperatures [2].

INTECH
open science | open minds

© 2013 The Author(s). Licensee InTech. This chapter is distributed under the terms of the Creative Commons Attribution License http://creativecommons.org/licenses/by/3.0), which permits unrestricted use, distribution, and reproduction in any medium, provided the original work is properly cited.

2. Nitridation of an oxygen containing boron compound with a nitrogen containing compound such as ammonia or ammonium compound. Borates such as boric acid (H_3BO_3) or borax ($Na_2B_4O_7$) are heated in the presence of a nitrogen-containing compound such as ammonia or urea ($(NH_2)_2CO$) to form a B-N bond. Such reactions are usually carried out at temperatures 800-1200ºC and due to the low melting of boron compounds, inert fillers such as tri-calcium phosphate are used. Carbon addition might be required to tide the excess oxygen. Since reactants and fillers create a large volume, the flow through the porous media is dominant in synthesis. Ingles and Popper tried re-nitrided BN as solid filler to eliminate the further processes to purify the powder but obtained lower yields probably due to the fact that BN filler is not wetted by boric oxide [2,3].

   Although this route can yield cheap product, the reaction kinetics and the state-of-art is not given in the scientific literature. Ingles and Popper employed this reaction; however, together with a complicated and excessive flow chart. Kuhn et. al, also employed this reaction in an arc plasma reactor. Their work consisted of a vapor-phase reaction between boric oxide and ammonia at 800-2400ºC. The product was ultrafine boron nitride powder having a thickness in the c-direction of between 20-50Aº. It was noted that the boric oxide vapor should not be allowed to condense into large particles susceptible to boron nitride barrier film formation and that the ammonia should not be allowed to fully dissociate before contacting the boric oxide vapor [2].

3. Reacting alkali metal borides or alkaline earth metal borides with silicon and/or aluminum or their alloys while mixtures of an alkali metal oxide and alkali metal borates or boron trioxide are present in a nitrogen containing atmosphere at a temperature between 200-1200ºC. The product is leached after the reaction to remove the water soluble alkali metal salts. This method however, requires the preparation of starting materials and a tedious leaching step in order to remove the difficultly soluble trimetal hydroxide impurities. Knorre et al., patented a process which employs such a reaction [2].

4. Through reactions involving the formation of elemento-organic BN compounds as intermediate-reaction products, the carbothermic decomposition of which would yield BN. By this method, high purity BN powder at temperatures below those of conventional techniques has been obtained at laboratory scale. Such elemento-organic BN compounds could be boranes and borazines. Among boranes, both amine boranes and amino boranes are potential candidates. Amine boranes are compounds of amines and the borane unit or substituted boranes.These compounds can be derived from halides through direct combination of reactants or by indirect elmination and displacement reactions.Monomeric aminoboranes are characterized by a structure in which boron and nitrogen are coordinately unsaturated. The disadvantage is that all of these starting materials are oxygen and moisture sensitive and best prepared prior to being used [2].

5. Pressure pyrolysis of borazine ($B_3N_3H_6$) can also yield to BN, namely in amorphous form. The yield in such a route is reported to be as high as 60% by Hirano et al. [2].

http://dx.doi.org/10.5772/53325

3. Properties of hexagonal boron nitride

Key boron nitride properties are high thermal conductivity, low thermal expansion, good thermal shock resistance, high electrical resistance, low dielectric constant and loss tangent, microwave transparency, non toxicity, easily machinability- non-abrasive and lubricious, chemical inertness, non-wetting by most molten metals [4].

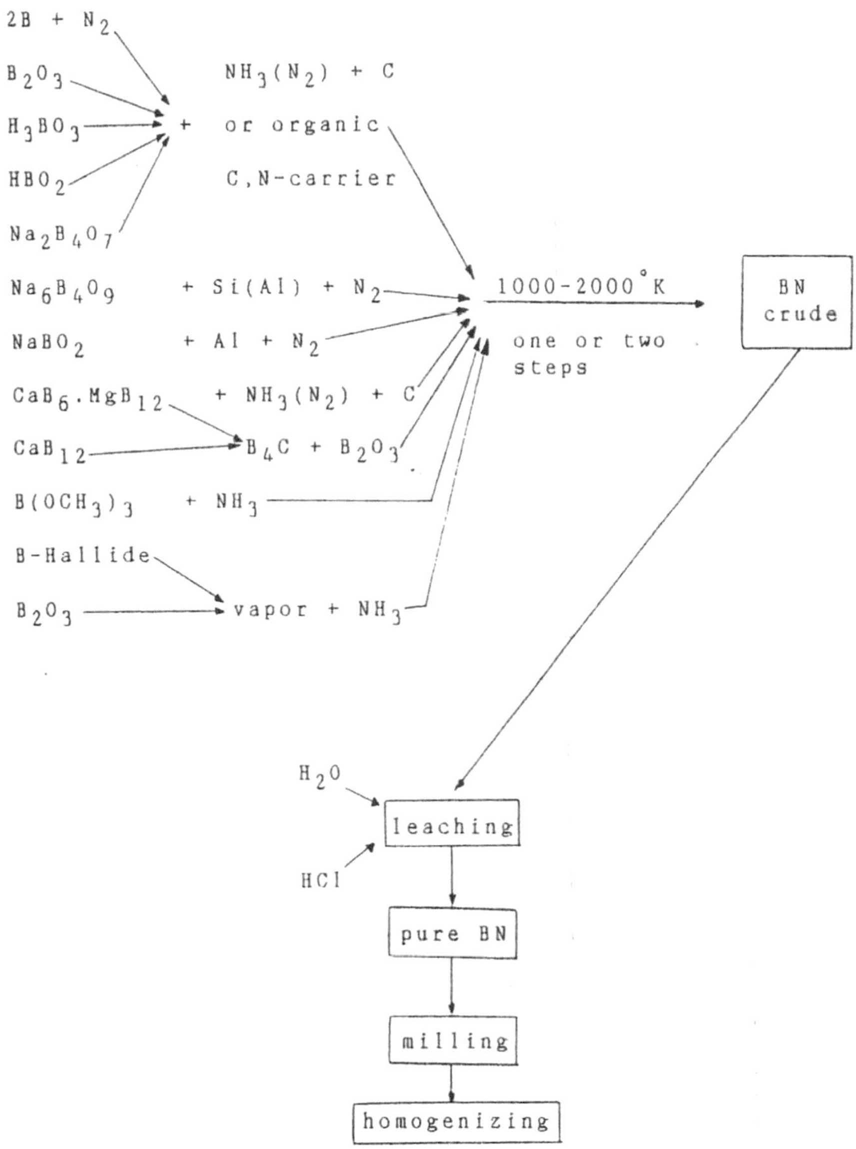

Figure 1. Possible routes to produce boron nitride [2].

Hexagonal boron nitride is being used increasingly because of its unique combination of properties which include a low density (2.27 $g.cm^{-3}$ theoretical density), high temperature stability (melting point near 2600ºC), chemical inertness (corrosion resistance against acids and molten metals), stability in air up to 1000ºC (in argon gas atmosphere up to 2200ºC and in nitrogen up to 2400ºC), stability to thermal shock, easy workability of hot-pressed shapes, excellent electrical insulating character as well as very high thermal conductivity. As a thermal conductor, BN ranks with stainless steel at cryogenic temperatures and with beryllium oxide, BeO, at elevated conditions; above 700ºC, the thermal conductivity of hexagonal boron nitride exceeds that of toxic BeO [5].

Of particular interest are its good dielectric properties (dielectric constant is 4, i.e half of that of α-Al_2O_3), also high dielectric strength and its ability to lubricate over a wide range of temperatures. Its small coefficient of friction is retained up to 900ºC, whereas other solid lubricants like graphite and molybdenum disulphide are burnt away at lower temperatures. Because of its high temperature stability and inertness against carbon and carbon monoxide up to 1800ºC it is as a refractory ceramic superior to the nitride ceramics Si_3N_4 and AlN and the oxide ceramics MgO, CaO, ZrO_2. Due to its non-wetting properties it is stable to attack by molten glass, molten silicon, boron, nonoxidizing slags, molten salts (borax, cryolite) and reactive metal melts (e.g Al, Fe, Cu, Zn). Because of its poor sinterability, dense shapes of hexagonal boron nitride are obtained almost exclusively by hot-pressing [5].

It must be recognized that the most chemical and physical properties of axial hot-pressed BN shapes depend on the nature and the amount of additives used for densification (up to 6 wt. % of B_2O_3, metal borates or SiO_2). Further some thermal (coefficient of expansion, thermal conductivity) and mechanical (flexural strength, Young's modulus) property values vary according to the direction of hot-pressing, BN being similar to graphite in respect of anisotropy. By hot-pressing isostatic of canned BN-powder, theoretically dense and pure hexagonal boron nitride shapes without texture (isotropic) and with improved properties can nowadays be obtained [5].

4. Applications of boron nitride

As can be seen from Figure 2, hexagonal boron nitride has established itself in the various field of chemistry, metallurgy, high temperature technology, electrotechnology and electronics [5].

Typical boron nitride uses are electronic parts (heat sinks, substrates, coil forms, prototypes), boron doping wafers in silicon semiconductor processing, vacuum melting crucibles, CVD crucibles, microcircuit packaging, sputtering targets, high precision sealing, brazing and metallizing fixtures, microwave tubes, horiziontal caster break rings, low friction seals, plasma arc insulators, high temperature furnace fixtures and supports [4].

In the following, a few examples of the uses for hexagonal boron nitride hot-pressed shapes are described in detail.

http://dx.doi.org/10.5772/53325

Application	Property Desired R	T	E	I	L	M
Hexagonal boron nitride powder						
* Solid lubricant for high temperature bearings	x				x	
* Mould release agent for die casting of glass and metals	x			x	x	
* Active filler for rubber, resins and plastics	x		x	x	x	
* Additive to oils and high-temperature grease				x	x	
* Ultrahigh-pressure transmitting agent	x			x	x	
* Coatings for evaporation plants	x			x		
* Coating for graphite hot pressing moulds	x			x	x	
* Embedding medium for heating wires	x		x	x		
* Boron source for preparation of c-BN and ceramic composites	x					
Hexagonal boron nitride hot pressed shapes						
* Crucibles for melting glass and metals	x	x		x		x
* Break rings for horizontal continuous castings	x			x	x	x
* Compounds for high-temperature electric furnaces	x	x	x	x		x
* Structural parts for magnetohydrodynamic devices	x		x	x		x
* Dielectric for radar antennas and windows			x			x
* Insulators for low- and high-frequency equipment	x		x			x
* Insulators for plasma-jet furnaces, arc pulse generators and ion engines	x		x			x
* Holders, mounting plates, substrates and heat sinks in valve and transistor circuits		x	x			x
* Pump components, pipes and nozzles for handling liquid metals	x			x		x
* Protective tubes and insulating sleeves for thermocouples	x	x	x	x		x
* Protective sleeves for electrodes in automatic welding	x		x	x		x
* Wafers for boron-doping of semiconductors	x					x
* Moulds for hot pressing of ceramics like B_6O or $B_{6.5}C$	x	x		x	x	x
* Neutron absorbers and shields for nuclear reactors	x	x		x		x

Figure 2. The applications of boron nitride [1].

4.1. Insulators for high temperature furnaces

The heating elements of resistance-heated electric furnaces, like graphite, molybdenum or tungsten-tube high temperature furnaces, require fixing devices made from ceramic insulating materials [5].

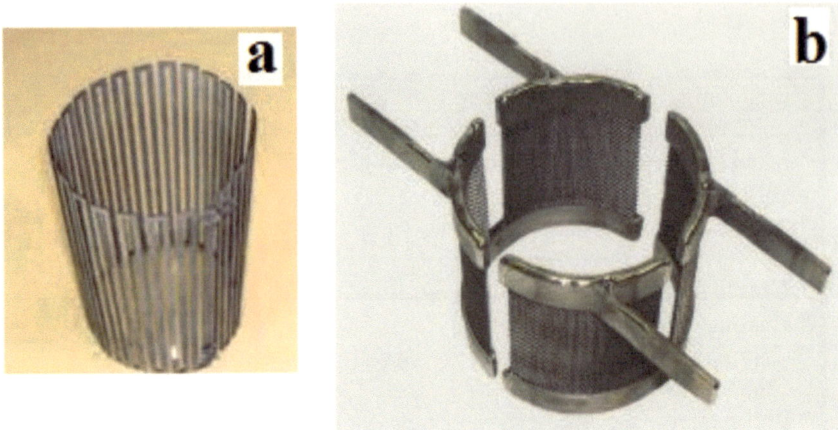

Figure 3. a.tunsten heater [6] and b. tunsten/molybdenum weave and mesh furnace elements [7].

Today due to frequent breakdown of oxide ceramic insulators at temperatures higher than 2000ºC, hexagonal boron nitride insulators are almost exclusively in use [5].

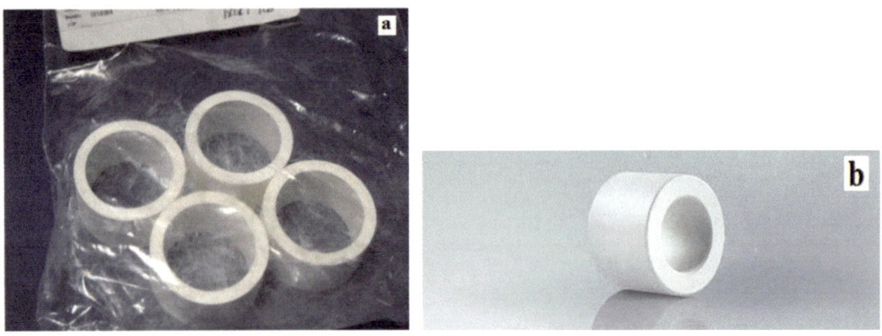

Figure 4. a. boron nitride insulator [8], b.insulator for furnace construction [9].

Therefore the BN material should contain only minimum amounts of boric oxide since B_2O_3 exudes at high temperatures and finally forms a conductive surface layer [5].

4.2. Protective tubes and insulating sleeves for thermocouples

The high temperature refractoriness, high thermal conductivity and high electrical resisivity of dense hexagonal boron nitride are utilized as thermocouple protection tubes and insulating sleeves. Preferably for the new B_4C/C thermocouple which can be used up to 2200ºC, hexagonal boron nitride sleeves are used. The B_4C/C couple consists of a graphite tube and a B_4C rod connected to ecah other by a conical fitting. This end is exposed to the heated area. Along its entire length the B_4C rod is insulated from the graphite tube by a hexagonal boron nitride sleeve of segmented construction [5].

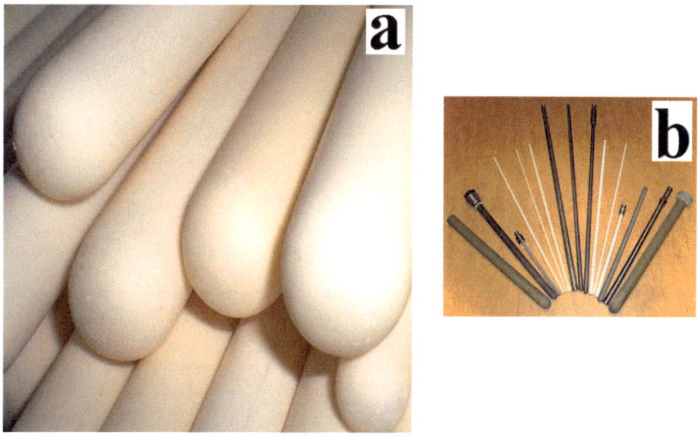

Figure 5. a. thermocuple protection tubes [10], b. tubes made of several materials [11].

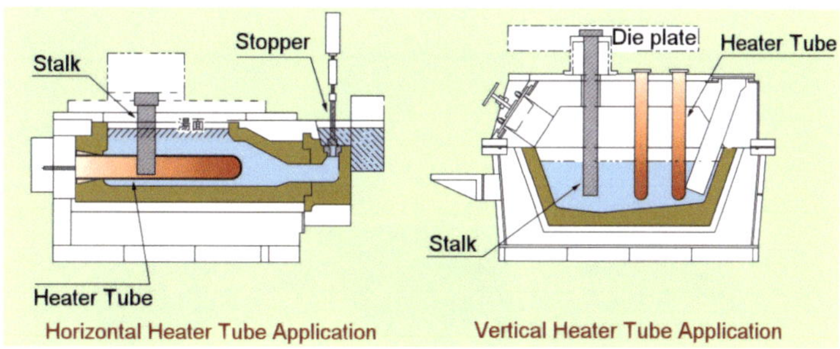

Figure 6. Heater protection tubes [12].

Typical fields of application are sintering furnaces, hot presses and hot isostatic pressing equipment working in the temperature range 1800-2200ºC. The thermocouple should only be used in a rare gas atmosphere or under vacuum. If the inert gas used contains oxidizing gases or nitrogen, reaction with the graphite tube and the B_4C rod will occurr, which can strongly influence the life of the thermocouple [5].

4.3. Break rings for horizontal continous casting of steel

The high chemical and thermal stability combined with the non-wetting property makes hexagonal boron nitride a useful crucible and structural material in metallurgical applications. For horizontal continuous casting of non-ferrous metals, graphite can be used as mould. However, the higher melting temperatures of steel alloys and their aggressiveness towads graphite and other conventional refractory metarials results in the need for special materials at the point of connection between tundish and the mould. This connection consists of a casting nozzle made of a high quality refractory such as zirconium oxide and a break ring made of boron nitride or a BN-composite ceramic. The sizes and shapes of the break rings and geometry are generally dependent on the type of equipment and type of steel (high or low alloy steels) [5].

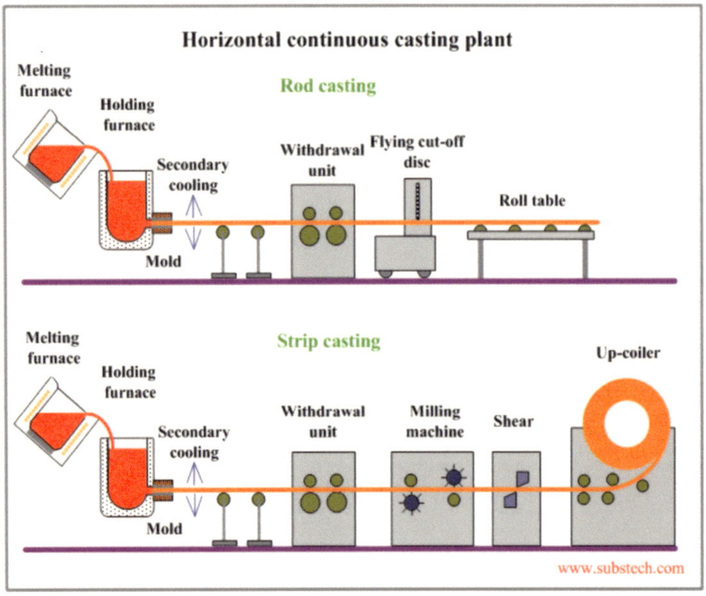

Figure 7. Horizontal continous casting [13].

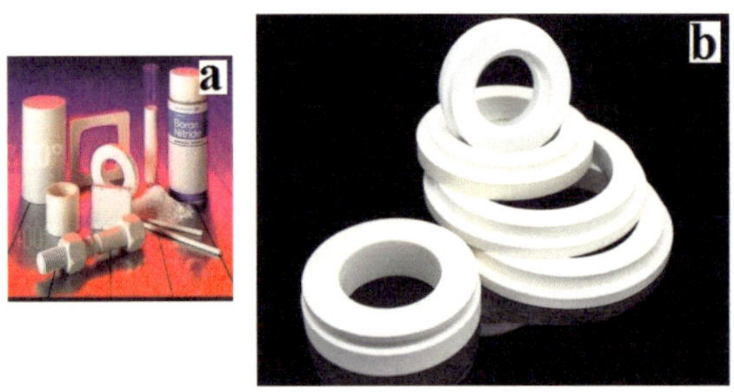

Figure 8. a. h-BN parts including break rings [14], b. Boron Nitride Horizontal Continuous Casting Split Rings [15].

Cross-sections of 50-250 mm diameters are now in use. The break ring determines the zone in which the liquid steel forms a solid. In addition the break ring must act as a seal or gasket with the tundish and the mould itself. The reaction of the ceramic boron nitride ring with steel alloying agents such as nickel, chromium, tungsten, manganese, sulphur etc. can cause breakdown of the BN material during service life. Therefore particular emphasis has been placed upon BN-composite materials to prevent this corrosive attack. Until now the least wear of all BN composites has been shown by the SC-BN material ; today 100-200 tonnes of steel per break ring in low alloy steel and 40-100 tonnes in high alloy steel can be cast using this new SC-BN material. Other BN metallurgical processing applications include crucibles for molten sodium carbonate at 900-1150ºC and crucibles for production of single crystals in high vacuum (Al-Li and Cu-Ti alloys) [5].

4.4. Pyrolytic boron nitride crucibles

A low-volume but high-value use for pyrolytic BN is for crucibles for growing GaAs single crystals by the liquid encapsulation technique [5].

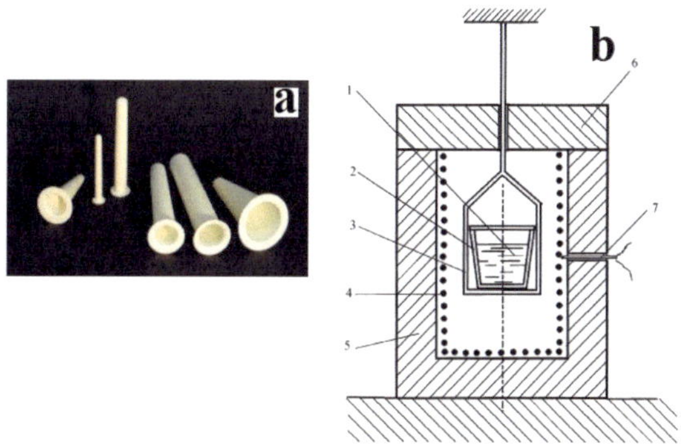

Figure 9. a. Pyrolytic boron nitride crucibles [16], b. thin-wall PBN crucible [17].

5. The hot-pressing and the microstructural properties of hexagonal boron nitride

5.1. Experimental procedure

The starting powders used for preparing h-BN ceramics were BN (C grade, H.C.Starck, Germany). Powder characterization was carried out on starting h-BN powders. The powder composition was confirmed by chemical analysis. Particle size, surface area, tap density and powder morphology of the h-BN powders were determined by Malvern Laser Particle Analyzer, BET analysis, Quantachrome pycnometer and SEM analysis, respectively. Hexagonal boron nitride (h-BN) based ceramics were fabricated by reactive hot-pressing in a graphite die with BN coating at 1900ºC under a pressure of 50 MPa for 60 min. with B_2O_3 as sintering additive in a nitrogen atmosphere. The obtained products had dimensions of 80 mm x 80 mm x 10 mm. Hot-pressed ceramics were cut into cube-shaped samples with dimensions approximately 10 mm x 10 mm x10mm. using a precision diamond saw. The sintered densities were determined on cube-shaped samples by Archimedes' method. The microstructural properties of h-BN powder and hot-pressed surfaces were evaluated by SEM-EDS analysis. Prior to SEM analysis, agglomerated BN particles were dispersed in acetone and adhered on a carbon tape and then coated with carbon. For the sintered bodies, the formation of the secondary phase was also investigated with particular interest on grain boundaries via SEM-EDS analysis [18].

5.2. Results and discussion

Powder morphology of the h-BN powders was shown in SEM images. Unlike the typical platelet shaped h-BN powder particles, near-spherical shaped particles can be observed in the

SEM images in Fig.10. These spherical particles could be explained as molten B_2O_3 particles that were not converted to BN during synthesis [18].

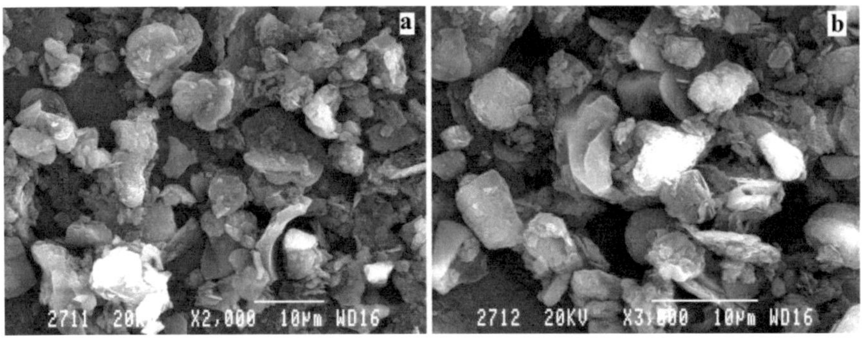

Figure 10. h-BN powder SEM micrograph (secondary-electron image) at (a)2000x and (b)3000x [18].

An average value of 92 % of the theoretical density was achieved by the hot pressing process. Apparent density values of 81-96 % of the theoretical density were obtained in related literature studies carried out on h-BN based ceramics formed by hot pressing depending on the process temperature and the composition [18].

Microstructural observations by SEM revealed an orientation of the c-axis of the crystal, parallel to the hot pressing direction as shown in Fig. 11. SEM observations indicated a fine platelet evolution in the hexagonal boron nitride grains. It is known that regardless of the type of sintering aids used, products manufactured from hexagonal boron nitride powder by hot-pressing exhibit a directional grain growth and an anisotropic microstructure due to the biaxial application of pressure at a high temperature so that the properties of the articles are direction-dependent. The degree of anisotropy increases as higher degrees of densification are achieved. In other words, a high degree of orientation of the hexagonal boron nitride grains leads to mechanical anisotropy. Densification via hot-pressing occurs by three consecutive mechanisms: re-arrangement of the powders, removal of open porosity by plastic deformation and elimination of closed porosity by bulk diffusion processes [18].

In the re-arrangement process, as the low melting compound (550ºC), B_2O_3, as a glass network-forming oxide, forms an intergranular liquid phase at a relatively low temperature during the sintering process. Thus promotes the orientation of the hexagonal boron nitride grains. Therefore the presence of a limited content of boron oxide is anticipated to enhance the densification process and grain orientation regarding the high purity h-BN grades. As can be seen from Fig.11, hexagonal boron nitride grains aligned perpendicular to the hot pressing direction and platy grains had an average diameter of 5 μm and a thickness of 1 μm. However, a high degree of orientation was not observed therefore not all the grains in the microstructure aligned towards the hot-pressing direction. This particular microstructural feature could be expected to reduce the strong anisotropy of the hot-pressed bodies. The low degreee of orientation prevents efficient packing and lowers density besides basal planes near parallel to the hot pressing direction act as flaws on the microstructural scale and decreases flexure strength. Thus orientation and densification characteristics are important from the mechanical point of view [18].

Figure 11. Hot-pressed hexagonal boron nitride microstructure (a) Low magnification, (b) high magnification and (c) plastic deformation at the fracture surface [18].

The hot-pressing operation, which involves the hexagonal boron nitride grains containing boron oxide, is not just a physical process. The trapped-in h-BN between the pressing rams and the inner wall of the matrix undergoes a decomposition reaction with the thermally released water as follows:

$$nBN + 3H_2O = B_2O_3 + (3n-2)BN + 3H_2 + N_2$$

Regarding the equation above, one can expect that a higher oxygen content will be in the hot-pressed body than in the starting h-BN powder since the yield of boron oxide in the reaction product is reversally proportional to the crystallite size. Above explanation was confirmed by XRD analysis results in Fig.12. Thus in XRD patterns a single phase was observed. Despite the presence of boron oxide in the microstructure, no sign of B_2O_3 was observed due to the amorphous structure.The results of the SEM/EDS analysis demonstrated that nitrogen is distributed throughout the cross section, while oxygen is localized in the grain boundary structures. SEM images indicated that a glassy amorphous phase is present in the grain boundaries. B and O in the grain boundaries suggest an amorphous oxide phase as confirmed by XRD analysis, which denotes a preferential oxidation [18].

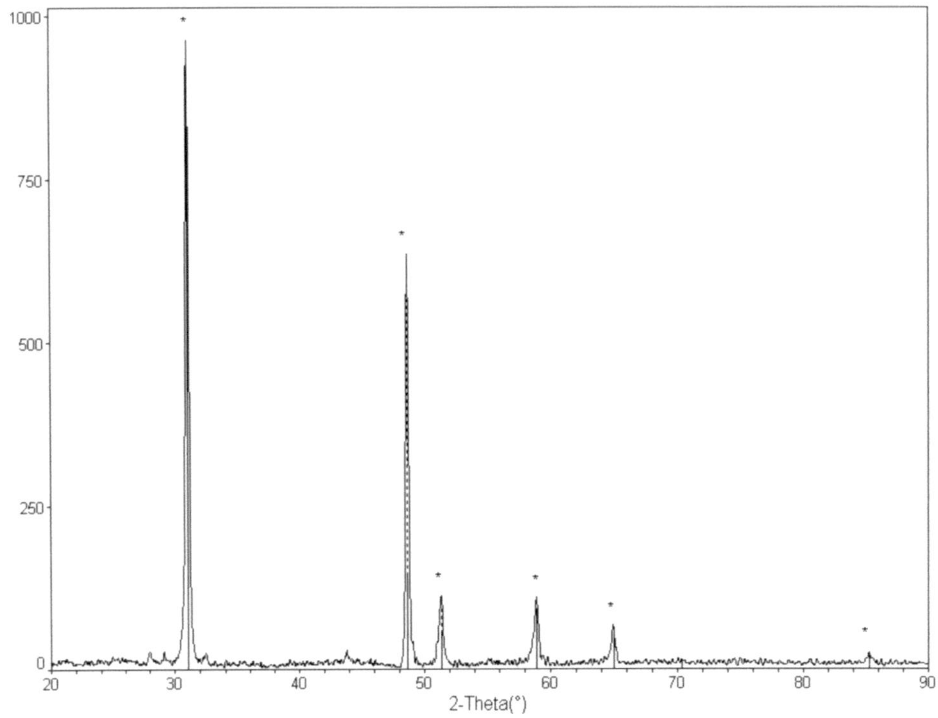

Figure 12. XRD pattern of hot-pressed h-BN ceramics [18].

6. The hot-pressing and the mechanical properties of hexagonal boron nitride

6.1. Experimental procedure

The overall mechanical characterization of hot-pressed hexagonal boron nitride (h-BN) based ceramics was performed. Hardness of the h-BN ceramics was tested on the polished samples by means of Vickers hardness tester, the applied load was 0.3 kg and the duration was 20 seconds. The indentation cycle was repeated at 15 points in each specimen with a step of 0.5 mm from one point of indentation to another. The length of the reciprocal diagonals of the indentation made by Vickers diamond pyramidal indenter tool was determined using optical microscope. The measurements were performed for parallel and perpendicular surfaces and then an average hardness value (HV) was determined for both surfaces. Specimens for strength measurements were cut from the hot-pressed plates into small rectangular blocks. The tensile surfaces of the samples were ground with silicon carbide papers and then polished with diamond paste along the longitudinal direction of the specimens until a scratch-free surface was obtained. The four edges of the specimens were chamfered in order to minimize the stress concentration induced by the machining flaw. The strength tests were carried out on Instron universal materials testing machine. Flexure strength (MOR) measurements were performed by three-point bending tests (test bars 4 mm x 3 mm x 45 mm) with a span of 40 mm and a crosshead speed of 0.5 mm/min. according to ASTM C1161-94. Five samples were tested for parallel and perpendicular surfaces and two average flexure strength values were obtained [19].

Fracture toughness (KIC) was evaluated by a single edge-V-notched beam (SEVNB) test, the dimensions of testing bars were 3 mm x 4 mm x 45 mm with a notch of 0.3 mm width and 2 mm depth. The fracture toughness tests were performed by four-point bending tests with a span of 40 mm. according to ASTM C1421. Five specimens were tested for each experimental condition. Prior to the testing, each specimen was mechanically polished to a mirror finish from both sides and five specimens were mounted parallel and side by side with their compression surface down on a plate and a straight notch to a depth of about 0.5 mm was tapered at the center of each specimen's tensile surface with a diamond wheel and then a second deeper notch was polished with a razor blade sprinkled with diamond paste as shown in Fig. 1. Following the four-point bending tests, the fracture toughness value, KIC was computed using the equation as follows:

$$KIC = F / B\sqrt{W} . S1\text{-}S2 / W . 3\sqrt{\alpha} / 2(1\text{-}\alpha)1.5 . Y^*$$

where, F, Sx, a, B are the fracture load, span, notch depth, specimen width, respectively and W, α, Y^* are specimen height, a/W, stress intensity shape factor, respectively. Finally the fracture characteristics of hot-pressed h-BN ceramics was investigated by analyzing fracture surfaces after bending tests using scanning electron microscope (SEM, JEOL JSM-T330 at 20 kV) [19].

6.2. Results and discussion

Several studies on hot-pressed boron nitride based ceramics denote a range of 15-285 Vickers hardness values for various compositions. On the other hand, particular work on hexagonal boron nitride ceramics point out a range of 0.15-0.24 GPa, which indicates that our results are agreeable within error. The average hardness value obtained, ranged from HV = 0.,20GPa to HV = 0.25GPa, which was higher in the surfaces parallel to the hot pressing direction due to the perpendicular alignment of the basal planes of the h-BN platelets to the hot pressing direction.

The results of the three-point bending tests carried out on surfaces parallel to the hot pressing direction indicated a relatively low average flexure strength (MOR) value of 0.14 GPa, which is in agreement with several studies on hexagonal boron nitride based composites confirming similar results regarding the flexure strength. A high degree of orientation of the hexagonal boron nitride grains was shown to lead to mechanical anisotropy. The flexure strength was found to be higher for the basal planes of the hexagonal boron nitride grains, which contain a covalent bonding of boron and nitrogen atoms. The average flexure strength value for the perpendicular direction was determined to be 0.17 GPa.

For the fracture toughness evaluation, a SEVNB test was conducted, where a linear stress profile occurs with tensile stresses below the neutral axis of the pre-cracked specimen and compressive stresses. This particular stress profile ensures the steady propagation of the crack initiated from the sharp notch root depending on the load. It is known that the anisotropy in the microstructure has also an influence on the fracture toughness. The cracks formed on the surface have a tendency to deviate in the direction of platelets due to the orientation and in some cases could not propagate at all. However, for the case that the direction of the crack propagation is not perpendicular to the basal planes of the platy grains, fracture toughness is not affected by the orientation of the grains. The average fracture toughness value was determined to be 2.6 MPa.m1/2. A single KIC value was obtained since a slight variation of the fracture toughness was observed with the orientation of h-BN due to aforesaid causes [19].

Finally the fracture characteristics of hot-pressed h-BN ceramics was investigated by analyzing fracture surfaces after bending tests using SEM using ASTM C1322. In Fig. 13, crack path of the fractured specimens after three-point bending test was indicated. No crack deflection was observed for each of the five bending tests, which is a sign of the high degree of grain orientation. Instead, crack followed a more straight path through the thickness of the specimen proving a low degree of grain orientation. The two sides of the fracture surface fit together very well after failure [19].

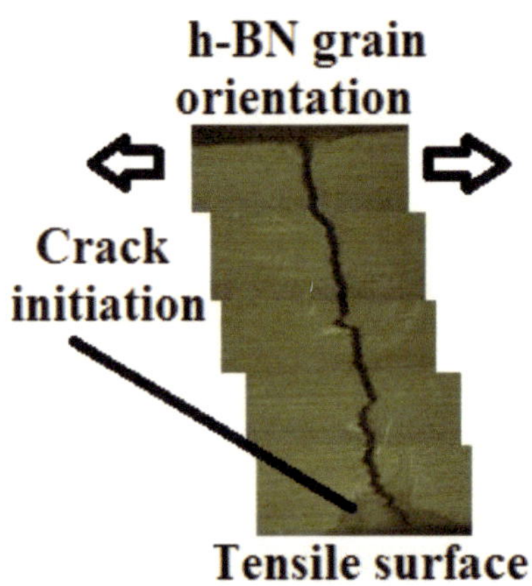

Figure 13. Crack path of the fractured specimens after three-point bending test [19].

The fracture halves were mounted back to back in order to reveal the crack path. Fig.11 shows the relatively flat fracture surface of the specimens. The stripped region at low magnification indicates the crack initiation pattern perpendicular to the applied stress on the fracture surfaces of three-point bending rods as shown in Fig.11.a. The crack formed at the maximum stress region, where typically occurs some plastic flow, easily propagates to reach the smooth surface of the specimen. Fig.11, shows a completely brittle type of fracture exhibiting small plastic deformation before failure. Thus the microstructural observations of h-BN grains indicated completely brittle fracture with intergranular voids and microcracks inside the individual grains revealing a mixed mode of fracture. A combination of a typical predominant transgranular fracture mechanism at room temperature and intergranular cracking due to the presence of grain boundary phase content was determined as a result of fracture tests [19].

7. The hot-pressing and the thermal and mechanical properties of h-BN/TiB2 composites

7.1. Experimental procedure

The starting powders used in this study were: TiB_2 powder (H. C. Starck), average size 0.7-2.0 μm and h-BN powder (H.C.Starck). In order to examine the effect of composition on mechanical and thermal properties, 5 series of TiB_2/h-BN compositions were prepared in this study. The powder mixtures were ball-milled in a YSZ vial (MSE Mill) using ethanol as a solvent for 3h and resulting slurry was then dried. The obtained powder mixtures were hot pressed in graphite dies at 1900 °C for 1 h under a pressure of 30MPa in vacuum to yield consolidated and sintered samples with dimensions of 40mm x 40mm x 40mm. The densities, ρ, of the hot pressed composite compacts were measured using Archimedes method with distilled water as medium. The theoretical densities of composites were calculated according to the rule of mixtures. Microstructure of thc composites was observed by scanning electron microscopy (SEM-Hitachi). The grain size, d, was determined by measuring the average linear intercept length, d_m, of the grains in SEM images of sintered BN ceramics [20].

The hardness values of the h-BN/TiB_2 composites was measured using a HMV™ Shimadzu Micro hardness tester (made in Japan). The surface of samples were prepared by using a Struers™ Tegrapol-15 polishing instrument to obtain the accurate hardness values of the composites and then the hardness tests were performed on the polished surface of the specimens by loading with a Vickers indenter for 10 s in ambient air at room temperature. The corresponding diagonals of the indentation were measured by using an optical microscope attached to the indenter. The indentation load of 490mN was used and five indentations were made for each measurement. In addition, X-ray diffraction (XRD) (Bruker™ D8 Advance) was used for crystalline phase identification of the composites. Moreover, to examine the effect of addition powders, thermal expansion measurement was done in argon (Ar) atmosphere and 10 °C/min. temperature by using Anter Unitherm™ Model 1181 Dilatometer instrument [20].

7.2. Results and discussion

The microstructural images of three different samples were given in Fig.14. Since c-axis of HPBN was aligned along hot-pressing direction, anisotropy could be anticipated in h-BN containing samples. As shown in Fig.14., basal plane of graphite-like plates (T0BN) was aligned perpendicular to hot-pressing direction. On the other hand, in B50T50 samples hexagonal grains did not indicate any sort of alignment despite hot-pressing process. The platelet thickness of HPBN sample (T0BN) was measured to be 1 μm after sintering at for 60 min. in vacuum atmosphere. The grain size obtained in the present study was nearly half of the densityof fully sintered TiB_2 in the literature. The grain sizes of 5 to 100 μm were reported for the TiB_2 sample which had theoreticaldensity of 4.5g.cm^{-3}. Since hot-pressing is a hot consolidation technique and stress-enhanced densification of powders, a higher density for HPBN (T0BN) was expected in comparison to those samples which contain lower amounts of BN, i.e B50T50. In the hot-pressing process, initially, compaction is by particle rearrangement and plastic flow (particle yielding at point contacts). As densification progresses, grain boundary and volume diffusion processes become controlling. Pressing before sintering as in pressureless sintering, reduces porosity and increases the dislocation population in powder. Higher disclocation density contributes to an initially faster sintering rate. However, in the present study direct hot-pressing was used and low density percentages were obtained. The microstructure of h-BN composition (T0BN) indicated a good agreement with literature microstructural features. The as-sintered microstructure consisted of hexagonal platelets with basal planes perpendicular to hot-pressing direction. However, as can be seen in Fig.14, the index of orientation preference or degree of orientation defined in the literature was low for our sample, T0BN [20].

Refining the microstructure of the starting material, i.e. ball-milling h-BN/TiB_2 powder to a defective, nanocrystalline or even amorphous state, enhances the reactivity. On the other hand, small grain sizes aid densification. That is why for five samples, ball-milled starting powder was employed. The main impurity in boron nitride powder was boric oxide, which is hygroscopic and is converted to boric acid. Since boron-oxygen compounds act as sintering aids during consolidation processes, 6 wt.% of B_2O_3 enhanced sintering behaviour (poor sinterability of h-BN) in all boron nitride containing samples except B0-TiB_2. Despite vertical die pressure of hot-pressing process at high temperatures, B_2O_3 also assisted sintering in HPBN. The amount of additives used for densification has a great effect on the chemical and physical properties of h-BN. h-BN powder utilized in the present study was a two phase system involving mixed powders of BN and low-melting boric oxide. Boric oxide provided for rapid transport and rapid sintering. That is why boric oxide content was kept constant for five different compositions. The resultant parts of five samples were easily machined with standard tools to bars for mechanical measurements [20].

Materials	Compositions %Vol.		Theoretical density g/cm^3	True density g/cm^3	Relative density %
	h-BN	TiB_2			
T0-BN	100	0	2.30	1.582	68.78
B75T25	75	25	2.86	1.977	69.11
B50T50	50	50	3.41	2.406	70.57
B25T75	25	75	3.97	2.845	71.66
B0-TiB_2	0	100	4.52	3.248	71.86

Table 1. Compositions, density and relative densities of the hot-pressed h-BN/TiB_2 composites [20].

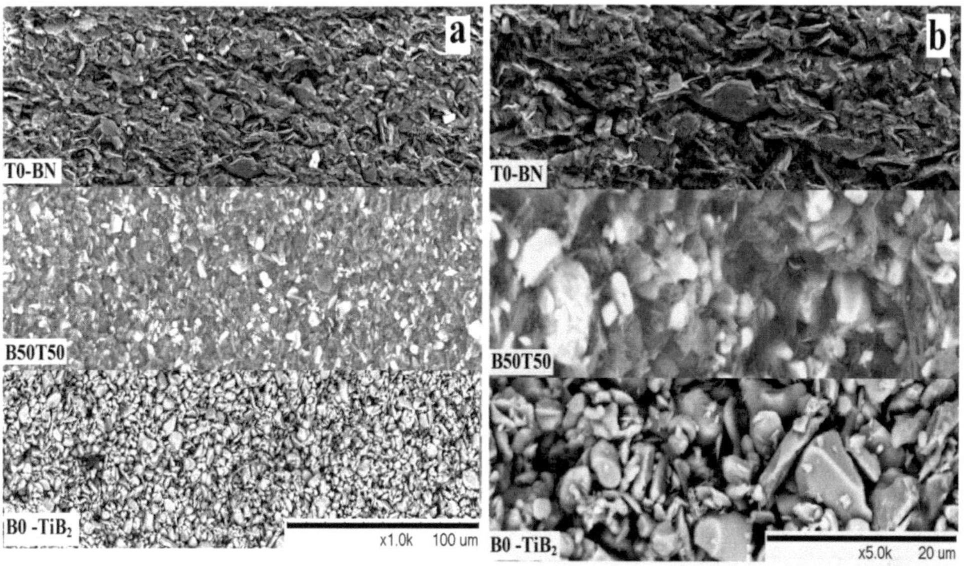

Figure 14. SEM images for 3 different samples at magnifications (a) 1kx and (b) 5kx. [20].

In general, micro-hardness is higher in the surfaces parallel to the hot pressing direction due to the perpendicular alignment of the basal planes of the h-BN platelets to the hot pressing direction. However, since h-BN content is lower than pure material in four of the samples, the anisotrophy in Vickers hardness was slight. The main difference between hot-pressed T0BN and B0-TiB_2 samples was the presence of high orientation of the hexagonal crystals and the grain growth, which was responsible for poor sinterability of pure h-BN sample. Due to the orientation of the crystals there could be different property data perpendicular and paralel to the pressing direction. B50T50 neither showed decrease in mechanical properties when measured at paralel to hot-pressing direction nor a dependence of microstructure on the direction [20].

In the Table 1, it is displayed that the densities of hot pressed samples decreased with the increase in the addition of boron nitride amount in the composition, the measured density for B25T75 is 2,845 g/cm^3, for B50T50 is 2,406 g/cm^3 and for B75T25 is 1,977 g/cm^3 and T0-BN is 1.582 g/cm^3 which is in accordance with the theoretical density for boron nitride (BN). The relative density for that is 68.78. The relative densities of the samples B0-TiB_2, B75T25 and B50T50 are 71.86 g/cm^3, 69.11 g/cm^3 and 70.57 g/cm^3 relatively, these values revealed that the measured densities of the samples much lower than the theoretical ones. This may be due to sintering conditions for TiB_2 containing samples [20].

The X-ray diffraction patterns of the sintered samples were shown in Fig.15 XRD analysis of the as-sintered samples revealed that main phases are titanium diboride (TiB_2, PDF card No. 35-0741) and boron nitride (BN, PDF card No. 34-0421). According to the X-ray patterns, in the composition as the ratio of boron nitride (BN) powder increased the peaks of TiB_2 phase gradually decrease while the peaks belong to the boron nitride (BN) phase getting stronger. The strongest peak for TiB_2 phase is about 52°, for boron nitride (BN) which is around 31° [20].

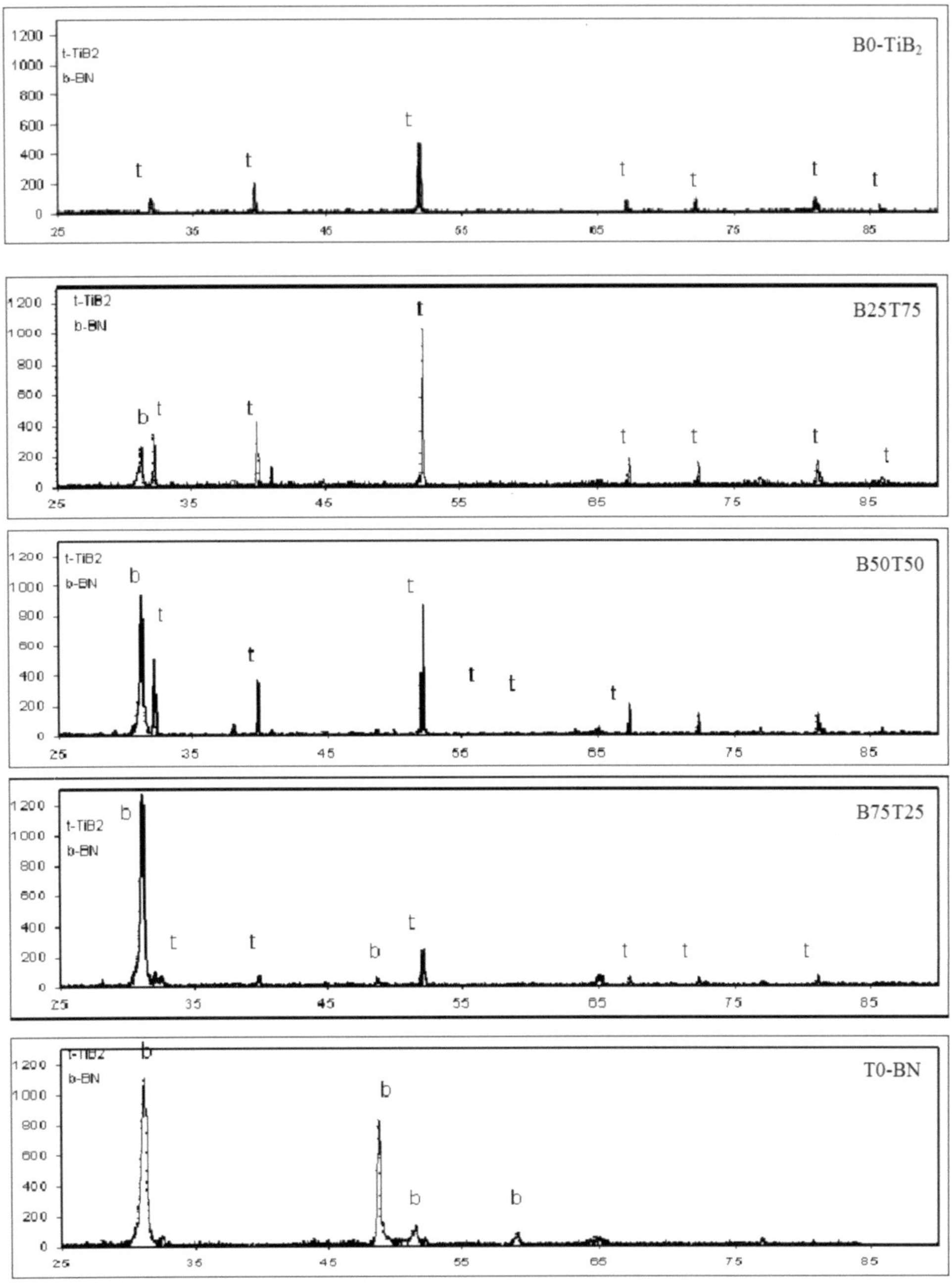

Figure 15. X-ray diffraction patterns of titanium diboride(B0-TiB_2) boron nitride(T0-BN), B25T75, B50T50 and B75T25 (t, titanium diboride; b, boron nitride) [20].

As expected, the content of TiB_2 increased, Vickers micro-hardness increased. The average Hv value of h-BN was measured to be 75kg/mm^2, which is good agreement with our previous study. As TiB_2 content was raised, micro-hardness increased gradually to 145kg/mm^2 when measured for the sample which contain 25% of TiB_2. The highest Vickers value was obtained for the sample which contain 100% of TiB_2. In a study by Munro, mechanical and thermal properties of nearly dense (98% of relative density and 4.5g.cm^{-3} of theoretical density) polycrystalline TiB_2 were examined [20]..

The diversity of the processing conditions is a significant factor in the often widely varying property values reported in the literature for polycrystalline TiB_2. Thus porosity greatly affected our Vickers hardness results. The cursory examination of the data for the Vickers hardness of TiB_2 has no immediately perceptible dependence on either density or grain size. However, our Vickers hardness, i.e 100% of TiB_2 (B0TiB_2 composition) was significantly affected by relative density and grain size (3-4 μm). The maximum Vickers hardness obtained for the present study in Fig.16, was that of B0T100 composition, which had a value of 450 kg.mm^{-2}. The higher values measured in the literature was the result of full densification of TiB_2 [20].

In Fig.17 the coefficient of thermal expansion curves as a function of temperature of h-BN, B50T50 and TiB_2 compositions are plotted. Hexagonal boron nitride is known to have low coefficient of thermal expansion, i.e high thermal shock resistance. The thermal expansion coefficients of hot-pressed h-BN in the literature were reported as 1.1x10^{-6} and 8.4x10^{-6} K^{-1} for perpendicular and parallel to hot-pressing directions, respectively. CTE of hot-isostatic pressed sample was measured to be 4.1x10 $^{-6}$ K^{-1}. In the samples studied, the lowest thermal expansion belonged to the pure h-BN composition, T0-BN. As TiB_2 content was raised, thermal expansion increased gradually as can be seen in the B50T50 curve. The coefficient of thermal expansion measured for 100% of TiB_2 (B0T100) in the present study (obtained at room temperature) well fitted with the literature data, which was 7.4x10 $^{-6}$ K^{-1} [20].

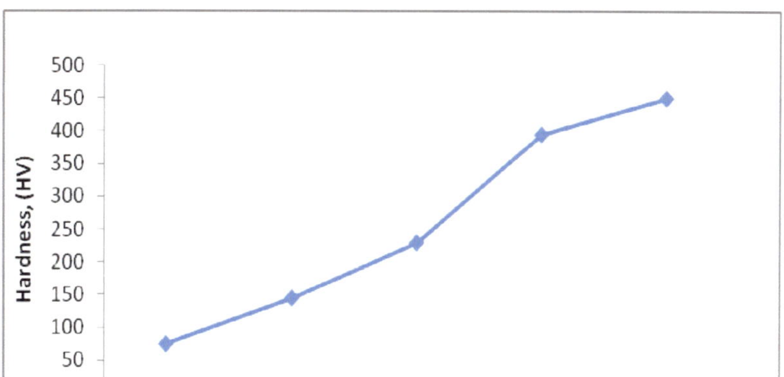

Figure 16. Plots of hardness of the hot-pressed h-BN/TiB_2 composites as a function of TiB_2 content [20].

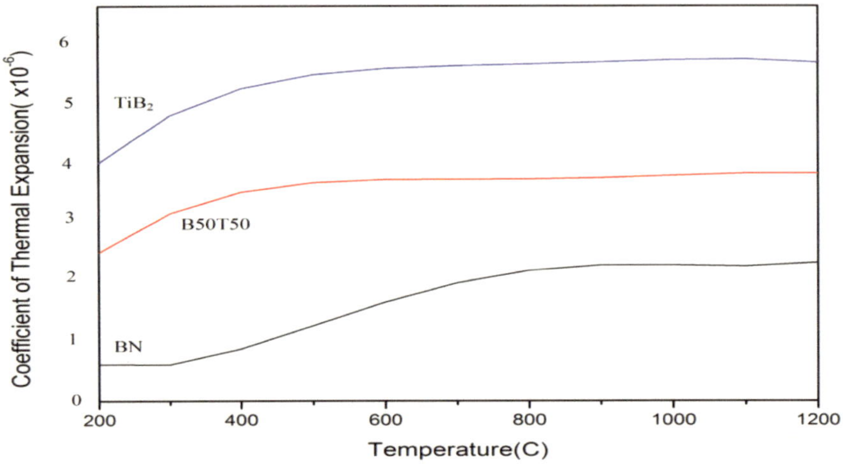

Figure 17. Plots of CTE of the hot-pressed h-BN/TiB_2 composites as a function of temperature [20].

8. Conclusion

1. The first synthesis method of boron nitride is the direct nitridation of boron. In another method, boric acid or borax are heated in the presence of a nitrogen-containing compound such as ammonia or urea. Reacting alkali metal borides or alkaline earth metal borides with silicon and/or aluminum is another method. However, there is a tedious leaching step after the synthesis. The carbothermic decomposition of elemento-organic BN compounds can also yield boron nitride. The disadvantage is that all starting materials are oxygen and moisture sensitive. Pressure pyrolysis of borazine ($B_3N_3H_6$) can also yield to BN, namely in amorphous form.

2. Key boron nitride properties are high thermal conductivity, low thermal expansion, good thermal shock resistance, high electrical resistance, low dielectric constant and loss tangent, microwave transparency, non toxicity, easily machinability- non-abrasive and lubricious, chemical inertness, non-wetting by most molten metals.

3. Typical boron nitride uses are electronic parts (heat sinks, substrates, coil forms, prototypes), boron doping wafers in silicon semiconductor processing, vacuum melting crucibles, CVD crucibles, microcircuit packaging, sputtering targets, high precision sealing, brazing and metallizing fixtures, microwave tubes, horiziontal caster break rings, low friction seals, plasma arc insulators, high temperature furnace fixtures and supports.

4. In the first experimental study descibed in the present work, hexagonal boron nitride ceramics with boron oxide content as sintering additive were fabricated by the hot-pressing technique. The major microstructural features, grain orientation and densification characteristics of the hot-pressed ceramics were influenced by the presence of B2O3. However, the enhancement of the sintering additive did not

cause a strong grain orientation or microstructural anisotropy as shown by SEM analysis. Only a slight orientation of the h-BN crystals towards the hot-pressing direction was observed. Thus it is anticipated that the mechanical properties of the h-BN samples will not show serious variations with the hot-pressing direction due to the low degree of orientation.

5. The overall mechanical characterization of hot-pressed hexagonal boron nitride (h-BN) based ceramics was conducted in the second experimental study. The effect of the anisotropy of hot-pressed h-BN based ceramics for various mechanical properties was determined. The mechanical characterization results were evaluated in terms of the orientation characteristics of hot-pressed ceramics. It was concluded that the hardness and the flexure strength properties in the hot-pressed h-BN ceramics are extremelydependent on the direction. Thus the average hardness and the flexure strength values are higher in the surfaces parallel to the hot-pressing direction. However, the measured fracture toughness (KIC) of the h-BN samples did not show serious variations with the hot-pressing direction. In addition, fracture characteristic of the hexagonal BN grains is determined to be completely brittle exhibiting a mixed mode of fracture. Crack path evaluations after the fracture tests indicated that no crack deflection occurred through the thickness of the specimen. Instead, the crack followed a straight path, demonstrating a low degree of grain orientation following the hot-pressing process.

6. In the final experimental study, the mechanical characterization of four hexagonal boron nitride containing and one titanium diboride samples, produced by hot pressing was conducted. It was concluded that the hardness and the fracture surface in the five samples were dependent on the h-BN content. On the other hand, fracture characteristic of the samples was brittle. There were both transgranular and intergranular cracks in the microstructures. After 3-point bending test, the crack followed a straight path, demonstrating a low degree of grain orientation. It was concluded that TiB2 content increased the thermal expansion property of h-BN ceramics.

Acknowledgements

This chapter would not have been possible without the guidance and the help of several individuals who in one way or another contributed and extended their valuable assistance in the preparation and completion of this study. First and foremost, my utmost gratitude to Professor Dr. A. Okan Addemir, my Ph.D thesis advisor, who gave me the idea of working on the interesting subject of hexagonal boron nitride. I would like to thank to Assistant Professor Dr. Tahsin Boyraz whose sincerity and encouragement I will never forget. Dr. Boyraz has been my inspiration as I hurdle all the obstacles in the completion this research work. Dr. Gülten Sadullahoğlu, has been a great friend and support for me during my post doctoral studies. Finally many thanks for my parents for their silent encouragement. But most of all a loving thank you to my sister, Deniz, for her understanding and gracious support.

Author details

Burcu Ertuğ*

Address all correspondence to: burcuertug@gmail.com

Istanbul Technical University, Department of Metallurgical and Materials Engineering, Maslak, Istanbul, Turkey

References

[1] S. Alkoy, Crystallization behaviour and characterization of turbostratic boron nitride, July 1994, Istanbul Technical University, Graduate School Of Science, Engineering and Technology.

[2] Ş. Pekin, The reaction rate controlling steps and turbostratic hexagonal structure transition in hBN synthesis, July 1992, Istanbul Technical University, Graduate School Of Science, Engineering and Technology.

[3] Development and application of high-purity hexagonal boron nitride (h-BN) powder, Kawasaki Steel Technical Report No:28, June 1993, 17-25.

[4] http://accuratus.com/boron.html

[5] A.Lipp, K.A. Schwetz and K. Hunold, Hexagonal boron nitride: fabrication, properties and applications, Journal of the European Ceramic Society, 5, 1989, 3-9.

[6] http://htemps.en.made-in-china.com/

[7] http://www.oxy-gon.com/

[8] http://buy.capitolareatechnology.com/

[9] http://www.esk.com/en.html

[10] http://www.diytrade.com/china/main.html

[11] http://www.rdccontrol.com/en/index.html

[12] http://www.crystec.com/tkralue.htm

[13] http://www.substech.com/dokuwiki/doku.php?id=main_page

[14] http://www.trademart.in/

[15] http://en.pensc.com/index.html

[16] http://www.junsun.url.tw/e-p2-1-1.html

[17] http://www.sciencedirect.com/science/article/pii/S0022024898006204

[18] B. Ertuğ, T. Boyraz, A.O. Addemir, Microstructural Aspects Of The Hot Pressed Hexagonal Boron Nitride Ceramics With Limited Content Of Boron Oxide, Materials Science Forum, 554, 197-200, 2007.

[19] B.Ertuğ, T. Boyraz, A.O. Addemir, An Investigation Of The Mechanical Properties And Fracture Characteristic Of Hot-Pressed Boron Nitride Ceramics, Materials Science Forum, 554, 201-205, 2007.

[20] A.C. Akarsu, H. Gökçe, T. Boyraz, B.Ertuğ, A.O. Addemir, M.L. Öveçoğlu, The Characterization of The Mechanical And Thermal Properties Of Hot-Pressed Hexagonal Boron Nitride-Titanium Diboride Composites, 3rd International Symposium on SiAlONs and Non-Oxides, June 1-4, 2010.

Chapter 3

High Pressure Sintering of Nano-Size γ-Al_2O_3

Nilgun Kuskonmaz

Additional information is available at the end of the chapter

http://dx.doi.org/10.5772/53324

1. Introduction

In recent years nano crystalline materials have been paid much attention because they have a variety of interesting and novel physical properties.

Research on the sintering of nano crystalline ceramics has focused on the problem of achieving high densities(> 95% of theoretical) without excessive grain growth.

Dense and fine grained alumina ceramics are widely used in practical applications, because of mechanical, electrical and optical properties. The mechanical strength, dielectric properties and transparency are strongly affected by the microstructure of alumina ceramics such as porosity, grain size and their distribution. (Fig.1)

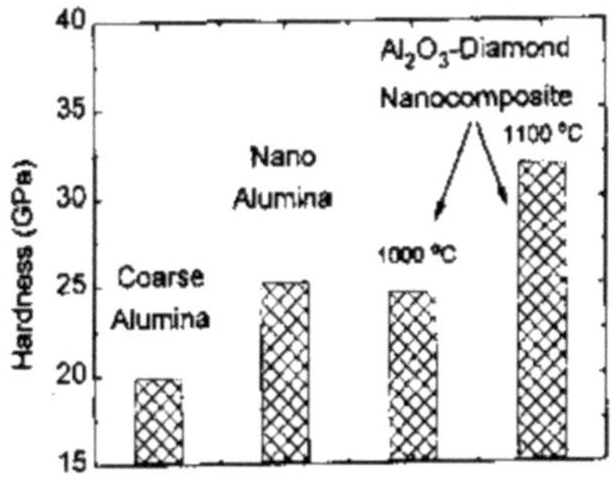

Figure 1. The variation of hardness in alumina-diamond nanocomposite with sintering temperature at 1GPa "in [1]"

For example, the transparency and mechanical strength of alumina were improved by decreasing the grain size, and the residul porosity less than 0.05% was required for obtaining a high transmission of light."in [2]"

INTECH
open science | open minds

© 2013 The Author(s). Licensee InTech. This chapter is distributed under the terms of the Creative Commons Attribution License http://creativecommons.org/licenses/by/3.0), which permits unrestricted use, distribution, and reproduction in any medium, provided the original work is properly cited.

Alumina ceramics with sub micrometer microstructure obtained by pressureless sintering have been widely studied.

It is difficult to obtain a fully dense ceramic with nanocrystalline grain size. Because phase transformation sequences that occur during the conventional sintering process at atmospheric (ambient) pressure:

$$\gamma\ Al_2O_3(750^0C) \rightarrow \delta\ Al_2O_3(900^0C) \rightarrow \theta\ Al_2O_3(>1000^0C) \rightarrow \alpha\ Al_2O_3$$

The transition paths and temperatures vary depending on the particle size, chemical homogeneity, heating rate, and water vapor pressure.

The transformation from θ –Al_2O_3 to α - Al_2O_3 involves a chance in the oxygen sublattice from cubic close packing to hexagonal close packing and generally requires temperatures above 1100 ^{0}C.

There is 10% decrease in specific volume during transformation because the density changes from 3.56 g/cm^3 (θ Al_2O_3) to 3.986 g/cm^3 (α - Al_2O_3).

As a result of the volume reduction and low real nucleation density (10^8-10^{11} nuclei /cm ^{3}) during the transformation, the α - Al_2O_3 colonies recede from the matrix and the microstructure develops into vermicular morphology containing larger scale interconnected porosity. (Fig. 2)

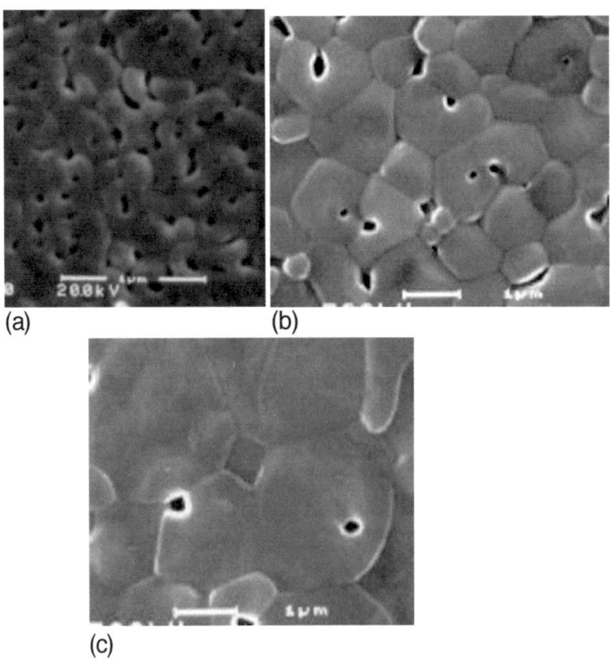

Figure 2. SEM images show the microstructure development of α - Al_2O_3 specimens pressureless sintered at (a) 1200 °C, (b) 1400 °C, (c) 1500 °C for 5 h. All specimens have fully transformed to α - Al_2O_3 at these sintering temperatures "in [3]".

http://dx.doi.org/10.5772/53324

The temperature required for the densification of this vermicular microstructure is over 1600^{0}C. to obtain dense, fine grained α- Al_2O_3 at low temperatures, the scale of the vermicular microstructure must be minimized "in [4]".

Pressure-assisted densification provide to obtain nanometric grains in fully sintered compacts.

The effect of pressure on nucleation and growth of the α- phase is the critical subject because the final grain size after sintering depends on these factors.

Low pressures create a smaller number of nucleation sites where the transformation begin.. These nuclei grow very fast forming vermicular clusters. The clusters stop growing when they touch on each other.

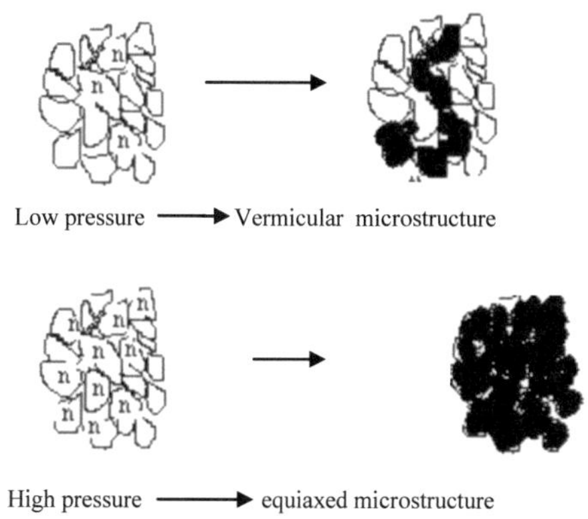

During High pressure each particle can act as an original nucleation site and transformation can occur within each particle, so that avoiding the formation of vermicular structure. In other words, high pressure can create a sufficient number of nucleation sites to prevent the formation of the vermicular structure.

Modifications in the structure, by the application of high pressure, have been investigated by several groups "in [5-7]". They have reported applying pressure decreases the thermodynamic energy barrier and kinetic energy barrier required for nucleation and causes the phase transformation to shift to a lower temperature.

Figure 3 shows that the start of transformation from γ- to α -Al_2O_3 temperature decreases from about 1075 ^{0}C at 1 atmosphere, to around 800^{0}C at 1 GPa, and to 640^{0}C at 2.5 GPa; 560 ^{0}C at 5.5 GPa and to 460 ^{0}C at 8 GPa.

Grain growth is limited by the low sintering temperature and variation of nucleation events in the γ phase at high pressure creates a nanoscale α grain size "in [7]".

In this work the required conditions were examined to produce high- density nano- Al_2O_3 ceramics using high pressure and different sintering temperatures. The effect of various sintering conditions on the properties of sintered sample such as microstructure and relative density were discussed.

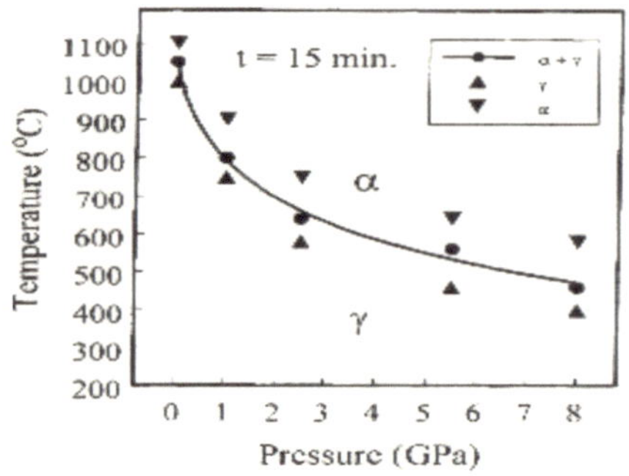

Figure 3. Pressure – Temperature phase diagram of α and γ phases of Al_2O_3 "in [7]".

2. Experimental procedure

The starting powder used consists of spherical γ- Al_2O_3 phase with an average particle size of 20 nm (Plasma & Ceramic Technologies Ltd.-Latvia) and specific surface area of 50 m^2/ g in the granulated state. (Fig.4)

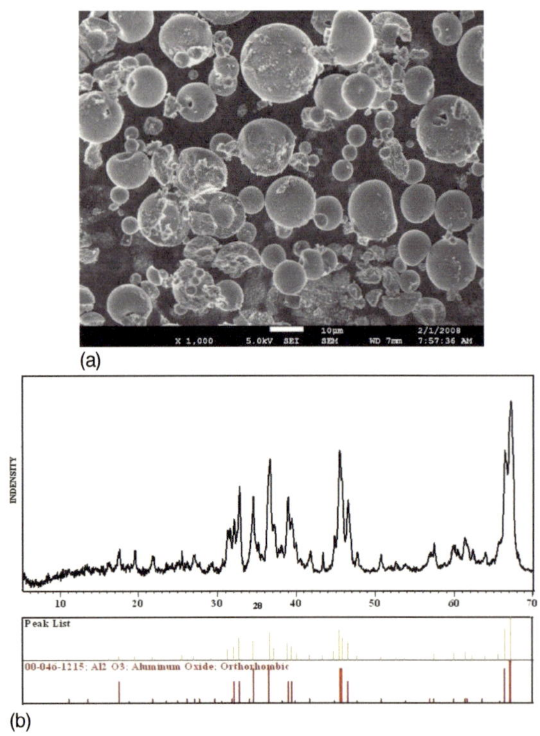

Figure 4. a) SEM micrograph of the starting granulated γ-Al_2O_3 powder. (b) XRD spectrum of the starting γ-Al_2O_3 powder (JSPDS Card No: 00-046-1215).

http://dx.doi.org/10.5772/53324

The impurity content of the initial powder was given in Table 1, in accordance with the analytical certificate supplied.

Chemical impurities: (ppm)	**Fe**	**Si**	**Na**
According to supplier's analytical certificate:	<1000	<200	< 1000

Table 1. The chemical composition of the starting γ -Al_2O_3 powder.

The powder was first preheated in the air at 700 ^{0}C for 3 h for the removal of the binder. After preheating, the pure γ-Al_2O_3 powder containing no additives was uniaxially cold pressed at 20 MPa into cylinders 20mm in diameter and 10mm in height. All the green compacts were pellets of 4 g. Green compact was encapsulated in a cube die made of pyrophyllite.

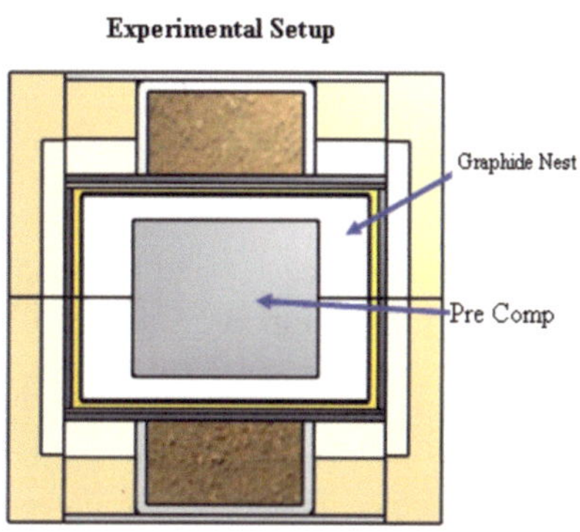

Figure 5. The sample assembly for Al_2O_3 ceramics sintered at high pressure.

All the high-pressure sintering experiments were carried out in a cubic anvil apparatus. High mechanical pressure was applied on all six faces of the die concurrently.

Additionaly 10 wt% TiO_2 was added by sol-gel to inhibited the grain growth.

TiO_2 doped alumina were prepared by adding titanium –isoproxide ($Ti(OC_3H_7)_4$ into alumina.

The TIP was hydrolyzed by addition of water. The ethanol was removed in a rotating evaporator at 65 ^{0}C and powders were dried in a furnace at 90 ^{0}C for 24 hour.After drying powder mixture was decomposed into oxide at 400 ^{0}C.

The Al_2O_3 bodies were fabricated in cubic anvil high pressure (2- 7 GPa) and varying temperature (600-1200 ^{0}C) for 1- 15 minutes.

Phase analysis of the sintered samples was carried out by X-ray diffraction (XRD). Grain sizes were estimated from high-resolution scanning electron micrographs taken from fracture surfaces and micro hardness was determined on the polished surfaces using an applied load of 500 g.

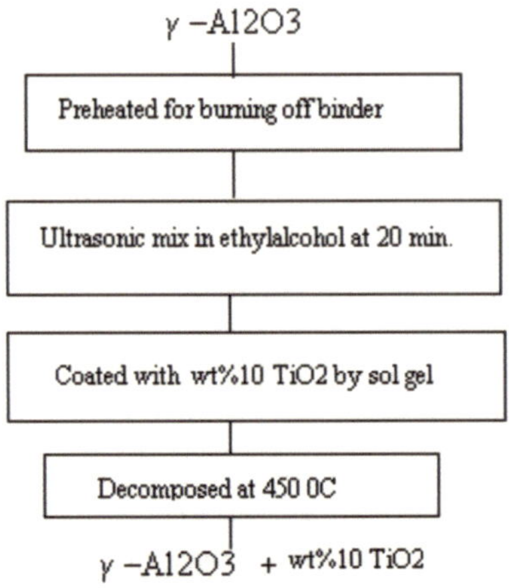

Figure 6. Processing flow chart for coating γ-Al_2O_3 particle with 10 wt %TiO_2 by sol-gel.

3. Results and discussion

3.1. X-ray difraction profile of sintered samples

Fig.7-8-9 clearly shows the significant effect of applied pressure time on phase content with the exception of the sample sintered at 5GPa and 500°C for only 5 min.(Fig a). All the sintered samples showed the presence of the α phase, with no evidence of any remaining γ phase. Because of pressure decreases the transformation temperature of γ to α phase of Al_2O_3

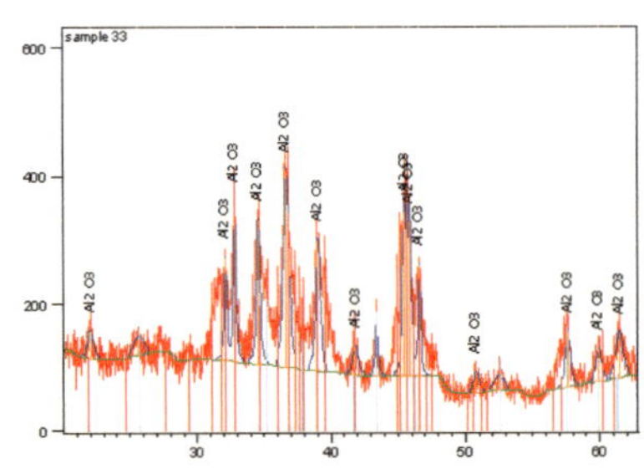

Figure 7. XRD patterns of Al_2O_3 sintered at 5GPa and 500 °C for 5 min.

Fig. 7-8 also shows that under the same applied pressure (5 GPa) and temperature (500 °C), the sintering time have a significant effect on the phase content, as the sintering time of 5 min. is not adequate for phase transformation from gamma to alpha alumina although both samples are translucent.

An alumina hydrate, AlO(OH), phase (Fig. 8-9) was found in the samples sintered at 5 and 7 GPa and 500 °C. This phase is caused by trapped water or surface OH groups which forms the hydrate phase during low temperature sintering.

http://dx.doi.org/10.5772/53324

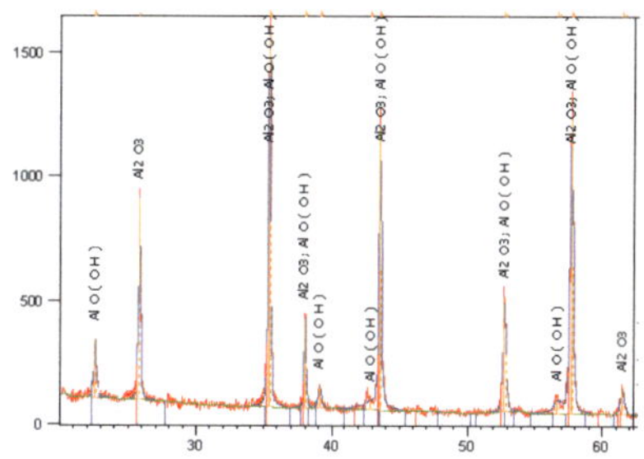

Figure 8. XRD patterns of Al_2O_3 sintered at 5GPa and 500 ^{0}C for 15 min.

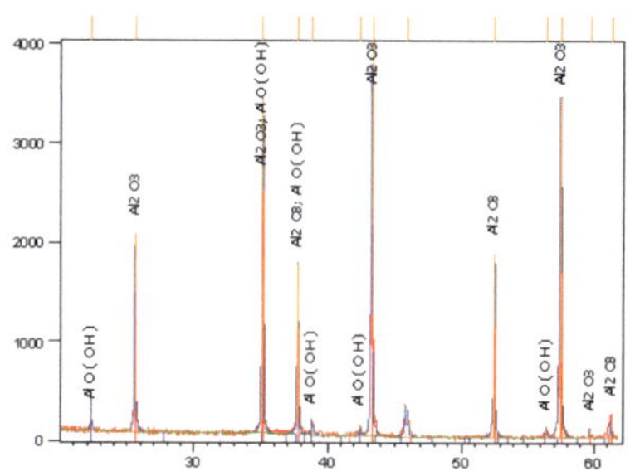

Figure 9. XRD patterns of Al_2O_3 sintered at 7GPa and 500 ^{0}C for 15 min.

3.2. Effect of temperature on the microstructure and density at 5 GPa

As can be seen in Fig 10, the morphology of the sintered samples at high temperatures were different from those sintered at lower temperatures.

The grain size of sintered sample increases with increasing sintering temperature and the grain size distribution was fairly wide at especially high sintering temperature.

Samples sintered at a low sintering temperature (500 ^{0}C-700 ^{0}C) showed fine (about 200 nm) equiaxed grains under 5 GPa pressures as shown in Fig. 10 (a), (b).

Grain Growth was started at 900 ^{0}C and abnormal grain growth was observed at the higher sintering temperature (1200 ^{0}C) as shown in Fig. 10 (c), (d).

Previous experimental results indicate that abnormal grain growth in commercially pure alumina is strongly correlated with presence of impurities. The minimum concentration for AGG over 300 ppm silicon or 30 ppm calcium. "in [8]"

This impurities are believed to form glassy films in the grain boundaries and somehow to catalyze AGG.

The sintered density generally increases with the sintering temperature for all samples.

From Figure 11, it is noted that the relative density first increased rapidly, than climbed slowly from 1000 to 1500 ^{0}C.

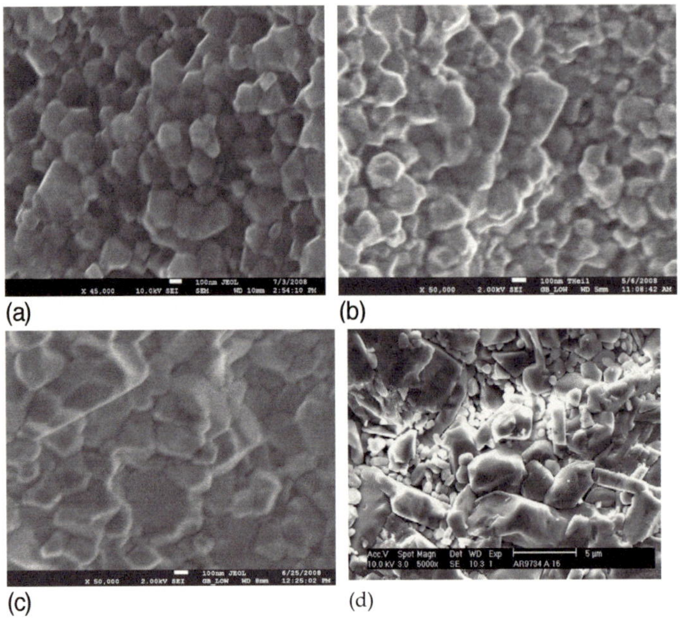

Figure 10. The fracture surfaces of Al_2O_3 samples sintered at different temperature (a) 500 ^{0}C - (b) 700 ^{0}C – (c) 900 ^{0}C - (d) 1200 ^{0}C at 5 GPa for 15 min. sintering time.

At 1500 ^{0}C. the relative density of sintered sample is reached a highest value of 98.1% when the pressure is 5GPa.

Due to significant grain growth, no complete densification could be reached with initial γ-Al_2O_3 powder

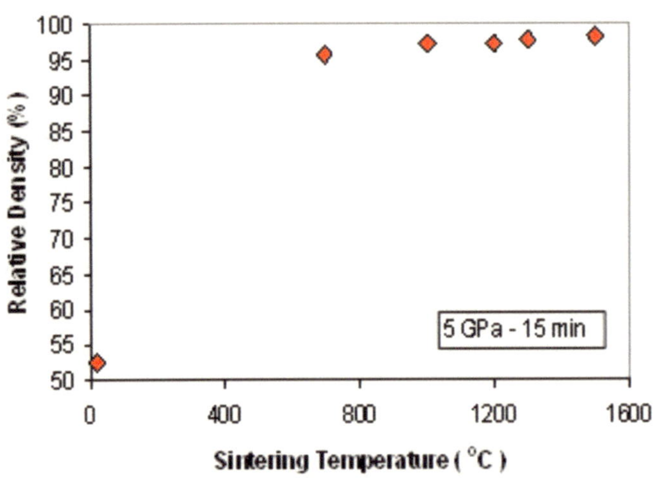

Figure 11. Relative density of sintered sample at 5 GPa as a function of sintering temperature.

3.3. Effect of pressure on the grain size of sintered samples

The microstructure of sintered samples obtained from γ -Al_2O_3 powders at 1000 ^{0}C and 5-7 GPa for 15 minute were examined. Representative SEM's of fracture surfaces are shown in figure 12.

(a) (b)

Figure 12. SEM micrograph of fracture surface for samples sintered at 1000 ^{0}C and (a) 5 GPa, (b) 7GPa for 15 minutes.

As shown in Fig. 12 the overall grain size for the sample sintered at 1000 ^{0}C at 7GPa appeared coarser than the sample sintered at the lower pressures (5 GPa) (for the same sintering temperature and time of 15 minutes).

Fig. 13 shows the SEM micrographs of the sample sintered at 1000 ^{0}C and 7 GPa for a shorter time (1 minute).

Figure 13. SEM micrograph of fracture surface for samples sintered at 1000 ^{0}C and 7GPa for 1 minute.

As shown in Fig. 13 the microstructure of the Al_2O_3 ceramics sintered at high pressure (7GPa) and high temperature (1000 ^{0}C) for 1 minute is obviously different from the ceramic microstructure sintered at the same pressure and temperature for 15 minutes (see in Fig. 12 (b)).

While Fig.13 contains much smaller fine grains, coarser grains are visible on the microstructure shown in Fig.12 (b), showing the considerable effect of sintering time on the sintered grain size.

This could be attributed to the higher input energy in the system at high pressure and high temperature conditions, thus the final stage in sintering was reached quickly as was the grain growth regime. Thus for high pressure high temperature conditions, either the sintering time or the temperature should be reduced to prevent grain growth.

The effect of the sintering time on the microstructure can also be seen from the pictures shown in Fig.14. Sintering γ-Al_2O_3 at 900 ^{0}C at 7GPa for 15 min. resulted in obtaining the opaque sample while a translucent sample was obtained with a sintering time of 5 min. as shown in Figs.14 (a and b, respectively.

The opaque structure is considered as a result of abnormal grain growth of alumina, as shown in Fig.14 (a). But when the sintering temperature is lowered to 500 ^{o}C, a translucent alumina with the finest grain size and the gamma form is evident using a sintering time of 5 min. as shown in Fig.14 (c).

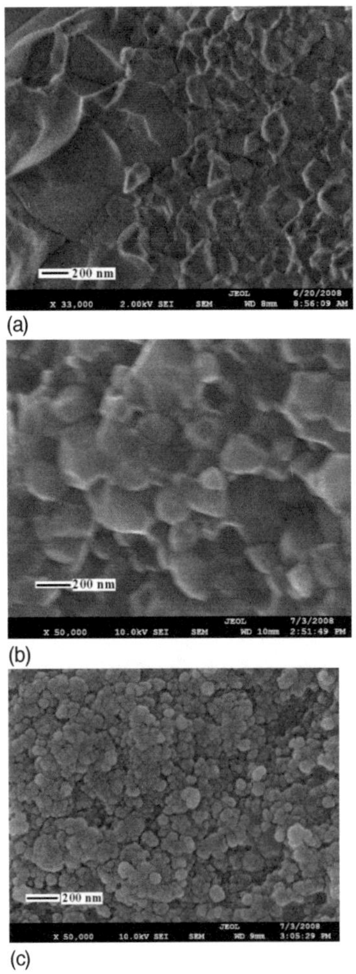

Figure 14. The fracture surfaces of Al_2O_3 samples sintered at 900 ^{0}C at 7 GPa for sintering time 15 min. (a), 5 min. (b) and 5GPa (c) at 500 ^{0}C for sintering time of 5 min.

http://dx.doi.org/10.5772/53324

Fig.15 clearly indicates the relationships between the translucency of alumina, temperature and sintering time. For example, a translucent alumina can be achieved either using a high sintering temperature of 900 °C for 5 min. or low sintering temperature of 500 °C for 15 min., as shown in Fig.15(a)

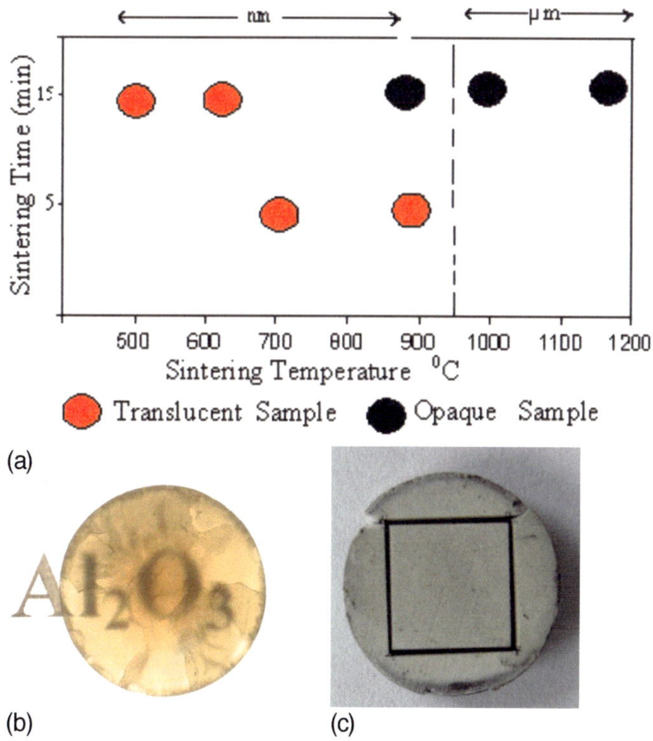

Figure 15. a) The effect of sintering temperature on the optical appearance of sintered sample at 7GPa (b) the optical micrograph of the translucent alumina sample sintered at 7 GPa, at 900 °C for 5 min. that was mechanically thinned to a 1mm in thicknes using lapping technique. (c) opaque alumina sample (at 7 GPa, 900 °C for 15 min.) after laser cut for characterization.

3.4. Influnce of TiO_2 additives on alumina microstructure

As you can see Fig. 16 The secondary phase precipitates mostly at along the grain boundaries (white phase)..The existence of this secondary phase reduces the driving force for grain growth by pinning effect.

It can be shown that TiO_2 additive leads to finer grain size after pressure assisted sintered at low and high temperature for same sintering pressure and time. (Fig 17 -18)

From figure 17 – 18 it is obvious that average grain size of TiO_2 doped sample is finer than that of undoped alumina sintered at the same temperature and pressure for same holding time.

However TiO_2 addition did not has significant effect on the density of sintered samples

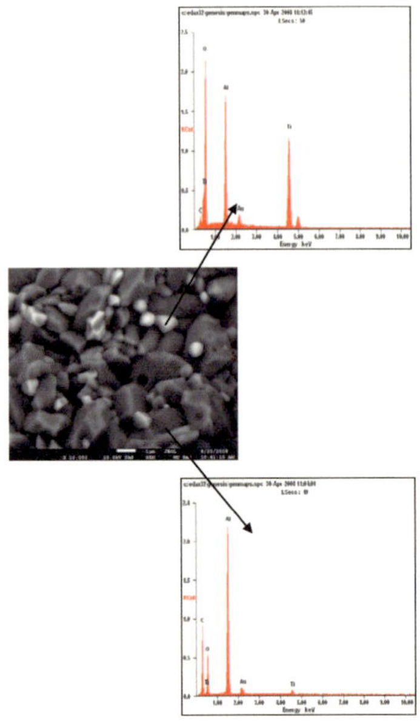

Figure 16. Typical microstructures and EDS spectrum of TiO_2 doped Al_2O_3 sintered sample.

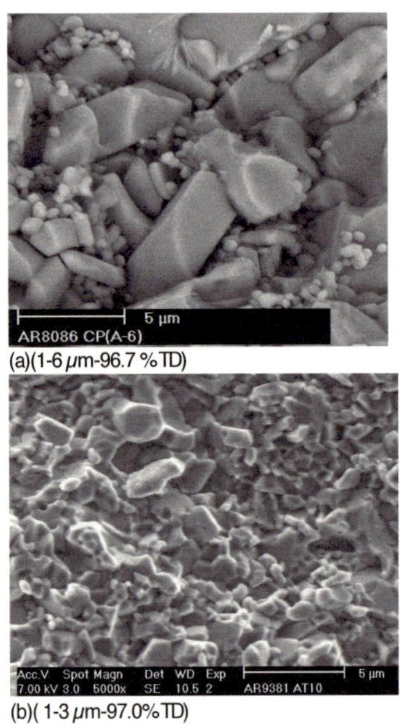

Figure 17. SEM micrographs of Al_2O_3 (a) and Al_2O_3 +10 %wt TiO_2 (b) sintered at 5GPa and 1200 ^{0}C for 15 min.

Figure 17 shows that Abnormal Grain Growth did not occur at 1200 ^{0}C in using 10 wt% TiO_2 additive

(a)(>100 nm-96.0 %TD)

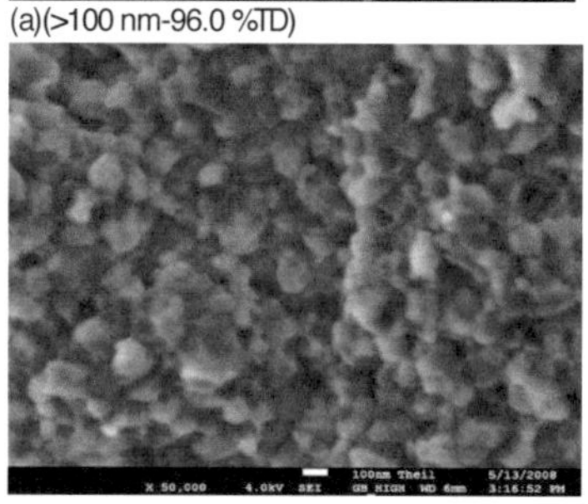

(b)(<100 nm-97.0 %TD.)

Figure 18. SEM micrographs of Al_2O_3 (a) and Al_2O_3 +10 wt % TiO_2 (b) sintered at 5GPa and 700 ^{0}C for 15 min.

Although Translucent Alumina is obtained with the alumina sample at 6GPa and 600 ^{o}C for 15 min (Fig 19 (a)). it is not possible to obtained translucent alumina with the alumina containing 10 wt% TiO_2 sample at same sintering conditions (Fig 19 (b))

(a)

(b)

Figure 19. Fracture surfaces of Al_2O_3 (a) and $Al_2O_3$10 wt % TiO_2 (b) samples sintered at 600 ^{0}C and 6 GPa for 15 min

3.5. Effect of pressure and sintering time on the hardness of sintered samples

Sintering time has a strong effect on the grain size and hence the hardness. For example, a sample sintered at 1000 °C (at 7 GPa) for 15 min. shows a hardness value of 8.30 GPa while a sample sintered at the same temperature for 1 min. provides a hardness value of 11.46 GPa.

The increment in hardness depending on the decrease in sintering time can be correlated with the grain size as shown by the SEM micrographs in Figs.12 b and 13. As shown in Fig. 12b, the sample sintered at 1000 °C for 15 min. (the applied pressure is at 7 GPa)contains much bigger alumina grains (the main grain size is about 5 microns) compared with sample in Fig.13 that was sintered at the same temperature and pressure for 1 min (the main grain size is about 200 nm).

Using applied pressures of 7 GPa and 5 GPa and lower sintering temperature of 600 °C, translucent α-alumina could be obtained with the hardness value of 13.89 and 13.35 GPa, respectively.

The highest hardness value of 20.31 GPa is achieved for the TiO_2 doped sample sintered at 700 °C for 15 min (the applied pressure is at 5 GPa) due to presence of very fine alumina grains with the main grain size of < 100 nm, as shown in Fig.18 b.

4. Conclusions

In the present work, the sintering behavior of a γ-Al_2O_3 powder subjected to different pressure, temperature and time conditions is examined.

Applying pressure drops γ- α Al_2O_3 transformation temperature and increase nucleation rate making it possible to obtain nano grain size sintered sample.

It was also found that, when using nano-size starting powder, the sintering time should be optimised in order to control the final sintered grain size under the same sintering temperature and applied pressure.

Chemical impurity in initial powder is very important to reduce grain growth during sintering.

To produce lower than 100 nm sintered sample it is very important to eliminate the hydrates & impurities before sintering.

TiO_2 additive in initial γ-Al_2O_3 leads to finer grain size after pressure assisted sintered

Using applied pressures of 5-7 GPa and lower sintering temperature of 500- 600 °C, translucent α-alumina could be obtained

Translucency can be controlled by increasing the applied pressure from 5 GPa to 7 GPa for a sample sintered at 700 °C for 15 min. or decreasing the sintering temperature and time. As the sintering pressure is decreased, the sintering temperature also needs to be decreased in order to obtain translucent samples.

http://dx.doi.org/10.5772/53324

The hardness of the sintered 10 wt % TiO_2 doped Al_2O_3 nanocomposites were higher than undoped Al_2O_3 at same sintering conditions.

The highest hardness value of 20.31 GPa is achieved for the TiO_2 doped sample sintered at 700 °C for 15 min. (the applied pressure is at 5 GPa)

Acknowledgements

Element Six (Production) (Pty) Ltd is greatly acknowledged for financial support and laboratory facilities used in the present work.

Author details

Nilgun Kuskonmaz

Yildiz Technical University, Metallurgy and Materials Engineering, Davutpasa Campus, Esenler, Istanbul, Turkey

References

[1] Mishra, R. S, & Mukherjee, A. K. Processing of high hardness- high toughness alumina matrix nanocomposites" Matererials Science and Enginnering A(2001). , 301(2001), 97-101.

[2] Krell, A, Blank, P, Ma, H, & Hutzler, T. J. Am. Ceramics Soc., 86 (1), (2003).

[3] Eduardo, J. Gonzalez, Grady White, and Lanhua Wei Ceramics Division National Institute of Standards and Technology Gaithersburg, MD 20899.

[4] Gonzalez, E. J, & Hockey, B. and G.J.Piermarini Materials and Manufacture Process, (1996). , 11(6)

[5] Mishra, S, Lesher, E, Mukherjee, K, & Am, J. Ceram.Soc, 79(11), (1996).

[6] Lioa, S. C, Chen, Y. J, Kear, B. H, & Mayo, W. E. Nanostructured Materials,10(6), (1998). , 1063-1079.

[7] Lioa, S. C, Chen, Y. J, Kear, B. H, & Mayo, W. E. High Pressure / Low Temperature Sintering of Nanocrystalline Alumina"Nano Structured Materials , 10(6), 1998.

[8] So Ik Bae and Sunggi Baik J. Am Ceramics Soc., 76 (4), (1993).

Chapter 4

Sintering Behavior of Vitrified Ceramic Tiles Incorporated with Petroleum Waste

A.J. Souza, B.C.A. Pinheiro and J.N.F. Holanda

Additional information is available at the end of the chapter

http://dx.doi.org/10.5772/53256

1. Introduction

Currently, the urgent need to preserve the environment has aroused great interest in academic and industrial areas for the reuse of pollutant wastes. In the XXI century one of the challenges of the environmental sustainability of the planet is to seek solution to the final disposal of huge volumes of pollutant wastes produced every year.

The petroleum industry generates various types of oily sludges in its production chain [1]: i) extraction of crude oil from the ground; ii) transportation to refineries and product distribution centers; iii) refining into finished products; and iv) marketing or sale of the products to consumers. These oily sludges are basically composed by hydrocarbons in the form of oil, water, solids, and traces of heavy metals in different proportions according to each area. For this reason, the oily sludges are considered as being hazardous waste materials. Thus, the petroleum industry is confronted with the environmental problem of the correct management of huge amounts of oily sludges produced worldwide.

Recently, the Brazilian petroleum industry has presented high growth in the production of crude oil. As a result, Brazilian oil industry generates huge volume of oily sludge during the oil extraction process. The traditional disposal methods used for management of this oily sludge include storage in ponds, dikes, and biodegradation (landfarming) [2,3]. More recently, the oily sludge has been treated with bentonite clay for disposal in sanitary sites, resulting in a new waste material that hereafter will be named as petroleum waste [4]. However, these waste management options are very limited, costing money and environmental impact.

Nowadays, the reuse of waste materials as an alternative raw material in the production of ceramic materials for civil construction has become a very attractive method [5-7]. The reuse approach is environmentally correct and can contribute to the environmental sustainability,

INTECH
open science | open minds

© 2013 The Author(s). Licensee InTech. This chapter is distributed under the terms of the Creative Commons Attribution License http://creativecommons.org/licenses/by/3.0), which permits unrestricted use, distribution, and reproduction in any medium, provided the original work is properly cited.

resulting in three main advantages: i) use of low cost raw material; ii) the conservation of the natural resources; and iii) the management of a pollutant waste.

The reuse of petroleum waste into ceramic materials for civil construction has already been investigated [2-4,8-11]. However, the sintering behavior of floor tiles containing petroleum waste during the firing step, which is a complex process, is not still thoroughly understood. Floor tiles are a multicomponent system primarily composed of clays, feldspar, and quartz, and is considered to be one of the most complex ceramic materials [12].

This chapter focuses on the sintering behavior of vitrified floor tiles containing petroleum waste. Emphasis is given on the effects of the petroleum waste additions on the sintering behavior, microstructural evolution, and physical properties of the vitrified floor tiles.

2. Experimental details

The raw materials used are commercial kaolin, Na-feldspar, quartz, and petroleum waste from petroleum extraction industry from south-eastern Brazil. The petroleum waste sample is a solid material in the form of granular powder. The raw materials were dried at 110 ºC, dry-ground, and then passed through a 325-mesh (45 μm ASTM) sieve. The chemical and mineralogical compositions are given in Table 1. The processing flow diagram is shown in Fig. 1.

Oxides	kaolin	Petroleum Waste	Na-feldspar	Quartz
SiO_2	49.07	41.73	69.55	09.97
Al_2O_3	33.74	10.93	18.82	0.41
Fe_2O_3	0.22	7.63	0.14	0.01
TiO_2	0.01	0.52	0.02	0.02
Na_2O	0.52	0.44	9.63	0.13
K_2O	1.97	0.95	1.47	0.18
CaO	0.30	7.76	0.17	0.01
MgO	0.06	5.87	0.09	0.01
MnO		0.02		
P_2O_5		0.09		
BaO		5.03		
SrO		0.29		
Loss on ignition	14.01	18.74	0.32	0.26
Mineral phases	kaolinite, quartz	quartz, kaolinite, barite, calcite, gypsum, hematite, galena, halite, calcium chloride, potassium chloride, montmorillonite	albite, quartz	quartz

Table 1. Chemical (wt.%) and mineralogical compositions of the raw materials.

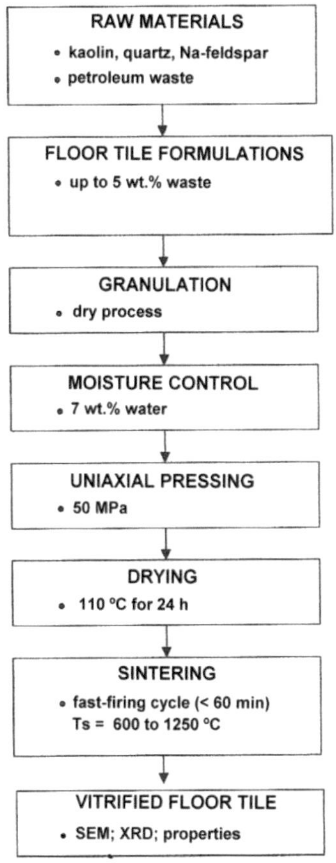

Figure 1. Process flow diagram of the vitrified floor tiles.

Several floor tile compositions were prepared using mixtures of kaolin, Na-feldspar, quartz, and petroleum waste (Table 2). The petroleum waste additions were up to 5 wt.% in gradual replacement of kaolin. The floor tile formulation used as a reference consisted of 40 wt.% kaolin, 47.5 wt.% Na-feldspar, and 12.5 wt.% quartz.

Raw materials	Tile formulations (wt.%)		
	MT1	MT2	MT3
Kaolin	40	37.5	35
Waste	0	2.5	5
Na-feldspar	47.5	46.5	47.5
Quartz	12.5	12.5	12.5

Table 2. Composition of the floor tile formulations containing petroleum waste.

The floor tile formulations (Table 2) were mixed, homogenized, and granulated via dry process using a high intensity mixer. After granulation step, the granules coarser than 2 mm were discarded. The moisture content (moisture mass/dry mass) was adjusted to 7 %.

Mineralogical analysis of the floor tile formulations was done using Cu-Ka radiation and 1.5º (2θ)/min scanning speed in a conventional diffractometer. Mineral phases were identified by comparing the intensities and positions of the Bragg peaks to those listed in the JCPDS/ICDD data files.

Thermogravimetric analysis (TGA and DrTGA) of the floor tile powder sample was performed in air between 25 ºC and 1200 ºC using a heating rate of 10 ºC/min. Dilatometric analysis of the floor tile samples was carried out on unfired test pieces within the 25 – 1200 ºC temperature range using a heating rate of 10 ºC/min under air atmosphere.

The tile powders were uniaxially pressed into test bars (11.5 x 2.54 cm2) under a load of 50 MPa, and then dried at 110ºC. The sintering step was carried out at soaking temperatures varying from 600 to 1250 ºC, using a fast firing-cooling cycle of total duration < 60 min. The heating and cooling rates used in this work were chosen to simulate actual sintering process used in the tile industry.

The following physical properties of the floor tile pieces have been determined: linear shrinkage, water absorption, apparent density, and apparent porosity.

The linear shrinkage values were obtained by measuring the length of the rectangular specimens before and after sintering step using a caliper with a precision of ± 0.01.

Water absorption values were determined from weight differences between the as-sintered and water saturated samples (immersed in boiling water for 2h) according to the ASTM C373 standardized procedures [13].

The apparent density of the floor tile pieces were determined by the Archimedes method according to the ASTM C373 standardized procedures [13].

The apparent porosity also was determined according to the ASTM C373 standardized procedures [13].

Scanning electron microscopy operating at 15 kV was used to examine the gold-coated fracture surfaces of the sintered floor tile pieces via secondary electron images (SEI/SEM).

The crystalline phases after sintering were identified between 2θ = 10º and 2θ = 80º via X-ray diffraction analysis with Cu-Ka radiation (40 kV, 40 mA). The mineral phases were identified using reference data from the JCPDS/ICDD data files.

3. Results and discussion

3.1. Characteristics of the floor tile formulations

The XRD pattern of the MT1 sample is shown in Fig. 2. The reference formulation exhibits peaks that are characteristics of kaolinite ($Al_2O_3.2SiO_2.2H_2O$), albite ($NaAlSi_3O_8$), and quartz

(SIO_2). XRD pattern of the petroleum waste containing formulation (MT3 sample) is shown in Fig. 3. In addition to kaolinite, albite, and quartz, peaks of barite ($BaSO_4$), hematite (Fe_2O_3), calcium sulphate ($CaSO_4$), and montmorillonite were identified. These results are in accordance with the raw materials data (Table 1).

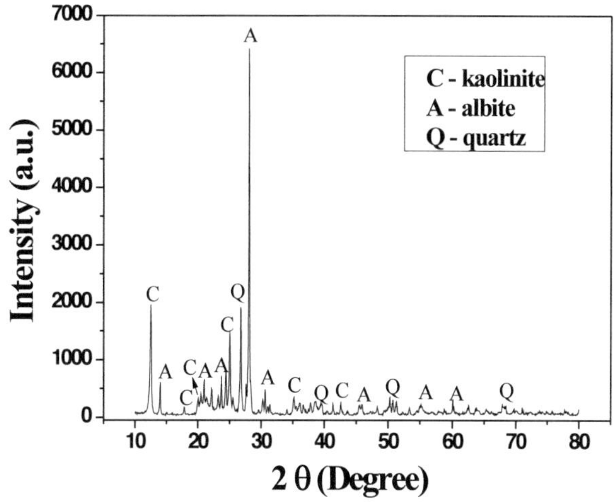

Figure 2. XRD pattern of the MT1 sample (waste-free formulation).

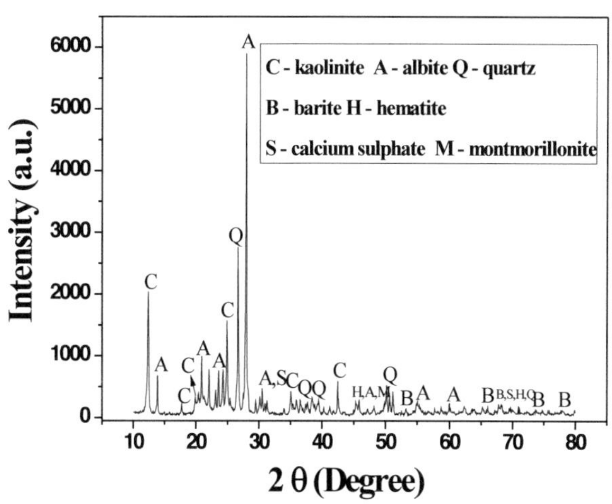

Figure 3. XRD pattern of the MT3 sample (with 5 wt.% petroleum waste).

The thermogravimetric curves (TGA and DrTGA) for the MT2 sample (with 2.5 wt.% petroleum waste) is shown in Fig. 4. DrTGa curve shows that the floor tile formulation exhibit one endothermic event around 573.5 ºC. This endothermic event is related mainly to the dehydroxylation of kaolinite. In addition, this endothermic reaction corresponds to an intense process of mass transfer in the tile sample as observed in the TG curve. The MT2 sample presented a total weight loss during sintering around 4.4 %. In addition the removal of chemically bound water of the kaolinite structure, the volatization of oil (hydrocarbons) of the petroleum waste should be also considered.

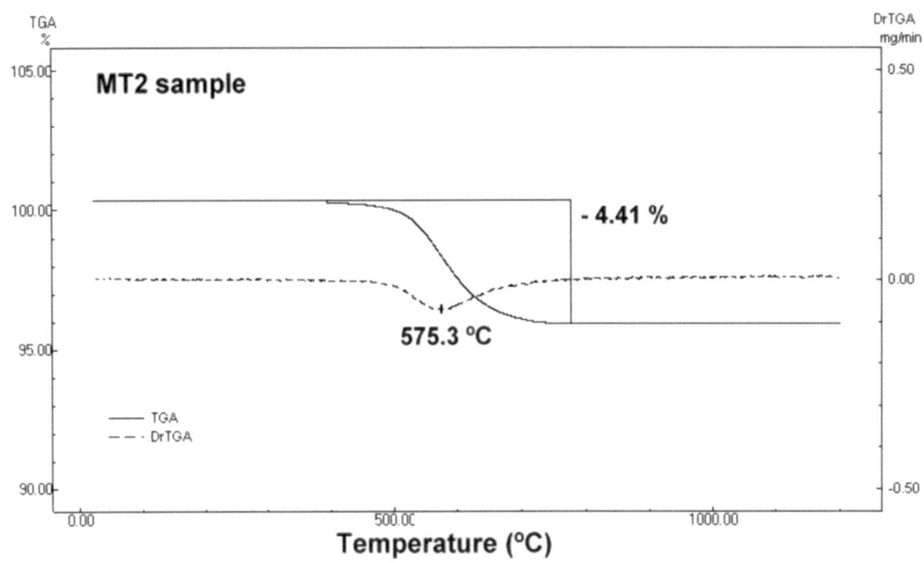

Figure 4. TGA/DrTGA curves of the MT2 sample.

Table 3 gives the particle size distribution of the raw materials used. The results revealed that the raw materials presented high concentration (83.2 – 96.8 %) of mineral particles < 63 μm. This means that the raw materials used in the floor tile formulations have good degree of comminution that could favor the reactivity of the particles during the sintering process. It also prevents the segregation of non-plastic and plastic components present in the tile formulations.

Particle size	Kaolin	Petroleum Waste	Albite	Quartz
< 2 μm	22.9	12	5	9.6
2 < x < 63 μm	73.9	83	89	73.6
63 < x < 200 μm	3.2	5	6	16.8

Table 3. Particle size distribution of the raw materials (wt.%).

3.2. Phase evolution during sintering

The XRD patterns of the MT1 samples sintered between 600 and 1250 ºC are presented in Fig. 5. Based on the X-ray diffraction patterns and previous studies [14-17] it is possible to describe the phase transformations underwent by the waste-free floor tile formulation (MT1 sample) at different temperatures.

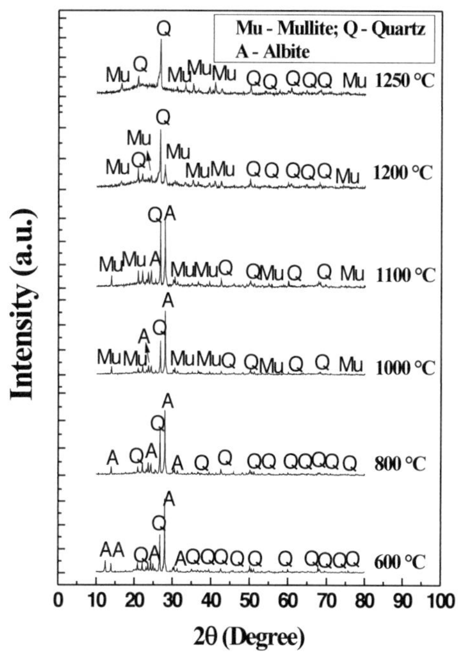

Figure 5. XRD patterns of the MT1 sample sintered at different temperatures.

The phase transformations involved in the waste-free floor tile body (MT1 sample) during sintering could be described as follow. The evolution of the physically adsorbed water by the mineral particles takes place until 100 ºC ($H_2O_{(l)} \rightarrow H_2O_{(g)}$). At 573 ºC, α-β quartz inversion of free silica occurs. At 600 ºC, the characteristic peaks of kaolinite have disappeared. In fact, between ~ 450 and 600 ºC, kaolinite loses the OH groups of the gibbsite sheet leading to the formation of amorphous metakaolinite according to:

$Al_2O_3.2SiO_2.2H_2O$ (kaolinite) $\rightarrow$ $Al_2O_3.2SiO_2$ (metakaolinite) + 2 $H_2O_{(g)}$

At 800 ºC, the peaks of quartz and albite are still seen. At 1000 ºC, peaks of mullite appear. In this temperature range, the silicate lattice totally collapse, followed by reorganization of the metakaolinite structure and the formation of amorphous silica. A spinel structure is formed and then quickly transformed to mullite according to:

$$2\,(Al_2O_3.2SiO_2)\ (\text{metakaolinite}) \rightarrow 2Al_2O_3.3SiO_2(\text{spinel}) + SiO_2(\text{amorphous})$$

$$2Al_2O_3.3SiO_2(\text{spinel}) \rightarrow 2(Al_2O_3.SiO_2)\ (\text{mullite}) + SiO_2$$

It can be seen that the mullite peaks increasing in intensity with the sintering temperature, but the quartz peaks decreasing slightly due its partial dissolution. Above 1100 ºC the albite peaks are not seen. At 1200 ºC, an amorphous band between 2θ = 15º and 2θ = 25º can be also observed. This is due to the fusion of albite to form a viscous liquid phase, which is then cooled to glass.

The XRD patterns of the MT3 formulation (with 5 wt.% petroleum waste) sintered between 600 and 1250 ºC are shown in Fig. 6. In addition the mullite and quartz, characteristic peaks of barite, hematite, and calcium sulphate were also identified. These results are in agreement with the chemical composition data (Table 1) and X-ray diffraction (Fig. 2). Thus, the partial replacement of kaolin with petroleum waste influenced the phase evolution of the vitrified floor tiles. This means that the petroleum waste addition can influence the sinterability of the floor tile formulations.

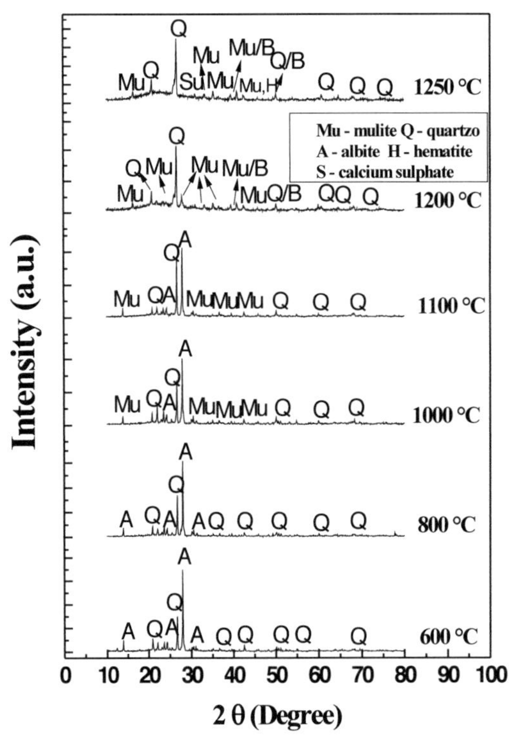

Figure 6. XRD patterns of the MT3 sample sintered at different temperatures.

3.3. Microstructural analysis of the sintered floor tiles

The microstructure of fractured surface of MT1 sample (waste-free sample) sintered al 1210 ºC obtained via SEI/SEM is shown in Fig. 7. One can clearly observe that the microstructure is composed mainly of dense zones (glassy phase) connected with rugous zones (open porosity). As indicated by the XRD analysis (Fig. 5), it consists of mullite, quartz, and glassy phase. In addition, the open pore volume is essentially formed by a narrow channels structure of irregular morphology. This means that the densification of the MT1 sample during sintering at 1210 ºC is incomplete.

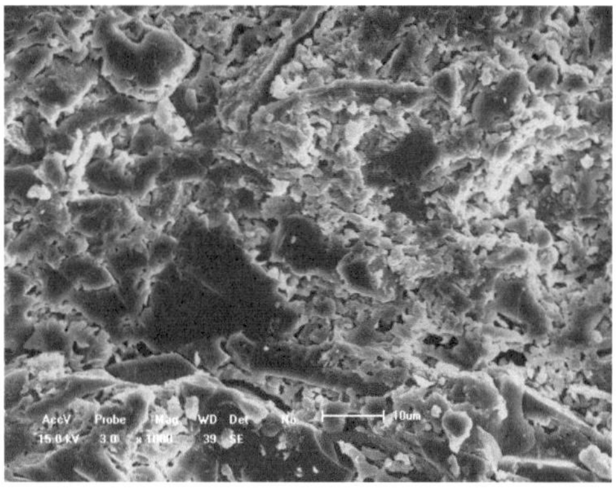

Figure 7. SEM micrograph of the MT1 sample after sintering at 1210 ºC.

Fig. 8 shows the fractured surface of the MT1 sample sintered at 1250 ºC. It also comprises kaolinite-derivate material (mullite), quartz, and glassy phase. However, the presence of few rounded and isolated pores indicates the consistent development of the liquid phase during sintering. As a consequence, the overall microstructure of the MT1 sintered at 1250 ºC is clearly denser.

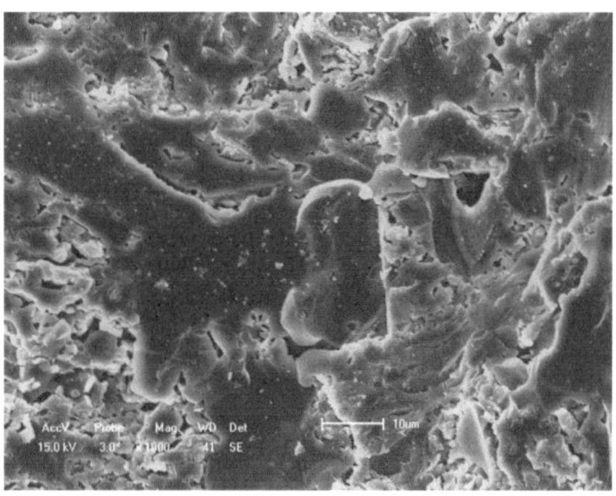

Figure 8. SEM micrograph of the MT1 sample after sintering at 1250 ºC.

SEM micrographs of the fractured surfaces of the MT3 sample are presented in Fig. 9a-c. The micrographs show the evolution of the microstructure of the MT3 sample as temperature increases. It may be noted that at 1210 ºC (Fig. 9a) and 1230 ºC (Fig. 9b), the microstructure is characterized by few nearly spherical isolated pores associated with high degree of vitrification. At 1250 ºC (Fig. 9c), however, a more porous fractured surface can be observed. Larger and more irregular pores can be seen. This effect suggests a gas evolution (gas trapped) for higher petroleum waste additions. Thus, the incorporation of petroleum waste into floor tile formulation brought about a relevant variation in the sintered microstructure.

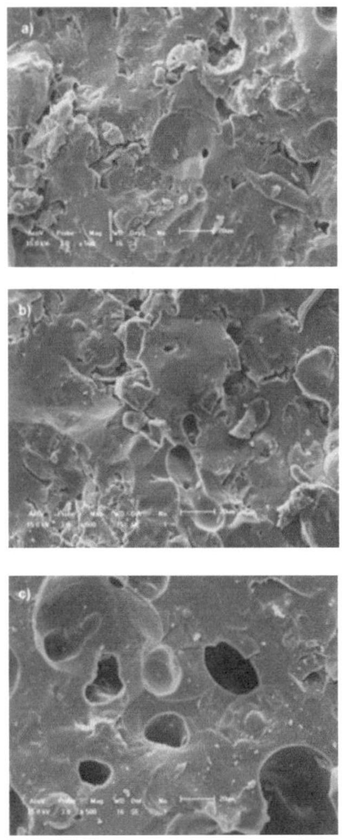

Figure 9. SEM micrographs of the MT3 sample after sintering: a) 1210 ºC; b) 1230 ºC; and c) 1250 ºC.

3.4. Sintering behavior

In the manufacturing process of floor tiles the sintering step is critical. On sintering, the floor tile materials undergo a series of physical and chemical reactions involving dehydroxylation of clay minerals, partial melting of feldspar-quartz eutectic compositions, collapse of the silicate structure to formation of mullite and progressive dissolution of quartz in the liquid phase. These processes play an important role in the sintering behavior of floor tile compositions.

The dilatometric curve of the MT1 sample is shown in Fig. 10. As it can be seen, the sintering behavior of the floor tile formulation is characterized by three main regions. The first region (~ up to 895 ºC) is characterized by a negligible shrinkage. The second region within the ~ 900 and 1100 ºC temperature range is characterized by a small shrinkage. In this temperature range, the reorganization of the metakaolinite structure and initial formation of the liquid phase occur. In addition, the vitrification of the tile piece is already in progress. The third region (above 1150 ºC) is characterized by a considerable shrinkage of the floor tile pieces. It is plausible to consider that an abundant amount of liquid phase was formed, resulting in high densification. This is consistent with the dense microstructure of the MT1 samples sintered at 1250 ºC.

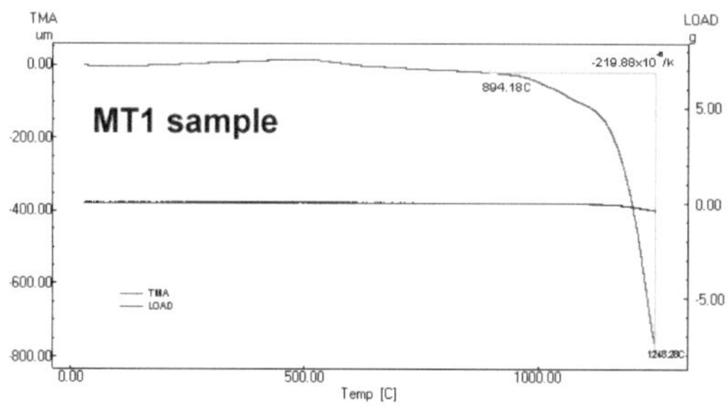

Figure 10. Dilatometric curve of the MT1 sample.

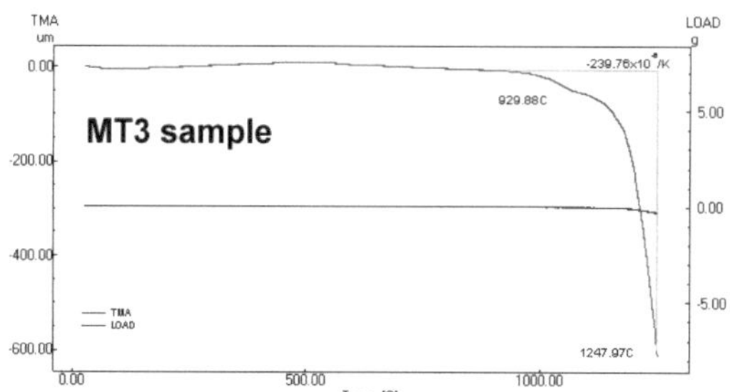

Figure 11. Dilatometric curve of the MT3 sample.

The dilatometric curve of the MT3 sample is shown in Fig. 11. The sintering behavior of the MT3 sample compared with that of the MI1 sample shows small but important differences. It seems that the petroleum waste tends to retard the densification of the tile pieces during sintering. This could be related to the complex composition of the petroleum waste. In fact, the petroleum waste used is composed of several mineral phases and oil (hydrocarbons).

The linear shrinkage of the floor tile pieces is shown in Fig. 12. The linear shrinkage indicates the degree of densification during sintering, and is a physical property very important for dimensional control of the finished tile products.

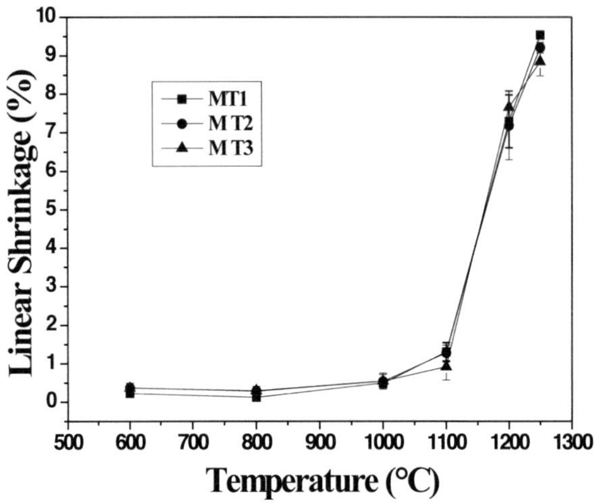

Figure 12. Linear shrinkage of the floor tile samples after sintering.

It is observed that the linear shrinkage presents different behaviors, depending on the sintering temperature range. This is consistent with the dilatometric curves (Figs. 10 and 11). This means that the predominance of distinct sintering mechanisms in their temperature range can occur. In the 600 and 950 ºC range, a small linear shrinkage (< 1%) was observed. This low linear shrinkage suggests that the dominant sintering mechanism is surface diffusion [18]. In this temperature range prevails mainly the growth of necks between the mineral particles coupled with significant specific surface area reduction. Moreover, the structural reorganization of matakaolinite and initial melting of quartz-feldspar eutectics are already underway. Between 1000 and 1100 ºC, a higher linear shrinkage can be observed. The formation of mullite occurs. The formation of a larger amount of liquid phase also occurs. Above, 1100 ºC, however, the linear shrinkage accelerates resulting in high densification of the tile pieces. In this case, the sintering is accompanied by the formation of a significant amount of liquid phase. This liquid phase acts to densify the structure by liquid phase sintering. Viscous flow closing open porosity is the dominant sintering mechanism [19]. It can also be seen in Fig. 12 that the shrinkage tends to be lightly lower for the samples containing petroleum waste. Appearance of the floor tile pieces after sintering is presented in Fig. 13.

The apparent density of the floor tile pieces is shown in Fig. 14. The results show that, in general, the density tends to lightly decrease with petroleum waste addition. Density presents only a small variation up to ~ 1100 ºC. This occurred due to the combined inverse effects of sintering and weight loss. As a matter of fact, the green tile pieces when heating up to ~1000 ºC underwent weight loss, as shown in Fig. 4. Above 1100 ºC, a substantial increase in density is observed. This behavior is in line with the formation of more abundant liquid phase during sintering that fills the open pores. This is consistent with the microstructure (Figs. 8 and 9) and linear shrinkage (Fig. 12).

http://dx.doi.org/10.5772/53256

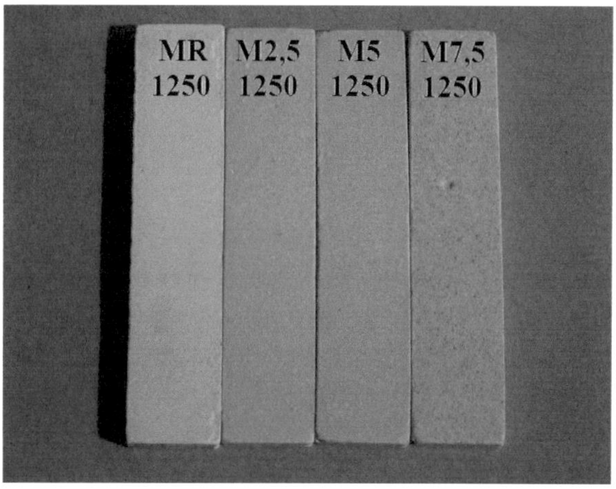

Figure 13. Appearance of the floor tile pieces after sintering.

The water absorption of the tile pieces is shown in Fig. 15. This property is related to the microstructure, and also determines the open porosity level of the pieces. For sintering temperatures up to ~ 1100 ºC, the water absorption remains practically constant. Above 1100 ºC, however, a strong decrease in water absorption is measured. This means that the sintering accelerates above 1100 ºC and causes densification. The results also showed that the addition of petroleum waste lightly increased the water absorption, except on temperatures above 1100 ºC. The variation of the apparent porosity (Fig. 16) is very similar to that observed for the water absorption.

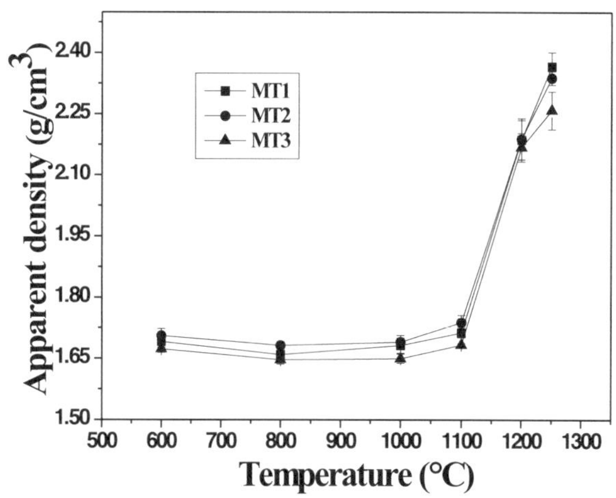

Figure 14. Apparent density of the floor tile samples after sintering.

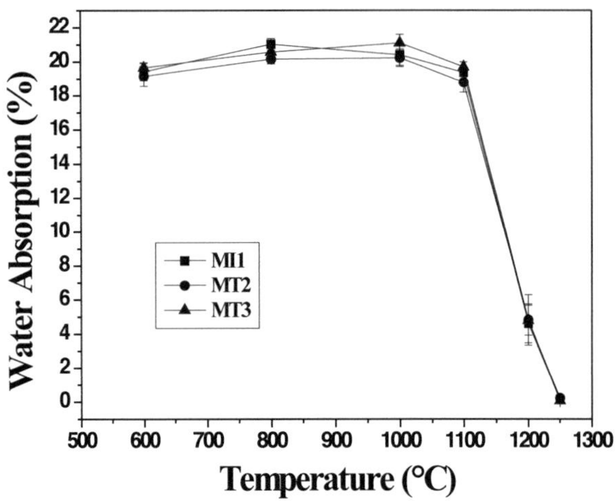

Figure 15. Water absorption of the floor tile samples after sintering.

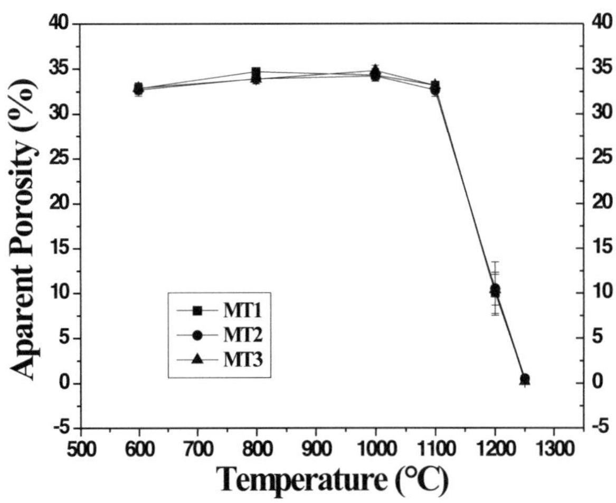

Figure 16. Apparent porosity of the floor tile samples after sintering.

4. Conclusion

In this chapter the sintering behavior of vitrified floor tiles bearing petroleum waste was investigated. It was found that the petroleum waste added influenced the chemical and mineralogical compositions of the floor tile formulations. It was also found that the sintering behavior was influenced by the petroleum waste and sintering temperature. Distinct sintering regions were observed. For the region above 1100 ºC, with predominance of the viscous flow sintering mechanism, important changes of the physical properties and sintered microstructure occurred. XRD analysis confirmed the mineralogical changes during sintering. Moreover, it was also observed that the petroleum waste tends retard the densification process of the floor tile pieces. It implies, therefore, that additions of very high petroleum waste amounts in the floor tile formulation should be avoided, because it affects the densification and technical properties of sintered floor tiles.

Acknowledgements

The authors acknowledge the FAPERJ and CNPq for supporting this work.

Author details

A.J. Souza, B.C.A. Pinheiro and J.N.F. Holanda

Northern Fluminense State University - UENF, Laboratory of Advanced Materials – LAMAV, Group of Ceramic Materials - GMCEr, Campos dos Goytacazes, RJ, Brazil

References

[1] Curran LM. Waste minimization practices in the petroleum refining industry. Journal of Hazardous Materials 1992; 29: 189-97.

[2] Amaral SP, Domingues GH. Application of oily sludge for manufacturing of ceramic materials. Proceedings of the 4th Brazilian Congress on Petroleum, Rio de Janeiro, Brazil, 1990.

[3] Li X, Lv Y, Ma B, Jian S, Tan H. Influence of sintering temperature on the characteristics of shale brick containing oil well-derived drilling waste. Environmental Science Pollution Research 2011; 18: 1617-22.

[4] Souza GP, Santos RS, Holanda JNF. Recycling of a petroleum waste in ceramic bodies. Materials Science Forum 2003; 416-418: 743-7.

[5] Segadâes AM. Use of phase diagrams to guide ceramic production from wastes. Advances in Applied Ceramics 2006; 105:46-54.

[6] Chiang K, Chien K, Hwang S. Study on the characteristics of building bricks produced from reservoir sediment. Journal of Hazardous Materials 2008; 159: 499-504.

[7] Souza AJ, Pinheiro BCA, Holanda JNF. Recycling of gneiss rock waste in the manufacturing of vitrified floor tiles. Journal of Environmental Management 2010; 91: 685-89.

[8] Sengupta P, Saikia N, Borthakur PC. Bricks from petroleum effluent treatment plant sludge: properties and environmental characteristics. Journal of Environmental Engineering 2002; 128: 1090-94.

[9] Acchar W, Rulf BM, Segadães AM. Effect of the incorporation of a spent catalyst reject from the petroleum industry in clay products. Applied Clay Science 2009; 42: 657-60.

[10] El-Mahllawy MS, Osman TA. Influence of oil well drilling waste on the engineering characteristics of clay bricks. Journal of American Science 2010; 6(7): 48-54.

[11] Souza AJ, Pinheiro BCA, Holanda JNF. Valorization of solid petroleum waste as a potential raw material for clay-based ceramics. Waste and Biomass Valorization 2011; 2:381-88.

[12] Manfredini T, Pellacani GC, Romagnali M. Porcelain stoneware tiles. Americam Ceramic Society Bukketin 1995 ; 74 : 76-79.

[13] ASTM, ASTM C373 – Test method for water abosrption, bukk density, apparent porosity, and apparent specific gravity of fired whiteware products, 1994.

[14] Brindley GW, Nakahira N. The kaolinite-mullite reactions:II, metakaolin. Journal of the American Ceramic Society 1959; 42: 314-18.

[15] Brindley GW, Nakahira N. The kaolinite-mullite reactions:III, hgh temperature phases. Journal of the American Ceramic Society 1959; 42: 319-23.

[16] Barba A, Béltran V, Fíleu C, García J, Ginés F, Sanchez E, Sanz V. Materias Primas Para la Fabricación de Soportes de BaldosasCerámicas. 2nd Ed. Castellón: ITC; 2002.

[17] Sousa SJG, Holanda JNF. Thermal transformations of red wall tiles. Journal of Thermal Analysis and Calorimetry 2007; 87: 423-8.

[18] Kingery WD, Berg M. Study of the initial stages of sintering solids by viscous flow, evaporation-condensation, and self-diffusion. Journal of Applied Physics 1955; 26 (10) 1205-12.

[19] Kingery WD, Bowen HK, Uhlmann DR. Introduction to Ceramics. 2dn ed. New York: Willey; 1976.

Chapter 5

Structural and Dielectric Properties of Glass – Ceramic Substrate with Varied Sintering Temperatures

Rosidah Alias

Additional information is available at the end of the chapter

http://dx.doi.org/10.5772/54037

1. Introduction

The main requirements of microwave circuit technology for wireless applications are a combination of high performance circuit, high integration density and frequency stability with a cost effective price [1-2]. Ceramic technology is particularly well suited to these requirements due to the superior performance of their electrical, electromechanical, dielectric and thermal properties in order to satisfy the many complex tradeoffs involved in electronic devices to meet many of today's and future needs [3]. However, as the demand for revolutionary changes in mobile phone and other communication systems using microwave as a carrier is increasing, the usage of ceramics may have reached at a certain limit. Thus, continuous research on this ceramic technology is urgently required in order to get an advanced improvement and innovation in the material properties in conjunction with the rapid change of required device characteristics. Although we may sometimes experience miserable insufficiency of accumulated knowledge in regard to basic research, it is believed that these research activities can contribute immeasurably to give scientific and technological understanding which is very important especially for the new engineer and researcher.

'Ceramic' is derived from the Greek word 'keramos' which means potters clay or pottery [4]. This material is usually used as ceramic-ware such as floor tiles. However the demand for ceramics is becoming more important in helping electronic devices to attain better performance for medical, global communication, military and other uses. Their introduction as engineering components in recent years has been based upon considerable scientific effort and has revolutionized engineering design practice. The development of engineering ceramics has been stimulated by the drive towards higher, more energy-efficient, process temperatures and foreseeable shortages of strategic minerals. Since 1750 ceramics materials have been used for the chemical industry, electrical insulation and especially for being heat resist-

INTECH
open science | open minds

© 2013 The Author(s). Licensee InTech. This chapter is distributed under the terms of the Creative Commons Attribution License http://creativecommons.org/licenses/by/3.0), which permits unrestricted use, distribution, and reproduction in any medium, provided the original work is properly cited.

ant to withstand the heating effects of high frequency electromagnetic fields. The last-mentioned property is crucial for radio and television broadcasting. The electro ceramics such as magnetic ceramics (Ferrites) and electrochemical ceramics (piezoelectric ceramics) were subsequently developed [5]. Presently, the ceramic packaging and multilayer substrate process have received much worldwide attention in response to an increasing demand for circuit miniaturization and higher performance device have led to the development of the Low Temperature Co-fired Ceramic (LTCC) technology.

1.1. An overview of low temperature co-fired ceramic (LTCC) technology

LTCC is a multilayer substrate technology for device integration. This technology has been growing continuously since the appearance of the first commercial cofired ceramic product for robust capacitors in the early 60's [6]. In the standard LTCC technology ceramic green tapes are processed by punching and screen printed to form vertical interconnect and planar conductor patterns, laminated and finally fired at 850 ºC to form a highly integrated substrate. The low sintering temperature provided by the LTCC technology is the key factor enabling its advantageous utilization for today's packaging concepts in microwave modules [2].The main motivation for the use of the LTCC technology is the possibility of fabricating three dimensional circuits using multiple ceramic layers allowing more complex design circuits and device structures. A simple diagram of an LTCC substrate is presented in Figure 1. Through this technology a highly integrated package which contains embedded passive parts which increase the level of circuit function can be realized [7-11]. Besides, LTCC modules or packages provide a hermetic multilayer interconnection structure with very uniform and stable frequency and temperature dependent properties. Furthermore they also allow high density of lines throughout the part, are able to construct various geometries of interconnects by layer cut outs and have good heat transfer ability [12]. In contrast to the conventional ceramic technology, this technology which uses the ceramic material as the main ingredient is growing continuously especially for industrial and telecommunication area due to the low investment and short process development in addition to the flexibility of the technology including obtaining interesting properties of the ceramic material itself by controlled processing methodology [13-14]. Moreover, high fired density with repeatable shrinkage and frequency characteristics are necessary for high performance and low cost modules.

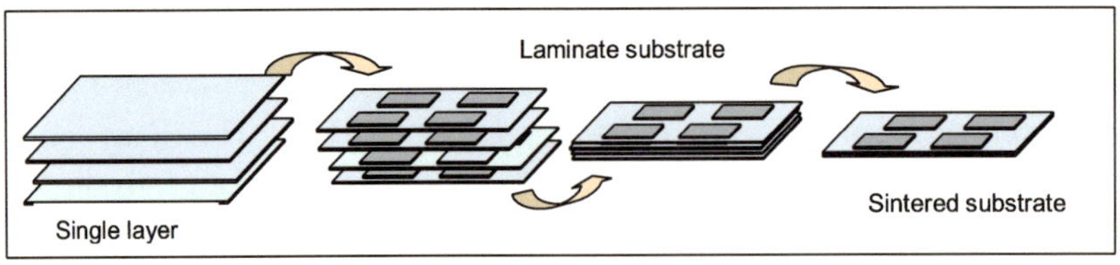

Figure 1. Multilayer ceramic substrates

LTCC is a system consisting of a glass ceramic dielectric composition which can be described as a polycrystalline material formed by controlled crystallization of glass [15-17]. This composite material consists of multi-components inside the multilayer structure involving more than one component, more than a single phase and complex morphology. The preparation of this composite system is not an easy task because its properties are known to depend significantly on the chemical and physical properties of the starting raw materials [17], chemical composition, chemical purity, homogeneity of particle size distribution and the nature of grains and pores (size, shape, orientation) in the conductor surface and the glass-ceramic composite. Consequently, accurate characterization of the starting powders is essential to achieve high quality and reproducible production of current materials and to development of new materials with optimized or designed properties [18]. Furthermore, the variation of sintering temperatures and chemical composition of the composite also strongly changes the properties of the substrate, so in order to obtain a high quality product, a special working knowledge of ceramic science is needed for a judicious and scientifically reasoned choice of starting constituents that could yield desired properties for required applications.

A multilayer component or microelectronic device requires a highly conductive metal to be used for internal circuitry pattern to obtain a high quality factor and should have a low melting point to co-fire together with the substrate [19]. Conductors are the carriers of electrical current and substrate are the carriers of electromagnetic waves that are related to the current being carried in the conductors. The conductors also must be match with other components in the LTCC substrate in terms of shrinkage and thermal coefficient expansion. So the right selection of the conductor is necessary because it will determine the overall system cost and will impact on the product reliability. A silver conductor is the most commonly used for this purpose due to its low resistivity, low cost, low melting point and being compatible with the high performance substrate, thus providing good electrical properties and good adhesion to the substrate [20-21].

The studies on glass-ceramic systems related to the dielectric and structural properties of $CaO-B_2O_3-SiO_2$ (CBS) for the LTCC technology have been previously reported by some researchers [22-24]. In 1992, the crystallization behavior of a non-alkali glass system was investigated by Ota and his-coworkers, 1992 [25]. They concluded that with an addition of wollastonite ($CaOSiO_2$) powder, the crystallization of the glass can be controlled. Jean et al., 1995 [26] reported a systematic study on the densification kinetic mechanism of glass-composite. They found that the addition of cordierite glass (CG) to the low binary glass (Borosilicate glass-BSG) significantly slowed down the densification kinetics but increased the activation energy of densification. It could be attributed to the strong coupling reaction which takes place between BSG and CG during sintering. More recently Wang et al., (2011) [27] investigated the structure evolution process for CSB glass ceramic powder for LTCC application. The prepared powder exhibits an amorphous network structure with steady Si–O and B–O bonds after heat treatment at 600 °C, in which Ca^{2+} is distributed. They also found that the main phases in the CBS glass ceramic are $CaSiO_3$ and CaB_2O_4, which made the as-sintered CBS glass ceramic to have excellent dielectric properties. Presently research on the CBS system has been carried out by Zhou et al., 2012 [28] on the effect of $La_2O_3–B_2O_3$ on the flexural strength and microwave dielectric properties of the low temperature co-fired $CaO–B_2O_3–SiO_2$ glass–ceramic. They found that LB addition promotes the crystallization of the

$CaSiO_3$ but a highly amount of the LB addition leads to the formation of more pores, which is due to the liquid phase formed during the sintering process.

The effect of sintering temperature variations on microstructure changes of the $CaO-B_2O_3-SiO_2$ glass-ceramic system has been carried out before [29]. In this chapter, an extension of some previous studies has been made to investigate the effect of sintering temperature on the structural properties and the dielectric properties in the range of 10 KHz to 10 MHz at room temperature. The work was constructed as follows: A low temperature co-fired ceramic substrate was prepared by laminating eight layers of ceramic-glass tape to study the relation of different sintering temperatures, 800ºC, 825ºC, 850 ºC, 875 ºC and 900ºC, on density and shrinkage of the whole substrate. The evolution of microstructural changes of the substrate was also studied and observed by using scanning electron microscopy. The density and the shrinkage of the substrate did not show any significant trend with increased sintering temperature. The main aim of this work is to analyze the changes of permittivity, dissipation factor and structural performance with varied sintering temperatures of laminated substrates. Thus these perhaps would provide good information for making high quality substrate materials comparable with those available in the market.

1.2. Sintering

Sintering forms the key stage of the fabrication route in which powder or sample is heated to a certain temperature and converted to the dense body by removal of pores between the starting particles (accompanied by shrinkage of the component) [30]. It involves heat treatment of powder compacts at elevated temperatures, where diffusional mass transport is appreciable which results in a dense polycrystalline solid [31]. The criteria that should be met before sintering can occur are (i) a mechanism for material transport must be present (ii) a source of energy to activate and sustain this material transport must be present. The primary mechanisms for transport are diffusion and viscous flow. Heat is the primary source of energy, with energy gradients due to particle contact and surface tension. A ceramist, physicist, chemist and metallurgist have to work together to produce a material with particular properties, must identify the required microstructure and must then design processing conditions that will produce this required microstructure. The objective of sintering studies is therefore normally to identify and understand how the processing variables such as temperature, particle size, applied pressure, particle packing, composition and sintering atmosphere influence the microstructure that is produced.

Thermodynamically, sintering is an irreversible process in which during this process the constituent atoms redistribute themselves in such a way as to minimize the free energy of the system. It involves consolidation of the powder compact by diffusion on an atomic scale resulting in reduction of surface area by grain boundary formation, neck growth between particles and densification of the system (Figure 2) [32-33]. The principal goal of sintering is the reduction of compact porosity. The effect of sintering temperature variations on the physical properties and microstructure of the finished product has always been a subject of great importance in ceramic literature since several decades ago [34-37]. The development of microstructure and densification during sintering is a direct consequence of mass transport through several possible paths and one of these paths is usually predominant at any given stage of sintering [38]. They are (i) evaporation/condensation (ii) solution/precipitation (iii) lattice diffusion (iv) bulk diffusion and (v) surface diffusion or grain boundary diffusion with neck surface diffusion and any combination of these mechanisms. Through the different diffusion mechanisms, matter moves from the particles into the void spaces between the particles, causing densification and resulting in shrinkage of the part and thus the microstructural developments will occur [39-40]. Several variables influence the rate of sintering. Some of them are initial density, material, particle size, sintering atmosphere, temperature, time and heating rate. The sintering phenomena are of two types: 1) Solid-state sintering, where all densification is achieved through changes in particle shape, without particle rearrangement and 2) liquid-phase sintering, where some liquid that is present at sintering temperatures aids compaction.

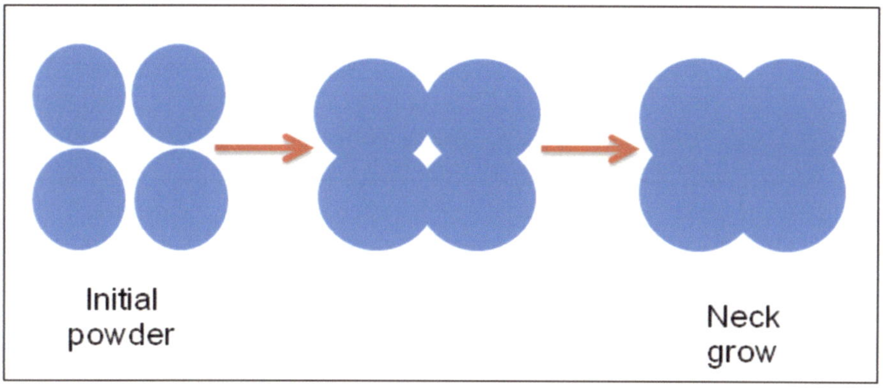

Figure 2. Diffusion process.

1.2.1. Sintering of LTCC substrate

In an LTCC technology, the sintering process of a glass-ceramic substrate is one of the most important steps from the green laminate to the final product. In this step, a ceramic and a metal conductor would densify simultaneously at the same firing profile and at similar densification and shrinkage rate. Any mismatch between the sintering shrinkage of the electrode and the dielectric substrate induces stress which later will affects the device performance [41]. During this process particles are bonded together by heating to form sturdy solid bodies [42]. A good sintering profile stage for a dense and crystallized glass-ceramic should consider several factors i.e. 1) the transition temperature of glass, T_g, which determines the starting sintering temperature. For all ceramic materials including LTCC substrates, it is important that no sintering occur before the complete binder burn-out. The binder burn-out stages is complex and historically least understood [43]. This is because mass/heat transport, chemical kinetics and binder distribution interact with each other to determine the binder burn-out characteristics. The volatiles are produced at a rate which depends on the amount of heat available. The binder distribution is also influenced by the thickness which means that the organics near the surface remove quicker than the inner organics. The binder decomposition temperature is usually in the range of 450 ºC - 750 ºC and recommended for a firing temperature of 850-950 ºC. A melting point below 450 ºC usually causes the organic material to be encapsulated resulting in blister forming in the dielectric layer as the organic decomposes [44]. The glass transition temperature determined by the composition of the glassy phase and is not affected by the presence of the crystal [45]. 2) Crystallization temperature (T_p) which determines the temperature that the crystal phase starts to grow. When the crystallization occurs the densification process rates is decreased. The crystallization of a glass-ceramic system usually occurs below 950 ºC. 3) Sintering rate; glass-ceramic substrates can be fully sintered before densification which depends on the sintering rate. Factor 1) and 2) contribute significantly in the temperature region for efficient densification.

According to Imanaka (2005) [46], the optimum firing temperature for an LTCC material is very important to achieve better performance of the LTCC module. So, the co-firing process of an LTCC substrate should be carefully controlled and monitored. The binder burn-out stage depends on the part size, the number of layers and the amount of metallization and should be considered as a starting point for the co-firing process of LTCC materials. At this stage the binder in the laminates must be completely removed after binder burn-out to make sure the joining of the layers are in good condition. The driving force for the joining of the tape is generally attributed to the surface tension of particles that exists during the burn-out process. The mechanism of binder removal transport are limited to the viscous flow, evaporation, condensation, volume diffusion and surface diffusion of binder phase which occurs at the interphase of the porous microstructure of the green tape and these causes capillary forces to arise [47]. These forces result in an approach of the ceramic layers towards to each other. Ceramic particles close enough after binder removal between ceramic layers, the particles at the edge can rearrange and move so that the ceramic layers interpenetrate each other and make a homogenous junction due to capillary force resulting in a laminate free of defects [48].

1.3. An overview of dielectric properties

The ceramic material provides the dielectric properties where the relative permittivity and dissipation factor of LTCC dielectrics become more important as frequency increases. It has been found that this parameter can be varied, therefore their values have to be determined before the design of any module or circuit can be started. In the microelectronics industry, the development of high speed interconnects for semiconductor devices demands new packaging-style technology to meet high-speed signal processing with high density and high frequency pulse requirements for multi-chip modules. To improve the performance of high speed systems, not only must high speed IC devices be exploited but also the signal delay in the substrate has to be reduced because the propagation delay time greatly depends upon the dielectric constant of the substrate [49-51]. The signal propagation delay time is given by [52]:

http://dx.doi.org/10.5772/54037

$$T = \frac{l\,x\sqrt{\varepsilon_r}}{C} \tag{1}$$

Where:

T_d: time delay

l: signal transmission length

ε_r dielectric constant of the substrate

c : speed of light

There are two ways to reduce the delay time by shortening the transmission length by:

1. forming fine patterns and many multilayers
2. lowering the dielectric constant

The relative permittivity values of commercial LTCCs which are usually measured at low frequency are in the range 3-10 as shown in Figure 3.

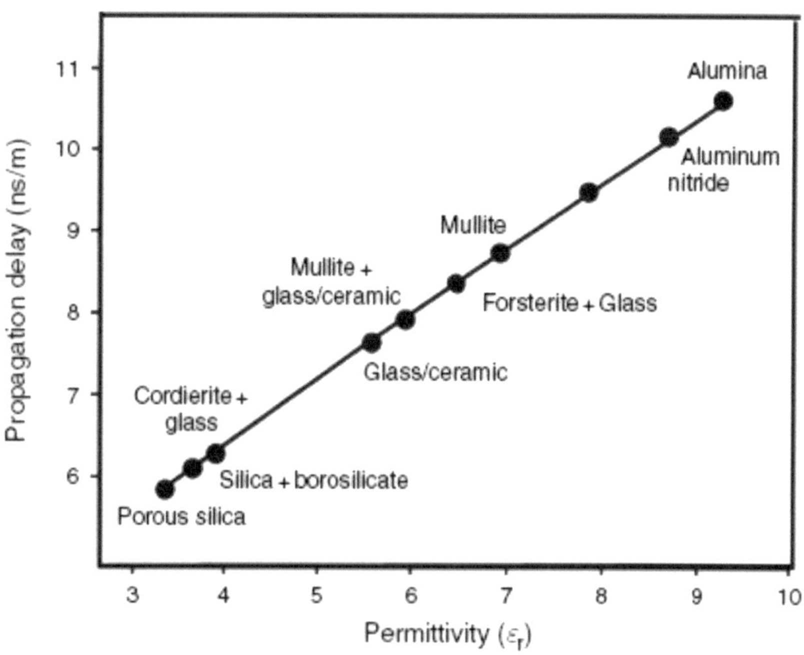

Figure 3. Variation of propagation delay as a function of permittivity of ceramic materials [50]

According to Wakino et al., (1987) [53], the permittivity is especially important because the length of a resonator is inversely proportional to the square root of the permittivity. Thus the footprint of RF components operating at low frequencies can increase to an inconvenient size if the ε_r is low. On the other hand, materials with too large a permittivity can also cause problems because they need narrow and accurate lines for proper impedance matching. In a later work by Sasaki and Shimada, 1992 and Kellerman et al., 1998, they noted that the low dielectric is necessary to shorten the delay per unit length and the total wiring length in order to reduce propagation delay for signal transmission speed with minimum attenuation [54,55]. Besides, the dielectric constant is also important to control the impedance of microwave and RF circuits to within a few percent of a nominal value so as to minimize energy reflections at interfaces between the circuit tracks and circuit components. In the RF world, circuit components, as well as input and output impedance, are generally 50Ω. Therefore circuit track impedance should be 50Ω as well [16]. The recent progress in microwave telecommunication, satellite broadcasting and intelligent transport system has resulted in an increasing demand for a dielectric resonator with low loss pucks which are used mainly in wireless communication devices [56]. NIL (Figure 3).

Besides appropriate dielectric constant, low dielectric loss is also an important property of the substrate material for microelectronic packaging with high propagation speed [57-58]. The dielectric constant (permittivity) describes the material behavior in the electric field and consists of a real part, ε', called the dielectric constant, and an imaginary part, ε'', called the dielectric loss factor. Thus, the permittivity is expressed as;

$$e^* = e' - je'' \tag{2}$$

where the dielectric constant represents the ability of a material to store electric energy and the dielectric loss factor describes the loss of electric field energy in the material. At microwave frequencies, ionic and electronic polarization mechanisms contribute predominatly to the net dipole moments and the permittivity as depicted in Figure 4 which basically have four basic mechanisms contributing to polarization [16]: 1) Electronic polarization, 2) Ionic/atomic polarization, 3) Molecular/dipole/orientation polarization and 4) Space charge polarization.

A low dissipation factor, DF is also required for low electrical losses in dielectric materials particularly at high frequency [59]. Dissipation factor or loss tangent (tan δ) is the ratio of the dielectric constant's imaginary part to the real part, given by:

$$Tan\,\delta = \frac{\varepsilon''_r}{\varepsilon'_r} \tag{3}$$

The loss tangent or loss angle may be used to determine how lossy a medium is. A medium is said to be a good (lossless or perfect) dielectric if tan δ is very small ($\sigma \ll \omega\varepsilon'$) or a good conductor if tan δ is very large ($\sigma \gg \omega\varepsilon'$). The characteristic behavior of a medium also depends on the operational frequency ($f = \omega/2\pi$). A medium that is regarded as a good conductor at low frequency may be a good dielectric at high frequencies. The simplest way to define dissipation factor (loss tangent) is the ratio of the energy dissipated to the energy stored in the dielectric material. The more energy that is dissipated into the material the less is going to make it to the final destination. This dissipated energy typically turns into heat or is radiated as RF (Radio Frequencies) into the air. The optimal goal is to have 100% of the signal pass through the interconnection network, and not be absorbed in the dielectric. With "high power" signals, a material with a large dissipation factor could result in the development of a tremendous amount of heat, possibly culminating in a fire (advanced dielectric heating). When the signals are very weak a high loss material means that little or no signal is left at the end of the transmission path. In order to retain maximum signal power, a low loss material should be used [61].

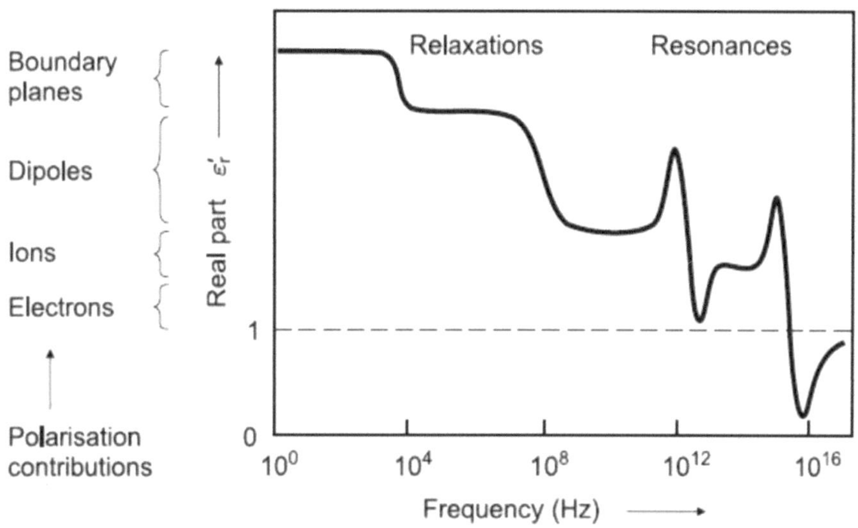

Figure 4. Polarization mechanisms [32].

As well known, the dielectric loss tangent of materials denotes quantitatively dissipation of the electrical energy to the different physical process such as electrical conduction, dielectric relaxation and loss from linear processes. The total dielectric loss is the sum of intrinsic and extrinsic loss. Intrinsic dielectric loss depends on the crystal structure and can be described by the interaction of the phonon system with the AC electric field. The ac electric field alters the equilibrium of the phonon system and the subsequent relaxation associated with energy dissipation. The extrinsic losses are associated with the imperfections in the crystal lattice such as impurities, microstructural defects, grain boundary, porosity, microcrack, random crystallite orientation, dislocations, vacancies, dopant etc. The extrinsic losses are caused by lattice defects and therefore can be in principle eliminated or reduce to the minimum by proper material processing [62].

2. Experimental procedure

2.1. Multilayer substrate process

The multilayer process was carried out using the LTCC process flow as shown in Figure 5. It started with cutting the Ferro A6S tape for the required dimension (204 mm x 204 mm) using a die cutting machine ATOM SE 20C to suit with the printing table dimension. Via hole filling with CN 33-407 was carried out using the manual filling process before printing the pattern onto the substrate using the suggested printing parameter with CN 33-391 for the surface and inner conductor on a KEKO P-200Avf Screen Printer Machine. The printed pattern and via hole filling were dried in an oven at 70 ºC for 10 minutes to avoid the printed pattern in one layer from adhering to each other during the stacking process. The printed pattern was then stacked using a manual stacker plate as shown in Figure 6. A simple and low cost mechanical fixture with alignment pins can be used to accomplish the stacking process layer by layer before insertion into an aluminum bag and vacuumed to remove the air in order to avoid the presence of warpage and delamination problems.

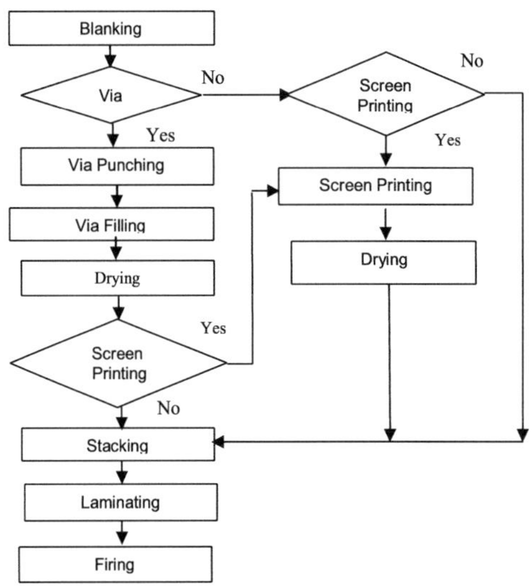

Figure 5. LTCC Multilayer Fabrication Process Flow.

Then the lamination process was performed by using an isostatic laminator system under pressure and temperature of 21 MPa and 70 ºC respectively to ensure that the layers of the stack were well adhered to each other to form a compact multilayer substrate. This process used heated water which acted as a medium to uniformly distribute the force of lamination. It minimized the delamination and gave more uniform shrinkage which improved the yield of complex parts. Finally, the laminated substrate was placed on the setter material in a Naberthem tube furnace and fired using a sintering profile as suggested by the tape manufacturer [63]. The setter material must be properly selected to suit the LTCC material set and fabrication process [64]. This sintering stage for cofiring the materials is very important in the process methodology and can affect the microstructure and properties of ceramics. Through the sintering process, a transformation from the original porous compact to a dense ceramic takes place.In this process a specific temperature time profile which depends of the glass-ceramic mixture has to be observed during the firing process. A typical practice for the sintering cycle is to heat up the furnace from room temperature to 450 ºC with a gradient of about 2 ºC/min to remove the organic solvent and 6-8 ºC/min from 450 ºC to 850 ºC at which the sintering process of the composite material starts. The temperature remains constant for 10 minutes to finish the sintering. A controlled cooling of the furnace which takes at least three to eight hours (depending on material; thick or large size) causes the need to modify the firing profile. The process was repeated for sintering at 800 ºC, 825ºC, 850 ºC, 875 ºC and 900ºC using the same profile as shown in Figure 7.

Figure 6. The manual stacker for collating and stacking process.

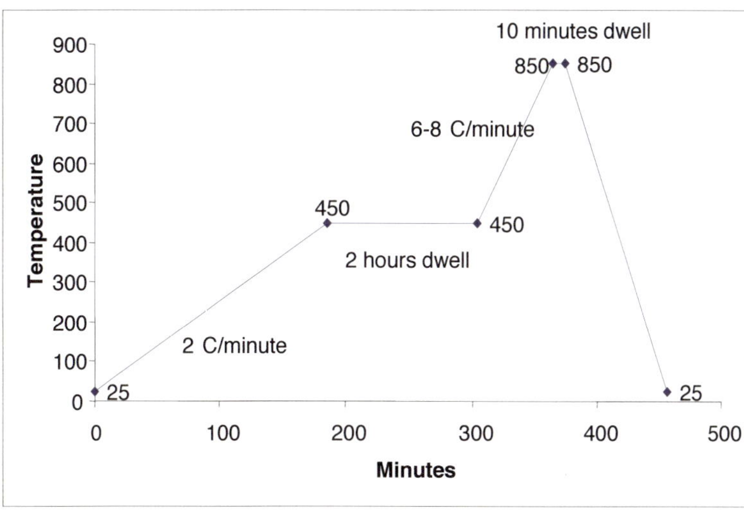

Figure 7. Firing profile for multilayer structure [63]

2.2. Sample characterizations

The density of the samples was measured using an electronic densitometer. The microstructure of the top surface of the ceramic dielectric tape was observed by FEI-Nova NanoSEM Field-Emission Scan (FE-SEM). Most of the samples were imaged several times, with at least three pictures in each case, from different areas of the sample holder. The average grain size was calculated using the line intercept method. The EDX spectrum was used to identify elements within a sample.

The shrinkage along the compaction direction and the diametrical shrinkage were measured from the geometry of the samples. The shrinkage was then calculated using equation.1 below.

$$\text{Shrinkage} = \left(\frac{\text{Length}_{\text{before fired}} - \text{Length}_{\text{after fired}}}{\text{Length}_{\text{before fired}}} \right) \text{x}100\% \tag{4}$$

The shrinkage during firing depends on the reactivity of material, binder content and the firing conditions (temperature, time and atmosphere). So the control of all the parameters is important to achieve acceptable values of shrinkage which will finally be used for circuit designing. The set of data for density and shrinkage percentage for all substrates will be discussed; this will take into account the composition of LTCC materials and the effect of sintering temperatures on microstructure.

Dielectric properties measurements (dielectric loss, dielectric constant and loss tangent) of the eight layers of the substrate materials were measured using Novocontrol Impedance/ Material Analyzer in the range of 10 KHz to 10 MHz at room temperature. The eight dummy layers were used to make sure an easy handling due to single layer tape is only about 10 micron in thickness. This instrument is very useful and has high accuracy for dielectric properties measurement. Before the substrate material was measured, the analyzer should be warmed up for about 30 minutes after switching on the analyzer for all specifications. The dielectric measurements were made using flat substrate surfaces. The top and bottom surface of the substrate were coated with silver conductive paint to make sure they had a good contact with the sample holder plates.

3. Results and discussion

3.1. Shrinkage and density

The properties of the final ceramic composite materials depend on the sintered density of the whole substrate. A stacked and laminated LTCC substrate before firing consists of a relatively porous compact of oxides in combination with a polymer solvent. During sintering the organic solvent evaporates and the oxides react to form crystallites, or grains of the required composition, the grains nucleating at discrete centers and growing outwards until the boundaries meet those of the neighboring crystallites. During this process, the density of the material rises; if this process were to yield perfect crystals meeting at perfect boundaries the density should rise to the theoretical maximum, i.e. the x-ray density, which is the material mass in a perfect unit crystal cell divided by the cell volume. In practice imperfections occur and the sintered mass has microscopic voids both within the grains and at the grain boundaries. The resulting density is referred to as the sintered density. The density of the sample was measured using the Archimedes principle shown in equation (5);

$$\rho = \left(\frac{W_a}{W_w} \right) \rho *_w \tag{5}$$

where W_a = weight of sample in air; W_w = weight of sample in water;

ϱ^*_w = density of water = 1 gcm^{-3}

Sintering Temperature (°C)	Green Density (g/cm)	Sintered Density (g/cm)³	Densification factor (D_f)	Shrinkage	% porosity
800	1.725	2.379	1.023	16.445	0.211
825	1.733	2.372	0.989	16.3	0.214
850	1.733	2.415	1.131	16.42	0.20
875	1.731	2.365	0.971	15.52	0.217
900	1.785	2.412	1.034	16.79	0.201

Table 1. Physical properties of LTCC tape samples fired at various sintering temperatures

The densification behavior is described as densification factor (DF) as a function of time. The DF which represents the porosity removed during densification is defined as

$$D_f = \frac{(D_s - D_g)}{(D_{TH} - D_g)} \tag{6}$$

Where Ds is the instantaneous sintered density, Dg is the green density and D_{TH} is the theoretical density calculated for the composite by the mixture rule. The data of green density, sintered density, shrinkage, percentage of porosity and the densification factor of the LTCC substrates fired at various temperatures are tabulated in Table 1. The density and linear shrinkage as a function of sintering temperature for all the substrates is plotted in Figure 8. The sintered density of the substrates material was between 2.365 -2.412 g/cm³ or about 79%

of the theoretical density of 3.018 g/cm^3. The density obtained is found to be lower than that produced by other researchers [65-66].The total pores are directly related to the density. Therefore the higher heat treatment is believed to have removed a fraction of the pores. In this work all the samples had roughly similar amounts of porosity.

The linear shrinkage of the substrate material also fluctuates with increased sintering temperature. However, it is still in the standard values within the XY shrinkage range of about 14-17 %. In commercial production, the designed shrinkage is generally between 12-16 % for the XY direction and 20-25 % for the Z direction. The shrinkage is the key factor and the major technical problems in the LTCC technology. The repeatability and consistency of the shrinkage percentage must be the top criteria when designing an LTCC product. This is because the shrinkage of the LTCC substrate depends on the reactivity of the co-fired material containing ceramic oxides, glass, metal, organic solvent and also the firing conditions such as temperature, time and ambient air [29]. Better reproducibility increases the uniformity of finished product characteristics and therefore increases the process yield. It is not an easy task because all process parameters (lamination, binder burnout, sintering, etc) and material properties (high temperature reactivity, thermal expansion, etc.) must be matched [67].

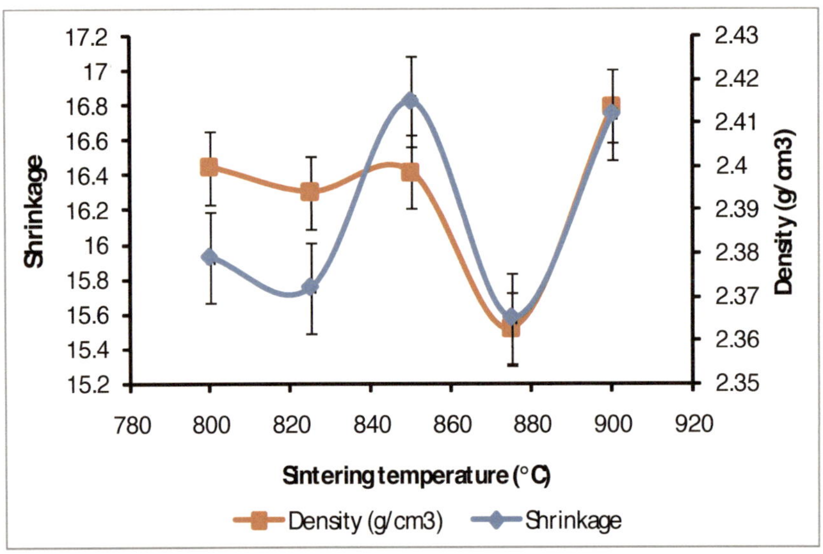

Figure 8. Relation of densification and shrinkage as a function of sintering temperature.

The density and shrinkage are highly microstructure sensitive. They are very much influenced by the way the samples are processed, grain size, chemical homogeneity and the presence of pores where all of these characteristics depend on the particle size, size distribution, particle packing density and particle orientation in ceramic powders and reactivity of the powder including the physical attributes of the green compacts [68]. Apart from the size distribution of the powder employed, the outcome is impacted by the other factors such as dispersion, solid loading and the heterogeneities in multicomponent systems [68-71]. Small

http://dx.doi.org/10.5772/54037

particle size and narrow size distribution may increase the shrinkage rate. A higher shrinkage rate means a good densification process which is due to a good particle distribution contributing to the good particle packing of the samples and giving high density of the tape. These results may be attributed to the preparation condition and the problem of starting raw material used in this work. It finally produced a bad substrate. Keizer and his co-workers in 1973 mentioned in their work that the electrical properties of a ceramic material of a given composition depend primarily on the chemical homogeneity and the average particle size. The value of these characteristics is considered to be the results of quality of starting raw materials as well as of the preparation condition since the density is considered as the most sensitive property [72]. Furthermore, the material quality of the finished product and process conditions also must be optimized in micro and macro structures in order to make sure the work in progress is highly repeatable at each process step [46]. All science and technology knowledge including technical skills regarding the LTCC fabrication process should be applied to achieve desired shrinkage data; the process engineer must have proper protocol to control the critical process variables.

3.2. Microstructure

Microstructural characterization has been used since many years ago in attempts to understand the crystallization mechanism of complex and multiphase ceramics which plays an important role in the improvement of the thermal, electrical and mechanical properties [73]. It covers by the microstructural features such as phases, crystalline morphology, crystallography and the chemistry of the interfaces. According to Yamaguchi, 1987, the desirable properties at high frequency cannot be determined only by chemical composition or crystal phase alone but also require the true characteristics of the microstructure which depend primarily on the sintering temperature, packing density and packing homogeneity of the green body which, in turn, are controlled by the powder characteristics and the method used to form the green body [42]. Generally, microstructure control during firing becomes less difficult when the homogeneity of the green body is improved since a sintered structural ceramic body should have a chemically and physically homogeneous microstructure [74]. Thus for the same green density, improved packing homogeneity leads to improved sinterability in the form of higher sintered density at a given sintering where fine particles tend to be active and sinter more readily at a given temperature especially in the presence of a liquid phase [75-76].

The effects of sintering temperature on microstructure of glass-ceramic tape are shown in Figure 9. It is clearly seen that there is inhomogeneous microstructure for the entire sample with some sample showing a big size pore meaning that probably only a few crystallite sizes are present in the bulk samples. This feature of the microstructure could be due to the presence agglomeration within the sample. The agglomerated tape however leads to lower densities with large shrinkage deviations in particular direction giving a poor quality [77]. NIL. In the LTCC process, some particles are densely packed and some particles are loosely packed due to the agglomeration of particles at some places. The presence of agglomeration is a common problem in ceramics processing and influences the microstructure behavior of the whole substrate. As mentioned by Lange (1984) and Hirata et al., (2009) [78-79], when a laminated substrate or powder compact is heated, the inhomogeneity of the packing provides the different densification rates producing a microstructure which is usually not uniform when agglomeration is severe. An inhomogeneous distribution of particles leads to an inhomogeneous liquid distribution such that there is no driving force for redistribution of the liquid, so the densification rate is not homogeneous and the microstructure development also becomes inhomogeneous. Agglomeration promotes uneven sintering which sometimes results in a mechanically weak and porous product. Thus, to achieve a high density material and good microstructure development, the agglomeration needs to be controlled [80].

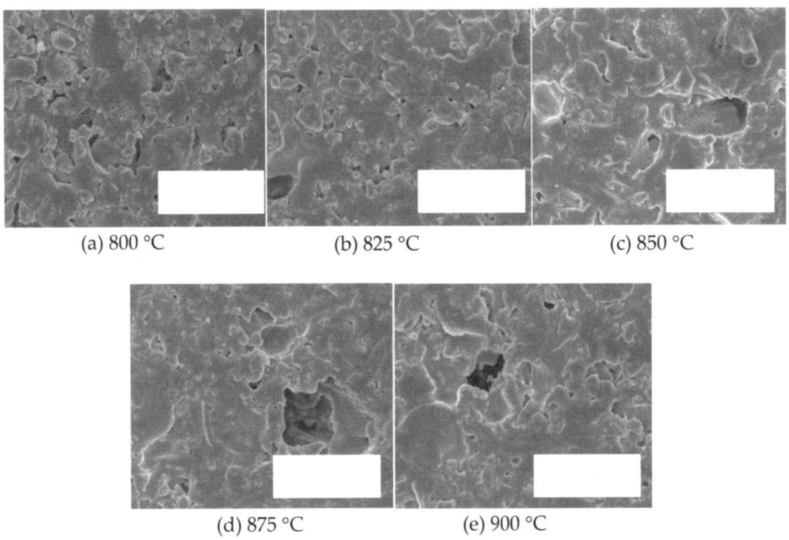

Figure 9. SEM micrograph of laminated sample sintered at a) 800 °C, b) 825 °C, c) 850 °C, d) 875 °C and e) 900 °C.

The densification process of the glass-ceramic composite in general can be described by the conventional three-stage liquid phase sintering as previously suggested in the early work [81-82]: particle rearrangement, dissolution and precipitation and solid state sintering. Compared to solid state sintering, each of the three stages is connected with the dominant microstructural changes; however, each stage in a liquid phase sintering is identified with that which occurs in the stages [18]. As previously mentioned the presence of glass phase ($CaSiO_3$) in this composite system produces liquid phase formation at a temperature lower than the sintering temperature and may considerably increase the rate of sintering [29]. The importance of the rearrangement of particles under the action of liquid surface tension and the solution precipitation mechanism has been considered essential features of the liquid phase sintering which finally affects the sintering. The redistribution of liquid during densification is more problematic since it depends on the initial particle packing which is difficult to identify [83]. The presence of this secondary

http://dx.doi.org/10.5772/54037

phase may probably restrain the growth of the grain. Compared to the solid state sintering, the effectiveness of liquid phase for enhancement of sintering rates depends on the introduction of new densification process. When a liquid is added which is present in sufficient amounts at a sintering temperature, it might completely wet and penetrate between solid particles; rapid densification can be obtained by the rearrangement of solid particles under capillary forces and by the solution at the contact points and precipitation elsewhere [84].

The whole densification rate at the final stage, is a summation of the developed local internal defect such as the presence of pores that are no longer large enough to prevent grain growth and this is the major process going on in addition to the final densification. With the increasing grain size, the densification rate decreases as the distance of the defects to the grain boundaries increases. Grain growth also gives pore coalescence where smaller pores are merged together into larger ones: this also reduces the densification rate and explains the density results obtained. In the case of the sintering process of glass ceramic material, if crystallization occurs before densification, the viscosity of samples will be increased. It is due to the contribution of glass composition into a crystalline phase structure, resulting in the reduction of viscous flow of the system. As a result, densification through viscous flow sintering will not occur properly and a porous body will be formed [85].

3.3. Dielectric properties

The ceramic substrate acts as a support material for active and passive devices and the interconnecting conductors that makes up the substantial subsystems. Discrete components integrated into the substrate also may be involved. Signal transmission from one chip pad to another chip pad is governed by the dielectric properties of the substrate or the interlayer and the electrical conductivity of the metal strip. If the total resistance is low, signals will be propagated with the speed of light, with a delay that is proportional to $n = \varepsilon^{1/2}$. Line termination is important to control reflections. If the resistance is large (greater than several ohms), transmission is slower and the effect of the dielectric constant exponent is even greater. Thus, the dielectric properties of the substrate play a crucial role in determining circuit speed.

Dielectric properties of substrates are dependent upon the several factors including the method of preparation, chemical composition and microstructure. Based on these factors, the dielectric materials have to have proper firing, so that a dense and nonporous product has to be obtained [86]. Besides, the dielectric properties are also affected by the bonding mechanism and crystallinity in the materials. The primary bonding mechanism in ceramic is ionic bonding while for the other material the bonding mechanism is covalent bond. In former times the dielectric constant and loss were measured at 1 MHz, the dependence of these data on frequencies up to GHz range is of high interest today, due to the dynamic development of mobile communications [87].

3.3.1. Effect density on the dielectric properties

The variation of dielectric constant and loss tangent with increased the sintering temperature are illustrated in Figure 10. The relation between dielectric properties and sintering did not show a similar trend to those the relation of density and sintering temperature. The decreasing trend of dielectric constant up to 850 ºC can be correlated to the presence of $CaSiO_3$ as a main phase in glass-ceramic system. It was reported in the previous work [29]. Chang and Jean (1999) [88] have noted from their studies and concluded that the presence of $CaSiO_3$ phases at temperature 850 ºC and above would cause some defects in microstructure due to the formation of liquid phase. This behavior will hinder the polarization which finally affects the dielectric constant of the sample sintered at 850 ºC. It was consistent with results by Lo and his team in 2002 and noted that the changes of permittivity attributed to the crystallinity presence in the NIL glass-ceramic samples which made the dielectric constant decline slightly [65]. It was later confirmed by Wang et al., 2009, for their research on the relation of dielectric properties and microstructure of $CaSiO_3$ in the presence of B_2O_3 additives. They found that the dielectric constant was decreased in the presence of $CaSiO_3$ at sintering temperature of 1100 ºC. They are the stable phases when crystallization is complete [89]. Thus, a decrease in the dielectric loss with an increase in the sintering temperature must have been caused by the phase reaction between the ceramic and the glass phase, resulting in an ionic exchange between Si2+, Ca2+ ions in the composite. This would cause a decrease in the electrical resistivity of the glass phase, which in turn would yield a decrease in dielectric loss.

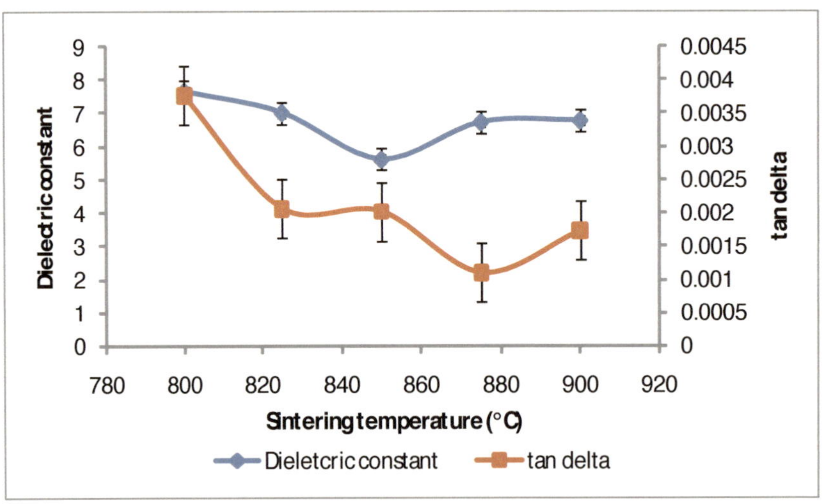

Figure 10. The relation of density and dielectric constant as a function of sintering temperatures.

Since LTCCs are basically composite structures of glass and crystals, controlling their dielectric constant depends largely on the combination of constituent materials of the composites and its material composition (i.e. volume fraction of the constituent materials)

[46]. Each constituent in the glass composition is crucial to the crystallization, densification and thus the microwave dielectric properties. Lo and Yang (1998) have studied the sintering characteristics of Bi_2O_3 added MgO-CaO-Al_2O_3-SiO_2 glass powder. They found that for a glass-ceramic, many factors affect the permittivity including of individual crystalline glass phases and the level of porosity or the degree of densification [90]. However, their effects are intercorrelated and sometimes unable to distinguish the contributions from an individual constituent.

Such a microstructural arrangement can affect the dielectric properties due to inhomogeneities phenomena which usually occur during sample preparation. Pores and grain boundary relaxation was found to be the major features that cause the dielectric loss in LTCC substrate in the low frequency region of dielectric where the response can be noted in the imaginary part of the dielectric constant [91]. Some factors such as the inclusion of the powder impurities, non-uniform density of green compact and controlled atmosphere is very important. The formation of second phases generated from the reaction between dielectric composition and the glass fritz does influence the dielectric constant of materials. As an example in the CaO-B_2O_3-SiO_2 system glass ceramic, the system with high SiO_2 content tend to possess lower dielectric constant at microwave frequency. It is because a higher SiO_2 content favors the formation of wollastonite ($CaSiO_3$) associated with the crystallization of cristobalite, trymidite or quartz depending on the sintering temperature and cooling rate. However after a certain weight percent addition of SiO_2 (65 wt%), the formation of wollastonite is prevented. Chiang et al reported that the formation of wollastonite will decrease the dielectric constant of a material. The reason is that wollostonite phase itself has a low dielectric constant which is ≈ 5. Thus one should determine the composition and the amount of glass fritz composition so that the second phase formation could be minimized, maintaining sufficient densification [92].

Figure 11 and 12 show the variation of dielectric constant and the dissipation factor with the variation of sintered density. The dielectric constant decreased up to the sintering temperature of 850 ºC and increased beyond that point. A closer look at the graph shows that only the sample sintered at 850 ºC has a reverse trend compared to others. The results are not consistent with the results found by Wang et al., 2009, Louh et al., 2005 and Zhang et al., 2005 which noted that the dielectric constant increases with a similar trend to the increase of density as a function of sintering temperature [89, 92, 93]. According to Yang et al., 2003, the dielectric constant increases as the sintered density increase. As the sintered substrate becomes denser, its dielectric constant tends to increase, which indicates that less air entrapment in the substrate favors the permeation of electromagnetic waves [94]. It was confirmed by Sun et al., 2009 [95] who's mentioned that the dielectric properties are affected by the density of the compact body. However in this work, the dielectric constant initially decreased up to sintering temperature of 850 ºC and then increased again beyond this turning point. The drop in dielectric constant value at slightly high sintered density could be attributed to the presence of pores within the substrate. Dielectric constant is linearly dependent on its density which is associated with the porosity; the greater the porosity, the smaller the dielectric constant [92].

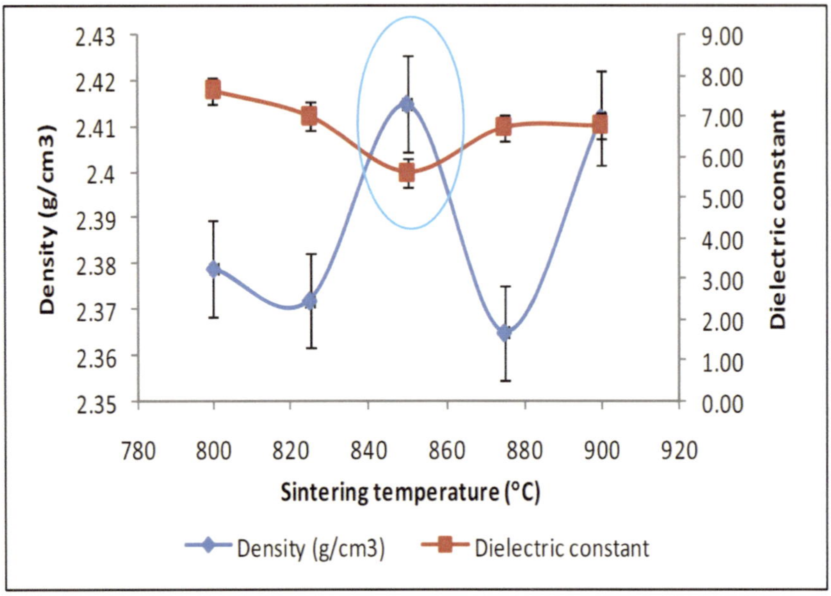

Figure 11. The relation of density and dielectric constant as a function of sintering temperatures.

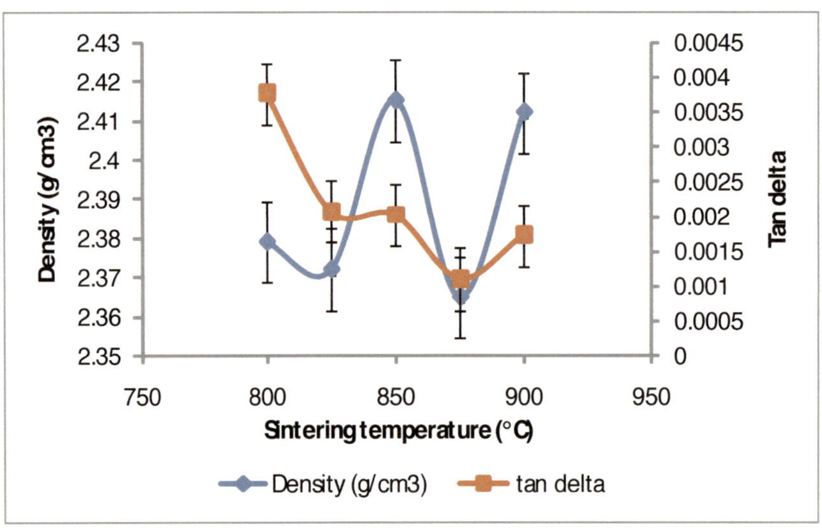

Figure 12. The variation of density and loss tangent as a function of sintering temperatures.

The dependence of the dielectric constant upon the properties of the material may be summarized in the statement that the dielectric constant depends upon the number and the mobility of the electron per cm^3 of material and when the molecules contain electric dipole, to much a greater extent upon the size and arrangement of the dipoles in the molecules, upon the freedom of the molecules to orient, and upon the number of molecules per cm^3 [96].

3.3.2. Dielectric properties with frequency

The dielectric properties of glass ceramic system were studied in the frequency range from 10 KHz to 10 MHz at room temperature. Figure 13 show the variation of the dielectric constant as a function of frequency for all samples. Obviously, the dielectric constant shows a decreasing trend for all the samples. The decrease is rapid at lower frequency and slower and stable at higher frequency. The decrease of dielectric constant with increasing frequency is a normal dielectric behavior which is also observed by other researchers [97-99]. A glass ceramic system is considered as heterogeneous material that can experience interfacial polarization as predicted by Maxwell and Wagner. They pointed out that at low frequency region (refer to Figure 3) the movement of charge carriers trapped at interfacial region which is caused by inhomogeneous dielectric structure. At high frequency, the dominant mechanism contributing to dielectric constant is the hopping mechanism in their respective interstice under the influence of alternating current. The frequency of hopping between ions could not follow the frequency of applied field and hence it lags behind, therefore the values of dielectric constant become reduced at higher frequency [100].

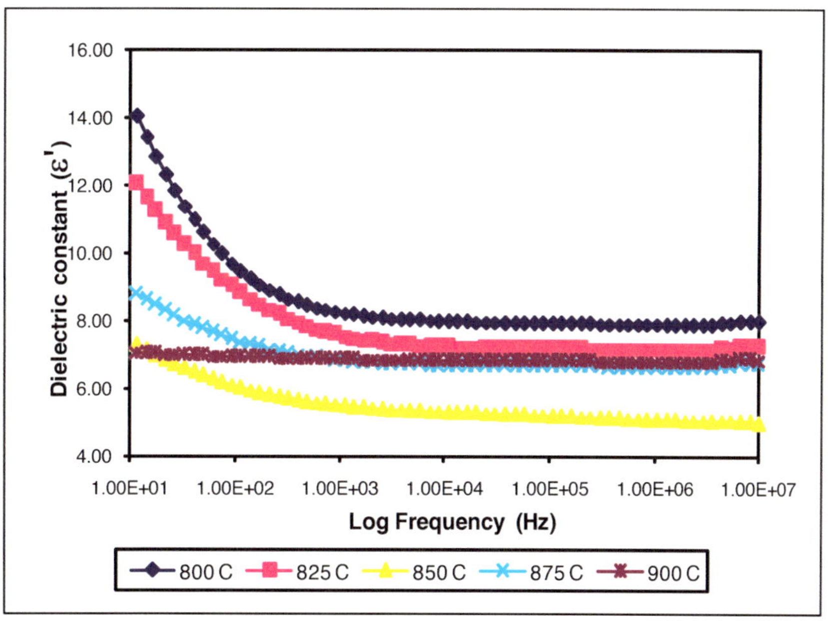

Figure 13. The variation of dielectric constant with frequency.

A low loss factor is desirable for a dielectric material so that the dissipated electric power to the insulator is minimized. This type of consideration is very important for high power circuits operating at high speed. Dissipation factor is a ratio of the energy dissipated to the energy stored in the dielectric material. The more energy that is dissipated into the material, the less is going to make it to the final destination. This dissipated energy typically turns into heat or is radiated as radio frequency (RF) into the air. The optimal goal is to have 100% of the signal pass through the interconnection network and not be absorbed in the dielectric. A high loss material means a little or no signal is left at the end of the transmission path. In other to retain maximum signal power, a low loss material should be used. The defect, space charge formation and lattice distortion is believed to produce an absorption current resulting in a loss factor and at high alternating frequency those could follow the field applied and later values the trend saturated [100].

Figure 14 show the variation of dissipation factor for all samples with frequency. The decreasing trend of loss tangent may be explained as follows: When the frequency of applied AC field is much larger than the hopping frequency of electron, the electron do not have an opportunity to jump at all and the energy loss is small. In general, the local displacements of electronic charge carriers (jumping electrons in case of ceramic substrate) cause (or determine) the dielectric polarization in materials. Since the dielectric polarization is similar to that for conduction and are mainly by the hopping conduction mechanism. Therefore the marked decrease in tan δ is due to the decreasing ability of the jumping electron (charge carrier) to follow the alternating frequency of AC electric field beyond certain critical frequency. This explains the decreasing trends for tan δ with increasingly frequency.

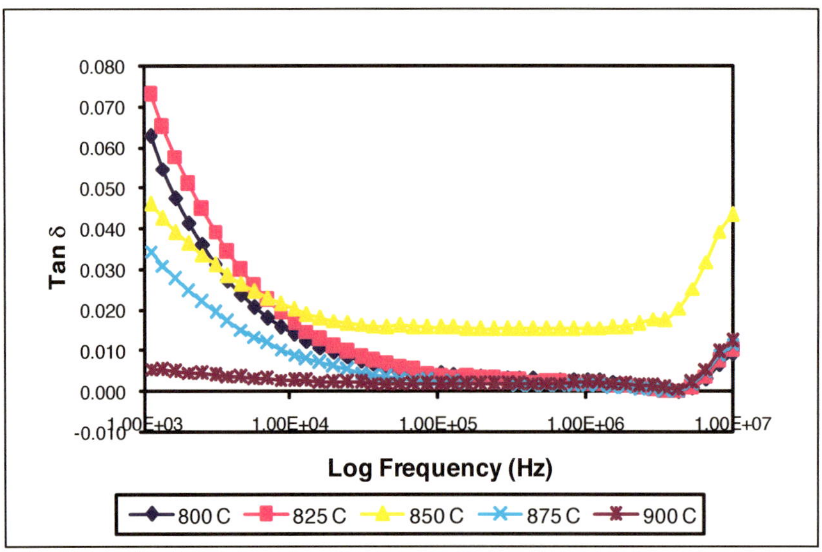

Figure 14. The variation of tan δ with frequency.

http://dx.doi.org/10.5772/54037

4. Conclusion

The effect of various sintering temperatures on some properties of laminated substrates has been successfully studied. It can be concluded that the sintering temperature strongly influenced the dielectric properties and some physical properties of the laminated substrate. The result of dielectric constant was decreased with increasing sintering temperature up to 850 °C but increased again beyond that point. It was also noticed that the dropped value of dielectric constant with higher sintered density at 850 °C is due to the presence of crystalline $CaSiO_3$ phases during the sintering process. The relation for the loss tangent and the density is generally a decrease with increased sintering temperature. It may be due to presence of crystalline phase and the porosity inside the substrate. The sintering process of glass-ceramic substrate is a complicated process because phase changes were involved. Thus the understanding of material behaviors such as binder burnout, densification of LTCC, grain growth behavior and the deformation of suspended LTCC is important in optimizing the fabrication process for multilayer LTCC substrate in order to achieve the required performance.

Acknowledgements

The author wish to thank Telekom Malaysia for their funding support under project IMPACT (RDTC/100745). Appreciation is also expressed to Assoc. Prof. Dr. Mansor Hashim, Mr. Mohamed Razman Yahya and Sabrina Mohd. Shapee for their guidance, technical support and for many useful discussions in this research work.

Author details

Rosidah Alias

Address all correspondence to: rosidah@tmrnd.com.my

Advanced Physical Technologies, TM Research & Development Sdn. Bhd., TMR&D Innovation Centre, Lingkaran Teknokrat Timur, Cyberjaya, Selangor, Malaysia

References

[1] Lecheminoux, L. "Advanced Design, Technology & Manufacturing for High Volume and Low Cost Production", IEEE Proceeding of International Electronic Manufacturing Technology Symposium, 2003.

[2] Joseph, T. And M. T. Sebastian (2010) "Effect of glass addition on the microwave dielectric properties of $CaMgSi_2O_6$ Ceramics" Int. J. Appl. Ceram. Technol. 7 [S1] E98-E106.

[3] El-Kheshen, A.A. and M.F. Zawrah (2003), 'Sinterability, microstructures and properties of glass/ceramic composites' *Ceramics International*, 29, 251-257.

[4] Smallman, R.E., R.J. BiShop (1999) "Ceramic and glass" Modern Physical Metallurgy and Materials Engineering (Sixth edition),1999, 320-350

[5] Haertling, G. H. (1999) "Ferroelectric Ceramics: History and Technology" J. Am. Ceram. Soc. 82(4) 797-818.

[6] Rodriguez, A. R., Franklinville and A. B. Wallace (Aerovox Corporation),1961. *Ceramic capacitor and method of making it.* United States Patent 3004197, 10/10/1961

[7] Modes, C., S. Malkmus and F. Gora (2002), "High-K low loss dielectric co-fireable with LTCC" Active and Passive Elec. Comp., 25, pp. 141-145.

[8] Kim, S.-H. and Koh, J.-H. (2008), 'ZnBo-doped $(Ba,Sr)TiO_3$ ceramics for the low-temperature sintering process', *J.of Eur. Ceram. Soc.*, 28, 2969-2973.

[9] Tok, A.I.Y., F.Y.C. Boey and K.A.Khor (1999), "Tape Casting Process of High Dielectric Ceramics Composite/Hybrid Substrate for Microelectronics Applications" Journal of material Processing Technology 89-90 pp 508-512.

[10] Roosen A (2000) "Ceramic Substrate: Trends in Materials Applications" Ceramics Transaction, 106, pp: 479-492.

[11] Stiegel Schmitt, A, A. Roosen, C. Ziegler, S. Martius and L. –P. Schmidt (2004) "Dielectric Data of Ceramic Substrates at High Frequency" J. European Ceramic Soc., 24, pp: 1463-1466.

[12] Morrison, D. G. (2000) "Low Temperature Co-fired Ceramic Fuel Growth of High Frequency" Technical Report. – http://www.elecdesign.com/article

[13] Schubert, H. and G. Petzow (1990) "Preparation and Characterization of Ceramic Powders in Advanced Ceramic III", S. Somiya, ed., Elsevier, pp. 45-56.

[14] Sigmund, W. M., N. S. Bell and L. Bergstrom (2000) "Novel Powder-Processing Methods for Advanced Ceramics" J. Am. Ceram. Soc. 83(7), 1557-1574.

[15] McMillan, P. W. G. PAartridge and F. R. Ward (1969) "Glass ceramics for the coating and bonding of silicon semiconductor material" *Microelectronics and Reliability,* Vol. 8, pp. 113-119.

[16] Charles A. Harper, "Handbook of Material for Product Design", Mc-Graw-Hill. Third Edition (2001).

[17] Chen, G.-H. and X.-Y. Liu (2007) "Sintering, crystallization and properties of MgO-Al2O3-SiO2 system glass-ceramics containing ZnO" J. Alloy & Comp. 431, 282-286.

[18] Rahaman, M.N. (1995) *Ceramic Processing and Sintering* Marcel Dekker, Inc. NY.

[19] Hamzawy, E.M., A.A. El-Kheshe, M.F. Zawrah (2005) "Densification and properties of glass/cordierite composites" Ceram. Int. 31, 383–389

[20] Lin, J. C. and C. Y. Wang (1996) "Effect of surface properties of silver powder on the sintering of its thick-film conductor" Matr. Chem. & Phys. 45, 253-261.

[21] Vu K., "Silver Migration – The Mechanism and Effects on Thick-Film Conductors" Materials Science Engineering, (2003).

[22] Wang, S.-H. and He.-P. Zhou (2003) "Densification and dielectric properties of CaO-B_2O_3-SiO_2 system glass ceramics" Mater. Sci. Eng. B99, 597-600.

[23] Chiang, C. C. S. F. Wang, Y. R. Wang, W.-C J. Wei, (2008) "Densification and Microwave Dielectric Properties of CaO-B_2O_3-SiO_2 System Glass Ceramics", Ceramics International 34, 599-604.

[24] Zhu, H., Liu, M., Zhou, H., Li, L. & Lv, A. (2007). Study on Properties of CaO-SiO_2-B_2O_3 System Glass-Ceramic. *Materials Research Bulletin*, Vol. 42, Issue 6, pp. 1137-1144.

[25] Ota, R., H. Kuribayashi, J. Fukunaga and T. Taguchi (1992) "Crystallization behavior of non-alkali glasses in the SiO_2-CaO-Al_2O_3-ZnO-TiO_2-B_2O_3 system" Journal of Non-Crystalline Solids 144 (1992) 81-88.

[26] Jean, J.-H. J.-I Shen, C.-R. Chang and C.H. Lin (1995) "Densification kinetics and mechanism of a low-dielectric glass composite" Materials Chemistry and Physics 41 (1995) 260-265

[27] Wang, M., R. Zu, J. Jin, S. Su, J. Zhai (2011) "Investigation of the structure evolution process in sol–gel derived CaO–B_2O_3–SiO_2 glass ceramics" Journal of Non-Crystalline Solids 357 (2011) 1160–1163

[28] Zhou, X., E. Li., S. Yang, B. Li, B. Tang, Y. Yuan and S. Zhang (2012) "Effects of La_2O_3–B_2O_3 on the flexural strength and microwave dielectric properties of low temperature co-fired CaO–B_2O_3–SiO_2 glass–ceramic" Ceramics International 38 (2012) 5551–5555.

[29] Alias, R (2012). The Effects of Sintering Temperature Variations on Microstructure Changes of LTCC Substrate, Sintering of Ceramics - New Emerging Techniques, Dr. Arunachalam Lakshmanan (Ed.), ISBN: 978-953-51-0017-1

[30] Kingery, W.D., Bowen, H. K., & Uhlman, D.R. (1976). *Introduction to Ceramics*, (2nd Ed.), John Wiley and Sons, New York

[31] Kang, S. J. L. (2005) Sintering; Densification, grain growth and microstructure Elsevier Butterworth-Heineman.

[32] Varela, J. A. and O. J. Whittemore (1983), 'Structural rearrangement during the sintering of MgO' *J. Am. Ceram. Soc.*, Vol. 66, No. 1, pp. 77-82.

[33] Thummler, F. and Thomma, W. (1967) Sintering processes, Metall. Review vol. 12, 69-108.

[34] Quirk, J. F. (1959) "Factors affecting sinterability of oxide powders": BeO and MgO" *J. Am. Ceram. Soc.*, Vol. 42, No. 4, pp. 178-181

[35] Gupta, T. K. and R.L. Coble (1968) "Sintering of ZnO: Densification and grain growth" *J. Am. Ceram. Soc.*, Vol. 51, No. 9, pp. 521-525.

[36] Kang, S.-J.L., Kim, K.-H. & Yoon, D.N. (1991). Densification and Shrinkage during Liquid Phase Sintering. *J. Am. Ceram. Soc.*, Vol. 74, No. 2, pp. 425-427.

[37] Huang, C.-L. & Tsai, J.-T. (2001). Effects of Sintering Temperature on CaO-Li_2O-Sm_2O_3-TiO_2 Microwave Dielectric Ceramic. *Proc. Natl. Sci. Counc. ROC (A)*, Vol. 25, No. 5, pp. 317-321.

[38] Coble, R.L. (1961). Sintering Crystalline Solids. I. Intermediate and Final State Diffusion. *J. Appl. Phys.*, Vol. 32, No. 5, pp. 787-792.

[39] Lance, D., F. Valdivieso and P. Goeuriot (2004), 'Correlation between densification rate and microstructural evolution for pure alpha alumina' J. of Europ. Ceram. Soc. 24, pp. 2749-2761

[40] De Jonghe, L. and M. N. Rahaman (2003) Handbook of Advanced Ceramics, vol 1, 187-264

[41] Chen, S. Y., C.-H. Chou and S. –Y. Cheng (2002) "Effect of metal oxide precurcor on sintering shrinkage, microstructure evolution and electrical properties of silver-based paste" J. Of Mater. Sci, 37, 169-175.

[42] Yamaguchi, T. (1987), 'Pores in ceramic processing in ceramic databook by Seihin, K. and G. Kyokai, Gordon and Breach Scientific Publishers, Switzerland

[43] Wersing W. and Oliver Dernovsek, "Multilayer Ceramic Technology" in Ceramic Materials for Electronics: Third Edition, Revised and Expanded, edited by Relva C. Buchanan, Marcel Dekker Inc., New York, (2004) 581-642

[44] Nebe, W.J. and T. R. Suess (2007), 'Photo-defined and photo-imaged films' in Ceramic Interconnect Technology Handbook, F. D. Barlow and A. Elshabi9ni, CRC Press, London.

[45] Higby, P.L. and J. E. Shelby (1984), 'Properties of some simple glass/ceramic system' J. Am. Ceram. Soc. 67 (7), pp. 445-449.

[46] Imanaka, Y. (2005). *Multilayered Low Temperature Co-fired Ceramic (LTCC) Technology*. Springer, New York.

[47] Wilder, D. R. and E. S. Fitzsimmons (1955) "Further study of sintering phenomena" J. Am. Ceram. Soc., 38 (2), 66-71.

[48] Cho, B.J, Park, E.-T. and Lee, J.-M. (2009), 'Novel structure ceramic tape for multilayer devices', *J. Eur. Ceram. Soc.*, 29, 451-456.

[49] Kata, K., Y. Shimada and H. Takamizawa (1990) "Low dielectric constant new materials for multilayer ceramic substrate" IEEE Trans. on Comp., Hybrid & Manuf. Tech. 13 (2), 448-451

[50] Tummala, R.R. (1991). Glass-Ceramic Packaging in the 1990's. *J. Am. Ceram. Soc.*, Vol. 74, pp. 895-908.

[51] Konsowski, G. and A. R. Helland (1997) *Electronic Packaging of High Speed Circuitry* McGraw-Hill, New York

[52] Schwartz, B. (1984), "Review of multilayer ceramics for microelectronic packaging" J. Phy.~. Chm Solid Vol. 45. No. I0 pp. 1051-1068

[53] Wakino K, Nishikawa T, Tamura H & Sudo T (1987) "Dielectric Resonator Materials and their Applications" Microwave Journal, 30(6): 133–159.

[54] Sasaki, A and Y. Shimada (1992) "Electrical Design Technology for Low Dielectric Constant Multilayer Ceramic Substrate" IEEE on Components, Vol 15(1) pp: 56-62.

[55] Kellerman, D., D. J. Nabatian, L. Retherford and A. Elshabini (1998) "Characterization and Processing of Low Dielectric Constant Thick-film Substrate for MCM-C Modules" International Conference on Multi-chip Modules and High Density Packaging.

[56] Sebastian, M. T. (2008) Dielectric Materials for wireless communication, Elsevier

[57] Shimada, Y., Y. Yamashita and H. Takamizawa (1988) "Low Dielectric Constant Multilayer Glass-Ceramic Substrate with Ag-Pd Wiring for VLSI Package" IEEE Transaction on Com… Vol. 11(1), pp: 163-170.

[58] Zhou, X., B. Li and H. Ning (2009), "Effect of Ca/Si ratio on the microstructure and properties of CaO-B2O3-SiO2 glass-ceramics" J. Mater. Sci.: Mater Elcetron 20: 262-266.

[59] Pecht, Michael G., Rakesh Agarwal,Patrick McCluskey, Terrance Dishongh, Sirus Javadpour, Rahul Mahajan. "Electronic Packaging:Material and Their Properties", CRC Press LLC, (1999).

[60] http://www.theeestory.com/topics/861

[61] Sen, D, Mahata. T, Patra A.K., Mazumder. S.and Sharma B.P, "Small angle neutron scattering investigation and the low frequency dielectric response of sintered ZrO_2-8 mol % Y_2O_3 ceramic compacts: the effect of pore characteristics," *Journal of Physics: Condensed Matter* 16, pp 6229-6242 (2004).

[62] Kume, S. M. Yasuoka, S.-K. Lee, A. Kan, H. Ogawa and K. Watari (2007) "Dielectric and thermal properties of AIN ceramics" J. European Ceram. 27, 2967-2971.

[63] Ferro Corporation; Ferro Design Guideline in www.ferro.com

[64] Sunappan, V., A. Periannan, C.K.meng and W.C.Khuean (2004) 'Process issues and characterization of LTCC substrates' Elect. Comp. & Tech. Conf. 1933-1937.

[65] Lo, C.-L.,J. G. Duh and B.-S. Chiou. (2002). Low Temperature Sintering and Microwave Dielectric Properties of Anorthite-Based Glass-Ceramic. *J. Am. Ceram. Soc.*, Vol. 85, No. 9, pp. 2230-2235.

[66] Erol, M., Kucukbayrak, S and Ersoy-Mericboyu, A. (2009). The Influence of the Binder on the Properties of Sintered Glass-Ceramics Produced from Industrial Wastes. *Ceramics International*, No. 35, pp. 2609-2617.

[67] Rabe, T., Schiller, W.A., Hochheimer, T., Modes, C. & Kipka, A. (2005). Zero Shrinkage of LTCC Self-Constrained Sintering. *Int. J. Appl. Ceram. Tech.*, Vol. 2, No. 5, pp. 374-382.

[68] Coble, R. L. and Kingery, W. D. (1956) "Effect of porosity on physical properties of sintered alumina"

[69] Subbanna, M., Kapur, P.C. and Pradip (2002), 'Role of powder size, packing, solid loading and dispersion in colloidal processing of ceramics', *Ceramics International*, 28, 401-405.

[70] Shui, A., L. Zeng and K. Uematsu (2006) " Relationship between sintering shrinkage anisotropy and particle orientation for alumina powder compacts" Scripta Materialia 55 pg; 831-834.

[71] Lee, D. Y., D.-J Kim, B.-Y. Kim and Y.-S Song (2003), "Effect of alumina particle size and distribution on filtration rate and fracture toughness of alumina-glass composite prepared by melt infiltration" Material Science and Engineering A341 pg: 98-105

[72] Keizer, K., E., H. Janssen, K. J. de Vries and A. J. Burggraaf (1973) "Inlfuence of particle size and structure of ZrO_2 on microstructure development and dielectric constant of $PbZr_{0.5}Ti_{0.5}O_3$" Mat. Res. Bull. 8, pp. 533-54.

[73] Lee, W.E., D.D. Jayaseelan and S. Zhang (2008) "Solid–liquid interactions: The key to microstructural evolution in ceramics" J. Eur. Ceram. Soc. 28,1517–1525

[74] Fortulan, C. A. and D. P.F.de Souza (1999), "Microstructural evolution of the Al_2O_3-ZrO_2 composite and its correlation with electrical conductivity" Mater. Res. 2(3), pp. 205-210

[75] Dey, D. and R. C. Bradt (1992) "Grain growth of ZnO during Bi2O3 liquid-phase sintering" J. Am. Ceram. Soc., 75 (9), 2529-2534.

[76] Valant, M. & Suvorov, D. (2003)."Microstructural phenomena in low firing .ceramics" Materials Chemistry and Physics 79, 104-110.

[77] Raj, P.M. & Cannon, R. (1999). Anisotropy Shrinkage in Tape-Cast Alumina: Role of Processing Parameter and Particle Shape. *J. Am. Ceram. Soc.*, Vol. 82, No. 10, pp. 2619-2625.

[78] Lange, F.F. (1984). Sinterability of Agglomerated Powders. *J. Am. Ceram. Soc.*, Vol. 67, No. 2, pp. 83-89

[79] Hirata, Y., Hara, A. & Aksay, A. (2009). Thermodynamics of Densification of Powder Compact. *Ceramics International*, Vol. 35, pp. 2667-2674

[80] Forrester, J.S., Goodshaw, H.J., Kisi, E.H., Suaning, G.J. & Zobec, J.S. (2008). Effect of Milling on the Sintering Behavior of Alumina. *J. Am. Ceram. Soc.*, Vol. 44, No. 1, pp. 47-52.

[81] Cole, Jr. A. S. (1972) "Sintering of Ag-Pd in the presence of a reactive glass" *J. Am. Ceram. Soc.*, Vol. 55, No. 6, pp. 296-299.

[82] Exner, H. E. and E. Arzt (1996) " Sintering processes" in R. W. Cahn and P. Haasen, Powder Metallurgy: fourth, revise and enhanced editin' Elsevier Science

[83] De Jonghe, L. and V. Srikatnth (1988) "Liquid-phase sintering of MgO-Bi2O3" J. Am. Ceram. Soc. 71(7), C-356-C-358.

[84] Kingery, W.D.,J. M. Woulbroun and F. R. Charvat (1963) "Effects of applied pressure on densification during sintering in the presence of liquid-phase" *J. Am. Ceram. Soc*, Vol. 46(3), pp. 391-395.

[85] Banijamali, S., Eftekkhari Yekta, B., Rezaie, H.R. & Marghussian, V.K. (2009). Crystallization and Sintering Characteristics of $CaO-Al_2O_3-SiO_2$ Glasses in the Presence of TiO_2, CaF_2 and ZrO_2. *Thermochimica Acta*, Vol. 488, No. 1-2, pp. 60-65.

[86] Jantunen, H. (2001) "A Novel Low Temperature Cofire Ceramic (LTCC) Material for Telecommunication Devices" Thesis Department of Electrical Engineering Oulu University

[87] McN Alford, N and Penn S. J. (1996) "Sintered Alumina With Low Dielectric Loss" J. Appl. Phys., 80, pp: 5895-5898.

[88] Chang, C.-R. & Jean, J.-H. (1999). Crystallization Kinetics and Mechanism of Low-Dielectric, Low Temperature, Cofirable $CaO-B_2O_3-SiO_2$ Glass Ceramics. *J. Am. Ceram. Soc.*, Vol. 82, No. 7, pp. 1725-1732.

[89] Wang, H. P, S.-Q. Xu, S.-Q. Lu, S. L. Zhao and B. L. Wang (2009) "Dielectric properties and microstructures of CaSiO3 ceramics with B2O3 addition" Ceram. Int. 35, 2715-2718.

[90] Lo, S.-H. and C.-F. Yang (1998) "Sintering characteristics of Bi_2O_3 added MgO-CaO-Al_2O_3-SiO2 glass powder" Ceramics International (24), 139-144.

[91] Maex, K., Baklanov, M. R, Shamiryan, D., Iacopi, F., Brongersma, S. H., Yanovitskaya, Z. S., (2003), Low dielectric constant materials for microelectronics, *J. Appl. Phys.*, 93:8793-8841

[92] Louh, S.P., I.C. Leu,, M.H. Hon (2005) "Effects of density and bonding structure on dielectric constantof plasma deposited a-C:H films" Diamond & Related Materials 14 (2005) 1000 – 1004

[93] Zhang, Q. L. H. Yang, J. L. Zhou and H.P.Wang (2005) "Sintering and microwave dielectric properties of LTCC-zinc titanate multilayers" Mater. Lett. 59, 880-884.

[94] Yang, T. C. K., Sea-Fue Wang, Chung-Chih Juan (2003) "Effect of lamination conditions on sintered properties of glass ceramic substrates for microelectronic packaging", Journal of Materials Processing Technology 148(2004) 165-17.

[95] Sun, H.-P., Q.-L. Zhang and H. Yang (2009) "Silver cofirable (Ca0.9Mg0.1) SiO3 microwave dielectric ceramics with Li0O3-Bi2O3 additive" Ceramic Internationals 35, 637-641.

[96] Smyth,C.P. (1929) "The properties of dielectric. I. Electric moment and molecular structure" J. Franklin Institute, 207 (6),813-824..

[97] Chenari H. Mahmoudi, Ali Hassanzadeh,*, M.M. Golzan, H. Sedghi, M. Talebian (2011) "Frequency dependence of ultrahigh dielectric constant of novel synthesized SnO2 nanoparticles thick films" Current Applied Physics 11 (2011) 409-413

[98] Venkataraman, B. H., K.B.R. Varma(2004) "Frequency-dependent dielectric characteristics of ferroelectric $SrBi_2Nb_2O_9$ ceramics" Solid State Ionics 167 (2004) 197–202

[99] Ravinder D., G. Ranga Mohan, Prankishan, Nitendarkishan, D.R. Sagar (2000) "High frequency dielectric behaviour of aluminium-substituted lithium ferrites "Materials Letters 44 2000 256–260

[100] Kumar, G. B. and Buddhudu, S. (2010) "Optical, thermal and dielectric properties of $Bi_4(TiO_4)$ ceramic powders" Ceram. Int. 36, pp. 1857-1861.

Chapter 6

Low Temperature Hybrid Processing Technology of Fine Electronic Ceramics

Hongfang Zhang, Chee-leung Mak,
Helen Lai-Wa Chan and Xi Yao

Additional information is available at the end of the chapter

http://dx.doi.org/10.5772/53255

1. Introduction

Low temperature sintering is of vital importance for commercial production of electronic ceramics, especially multi-layer devices [1, 2]. Due to superior mechanical strength, ferroelectric ceramics with fine-grained structures have attracted much attention and exhibited potential applications such as piezoelectric sensors and actuators [3, 4-5]. Together, composite materials are also widely applied for multilayer capacitors, piezoelectric transducers, packaging materials for integrated circuits, high voltage insulators and chemical sensors [6-10]. It is well known that materials' performances are closely related to the ways they are manufactured. Generally, ferroelectric ceramics fabrication process should achieve products of dense and homogeneous microstructure with well developed crystalline grains of uniform size in order to ensure the best physical properties. Using conventional ceramic process to achieve such ideal microstructure, the window of sintering temperature required is always high and narrow. Conversely, the sintering temperature of sol-gel derived ceramics is always lower. However, the drawbacks of sol–gel method are aggregation of ultrafine powders during fabrication and formation of secondary phases [11]. Thus, the physical properties of these sol-gel derived ceramics are deteriorated in comparison to those of their counterparts. Recently, dense, crack-free ferroelectric thick film integrated on silicon substrate has been the subject of considerable attention for potential application in micro-electro-mechanical systems (MEMS). Combining micro-machined surfaces of silicon wafers with ferroelectric films has resulted in novel devices such as micro-fluidic devices, micro-pumps, infrared sensors, dynamic random access memory, tunable microwave devices and several others [12-16]. Most of these ferroelectric thick film devices, such as multi-layer ceramic capacitors (MLCC), require the film thickness to be around 1 micrometer. In addition, these

INTECH
open science | open minds

© 2013 The Author(s). Licensee InTech. This chapter is distributed under the terms of the Creative Commons Attribution License http://creativecommons.org/licenses/by/3.0), which permits unrestricted use, distribution, and reproduction in any medium, provided the original work is properly cited.

devices need to be co-fired with less expensive base metal electrode at low temperature. Therefore, the existing technology can no longer meet such requirements. Many attempts have been devoted to develop new technologies to meet these new technological challenges. Indeed, a new hybrid ceramic processing technology has been developed [17-22]. The key characteristic of this new technology is to graft the sol-gel wet chemistry process onto the conventional mixed oxide ceramic process. With this new technology, the advantages of both processes can be superimposed together. The good performance of ceramics prepared by conventional ceramic process under high temperatures is mostly preserved, while the sintering temperature of the devices is effectively reduced down as inherited from the sol-gel process. The low sintering temperature of this hybrid process enables the technology better compatible to the thick film technology. Low sintering temperature also impedes the chemical reactions among different materials in the devices, such as ceramic/electrode, ceramic/substrate and ceramics/matrix. This new technology will be very useful in developing all-ceramics devices such as microwave barium-strontium titanate ($Ba_{0.6}Sr_{0.4}TiO_3$, BST) ceramics, thick films and composite materials. This chapter reviews our recent progress in the synthesis of various ferroelectric ceramic materials and composites via the proposed hybrid process. After the background introduction, a brief description of the experimental procedures is followed. In Section 3, synthesis processes and characterization of various ferroelectric ceramics and composites are presented. Finally, the chapter is ended with some concluding remarks.

2. Features of the low temperature hybrid processing route

Figure 1 is the schematic diagram illustrating the entire formation process of a representative microwave BST thick film.

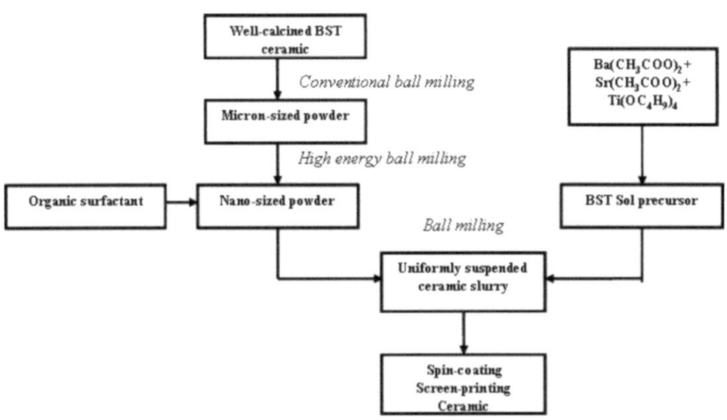

Figure 1. Flow chat of the hybrid processing of a typical microwave BST thick film.

As shown in Fig. 1, this processing route is quite flexible. Changing the process parameters of ball milling (i.e. conventional as well as high energy) such as revolution speed and ball

http://dx.doi.org/10.5772/53255

milling time, the particle size can be easily controlled. On the other hand, the thickness of the film can be increased by repeating the spin-coating or screen-printing process. In addition, the viscosity of the ceramic slurry can be adjusted by altering the powder/sol ratio as well as the amount of organic solvent. In fact, the viscosity of the slurry affects the film thickness for each spin-coating/screen-printing layer. For calcinated sol-gel precursor possessing the same composition as that of the nano-sized powders, the mixture obtained is a homogenous mixture, and the resulting composite has single composition. In contrast, if the precursor solution and the nano-powders are of different compositions, multiphase composite will be produced forming a heterogeneous composite. Using this method, uniformly distributed multiphase nanocrystalline composite can be obtained easily. The composite normally exhibits all the merits of the nano-powders and the sol-gel precursor. We can find different applications of this route by utilizing its respective merits. For example, by using uniformly dispersed nano-powders in sol-gel solution we demonstrated that homogeneous thick films were obtained by spin coating technique [20, 22]. Using sol-gel precursor as binder, the density of the compacted films was increased and the sintering temperature of the compacted films was reduced. Similar, for screen-printing films, low sintering temperature will be expected. This unique feature is particularly useful to low temperature co-firing of multilayer devices because cheap internal electrode paste can be used. We consider that the simplicity of this approach should make it useful to most of the ceramic processes.

3. Formation and characterization of fine ferroelectric and composite electroceramics via hybrid process

3.1. Microwave fine-grained BST and BST-MgO ceramics

3.1.1. Fine-grained BST ceramics

3.1.1.1. Sintering behavior

In Fig. 1, pre-calcined BST powders were firstly prepared by a conventional ceramic process using commercial $BaTiO_3$ (BT) powder of mean grain size smaller than 400 nm, and $SrTiO_3$ (ST) powder of size smaller than 100 nm. The powders were mixed together by using conventional ball milling for 2 h and then calcined at 1150°C for 2 h to form the BST ceramics. Then, the BST ceramics were ball milled by a normal planetary ball mill machine with zirconia balls as the milling medium. The size of the obtained powder particle was around few micrometers to sub-micrometers. Afterwards, a high-energy ball-milling machine (Frisch Pulverisette 5 planetary ball-milling machine, Germany) with a tungsten carbide milling jar and milling medium was used to prepare the nano-sized powders. The size of the final powder was in the range of 20–60 nm depending on the process parameters. During the course of high-energy ball milling, a selected dispersant was added to the nano-powder to prevent the agglomeration of the nano-sized powder. On the other hand, BST sol precursor in the same chemical composition as the powder was synthesized via a polymer-assistant sol-gel route. Here the starting materials for the sol precursor were $Ba(CH_3COO)_2$, $Sr(CH_3COO)_2$ and $Ti(OC_2H_5)_4$. The details of the fabrication process can be found elsewhere [20]. To graft the sol gel process onto the conventional ceramic process, pre-synthesized BST sol was mixed with BST nano-sized powder by conventional ball milling. Due to the effect of the organic surfactant used, the BST powder was well dispersed and suspended in the BST sol solution forming uniform paint-like ceramic slurry. In this hybrid process, the molar ratio of BST introduced in the form of powder and in the form of sol solution was very important. In the current investigation, an optimized mass ratio of ceramic powder/sol precursor equivalent to 72/28 was adopted. For preparing bulk ceramics, the as-prepared ceramic slurry was dried at 120~150°C and then calcined at 800~1000°C for 2 h. Then calcined powder was uniaxially pressed at 4 MPa to form a green BST compact which was heated to 1200°C for 2 h, with a heating rate of 2°C/min in air. For comparison, coarse-grained BST ceramics were prepared by a conventional solid-state reaction method from the pre-calcined micron-sized BST powders (i.e. the powders calcined at 1150°C) at a sintering temperature of 1400 °C for 2 h in air.

3.1.1.2. Microstructure and properties

Figure 2 shows the X-ray diffraction patterns of (a) commercial BT powder, (b) commercial ST powder, (c) coarse-grained BST ceramics fabricated by conventional solid-state reaction method and sintered at 1400°C, and (d) BST ceramics fabricated by hybrid process technology and sintered at 1200°C. In Figs. 2(a) and (b), the XRD patterns show that the BT and ST powders are in tetragonal and cubic phase, respectively. Figs. 2(c) and (d) indicate the presence of highly pure and crystalline pervoskite cubic structure. All the peaks were identified as BST phase and no impurity phase was observed, indicating the success in synthesizing BST ceramics by the two processing routes.

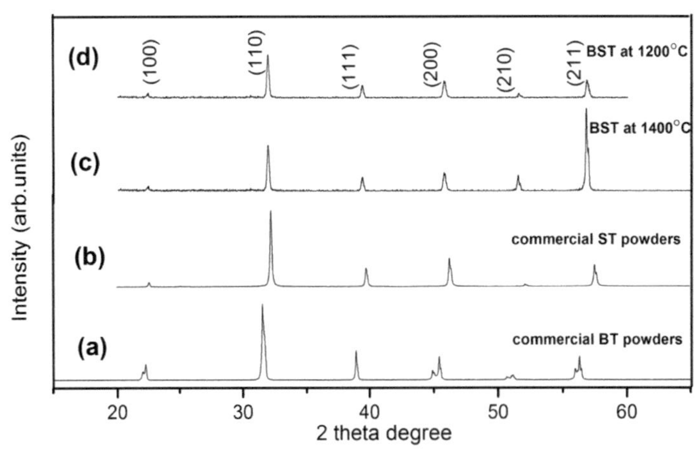

Figure 2. X-ray diffraction (XRD) patterns: (a) and (b) commercial BaTiO3 and SrTiO3 powders; and BST ceramics sintered at (c) 1400°C using conventional solid-state process, (d) 1200°C using hybrid process.

Figure 3 shows the SEM images of (a) coarse-grained BST ceramics fabricated by conventional solid-state reaction method, and (b) fine-grained BST ceramics fabricated by hybrid process technology. The images showed that both ceramics possessed well-densified grains of various size and shape. In general, the grains size in Fig. 3(a) was larger than those in Fig. 3(b) and the grain size ranges were 5-10 μm and 1-2 μm for Figs. 3(a) and (b), respectively. It is noticed that the fine-grained ceramic fabricated by the hybrid process was densified at a sintering temperature 200°C lower than that of the conventional solid-state process. The lower sintering temperature of the hybrid process is most inherited from the low crystallization and sintering temperature of the amorphous gel components of the hybrid BST powder. Sol-gel process is usually characterized with much lower sintering temperature because of higher chemical reactivity and shorter diffusion length [23]. On the other hand, the BST gel solution acts as binding agent to connect the nano-sized BST particles and fills the interstitials of the nano-sized particles, which is easy to achieve dramatic modification of the processing behaviors of ceramics, and useful to produce uniform and fine-grained high-quality ceramics at low sintering temperatures.

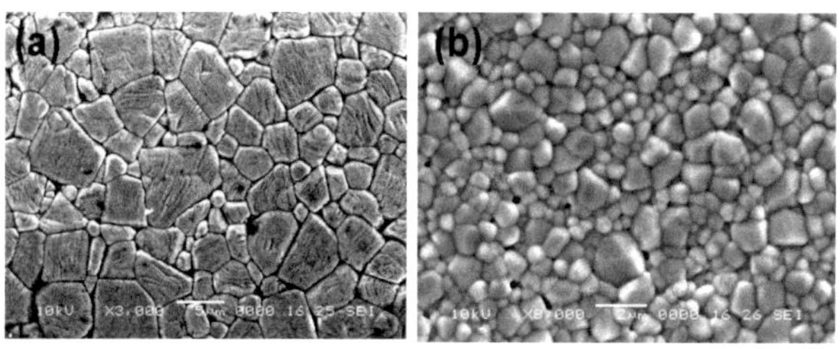

Figure 3. Surface morphology of BST ceramics by (a) the conventional solid-state process at 1400°C, and (b) by the hybrid process at 1200 °C.

The corresponding temperature dependency of dielectric constant and loss tangent for BST ceramic prepared by conventional solid-state process and hybrid process are shown in Figure 4. As shown in Fig. 4(a), for the coarse-grained BST ceramic, the peak dielectric constant at 10 kHz was around 6500 with a loss tangent about 0.02. Though the dielectric constant in coarse-grained ceramic was high due to the well-developed large grains; however, the grain size was too large and the grain size distribution was too large so that the microstructure was not very uniform as required in many applications. In Fig. 4(b), for the fine-grained BST ceramic, a broader and diffused peak was observed in both the dielectric constant and loss tangent spectra indicating the characteristic dispersion at the Curie point. The peak dielectric constant was about 2500 at 10 kHz with a loss tangent about 0.02. The peak dielectric constant is lower than that of the conventional solid-state process mostly due to the finer structure.

The dielectric tunability, defined as the change of dielectric constant in the presence of a dc electric field with respect to the dielectric constant in the absence of a dc electric field, is shown in Figure 5. The tunability of the fine-grained BST ceramic was found to be about 36% at 11 kV/cm at room temperature, the dielectric loss was in the range of 10^{-2}.

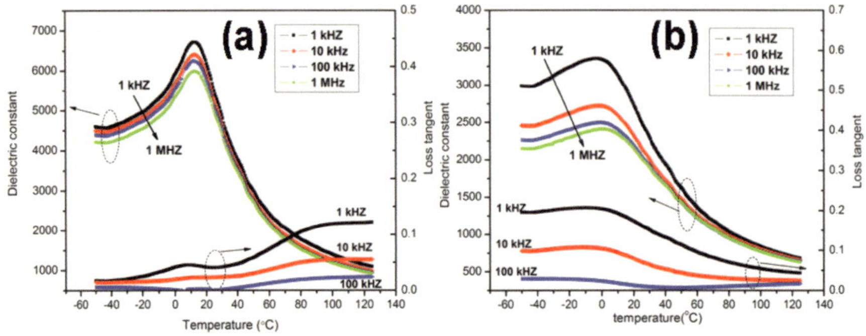

Figure 4. Temperature dependences of dielectric constant and loss tangent for BST ceramic via (a) conventional ceramic process, and (b) hybrid processing.

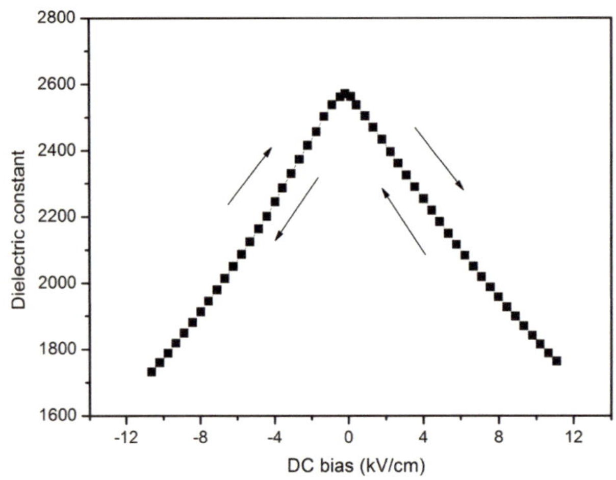

Figure 5. Tuning behavior of dielectric constant of fine-grained BST ceramic at 10 kHz and room temperature under DC bias voltage.

3.1.2. Fine-grained BST-MgO ceramics

3.1.2.1. Sintering behavior

To fabricate BST-MgO ceramics, the same BST ceramic slurry described in Section 3.1.1.1 was employed. Magnesium nitrate hexahydrate ($Mg(NO_3).6H_2O$) solution was added to the BST slurry, then subjected to a conventional ball-milling for 2h to obtain heterogenous BST-MgO ceramic slurry. The paint-like BST-MgO slurry was dried at 120°C, and then calcined at 800°C for 4h. The pre-calcined powder was uniaxially pressed into disk pellets at a pressure of 4 MPa in a stainless steel die. The pressed pellets were sintered in air at 1200–1300°C for 2h.

http://dx.doi.org/10.5772/53255

3.1.2.2. *Microstructure and properties*

Figure 6 shows the XRD patterns of (a) the nano-sized BST powders, (b) BST-MgO slurry dried at 120°C, (c) BST-MgO slurry calcined at 800°C for 4h, (d) BST-MgO ceramics sintered at 1200°C for 2h, and (e) BST-MgO ceramics sintered at 1300°C for 2h. The XRD pattern in Fig. 6(a) suggested the presence of BST powder of pure perovskite structure. The XRD pattern of the dried BST-MgO slurry in Fig. 6(b) only displayed the peaks of the BST powder and no MgO peaks were observed, indicating that the MgO was still in an amorphous phase and there was no chemical reaction between the BST and MgO phases. It is seen from Figs. 6(c)–(e) that the BST and MgO phases coexisted in both the calcined BST-MgO slurry and the sintered BST-MgO ceramics, and all the XRD peaks corresponding to both the BST and MgO phases were identified with no detection of intermediate or interfacial phases. The results indicate the success in synthesizing BST-MgO ceramics using the hybrid processing route.

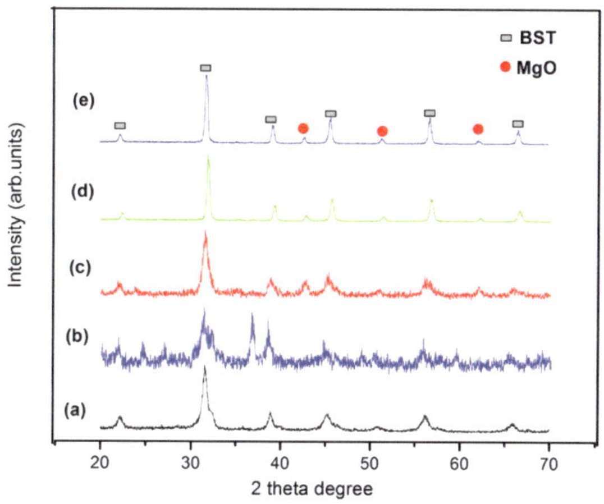

Figure 6. XRD patterns of (a) BST powders, (b) BST-MgO slurry dried at 120°C, (c) BST-MgO slurry calcined at 800°C for 4h, (d) BST-MgO ceramics sintered at 1200°C for 2h, and (e) BST-MgO ceramics sintered at 1300°C for 2h.

Figure 7 illustrates the SEM images of BST-MgO ceramics sintered at three different temperatures ranging between 1200 and 1300°C for 2 h via hybrid processing. It is noticed that densification of the ceramic was taken place at 1200°C with grain size of about 300 nm. With increasing sintering temperature to 1275~1300°C, the ceramics were further densified with grain size increased to about 1–2 μm, and no apparent grain growth was observed after 1275°C. Thus, dense, homogeneous and fine-crystalline BST-MgO ceramics were produced at sintering temperatures below 1300°C. It has been reported [24] that sintering temperature in the range of 1350–1500°C was indispensable in the fabrication of BST-MgO ceramics using conventional solid-state process. This high sintering temperature resulted in large grains as well as pores, and followed by the deterioration of microstructures as well as properties of the resulting ceramics. Fortunately, such phenomena were not evident in our BST-MgO ceramics synthesized by the hybrid processing route. As shown in Fig. 7, sintering temperature of 1200°C was good enough for the synthesis of BST-MgO ceramics with a reasonably high density of 4.50 g/cm^3. We attribute these promising results to the characteristics of the hybrid processing route as mentioned above. Besides, the sol–gel derived species can act as the buffer for another phase inhibiting the grain size in the sub-micron scale.

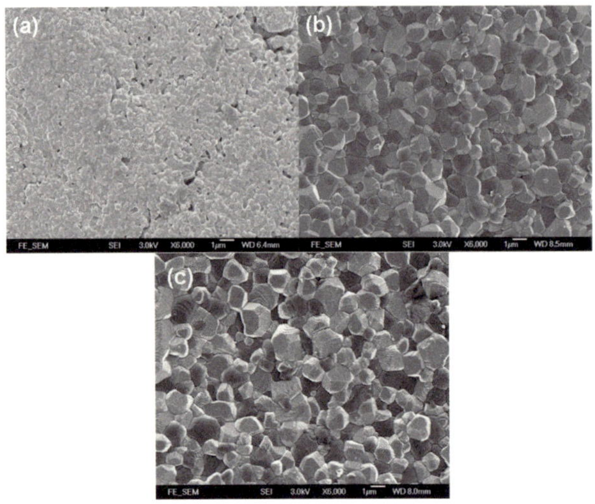

Figure 7. The surface morphologies of BST composite bulk ceramics sintered at (a) 1200°C ; (b) 1275°C; (b) 1300°C for 2 h via the hybrid process. Note: the EDX results suggested that grains with light color corresponded to the BST phase while grains with dark color were MgO phase.

Figure 8 shows the temperature dependency of the relative dielectric constant ε_r and loss tangent (tan δ) in the temperature range of -60°C to 80°C for three selected frequencies (1 kHz, 10 kHz, and 100 kHz). A relatively broad and diffused phase transition was clearly identified at about -20°C for both samples. The ceramics demonstrated very low room temperature loss tangent below 0.005 for frequency above 1 kHz. Reduced temperature variation of dielectric properties in these samples is desirable for application of the ceramics in electrically tunable microwave devices to improve their temperature and frequency stabilities [25].

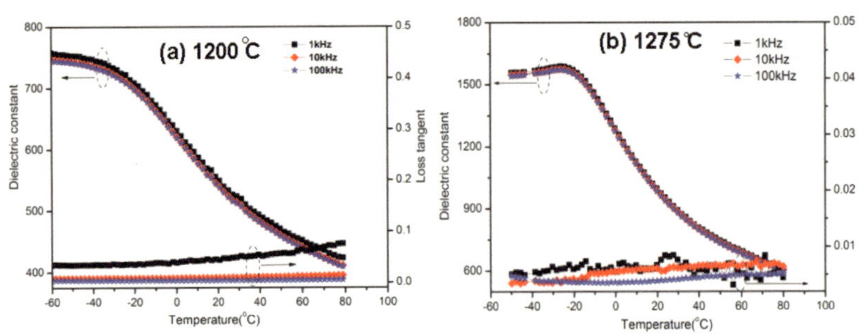

Figure 8. Temperature dependences of dielectric constant and loss tangent for the BST-MgO ceramics sintered at 1200 and 1275°C via hybrid process.

http://dx.doi.org/10.5772/53255

To explore the application of the ceramics in microwave devices, dc electric field was employed to change the dielectric constant of the BST-MgO ceramics. Figure 9 shows the room temperature dielectric constant as a function of dc electric field for BST-MgO ceramics sintered at 1200°C and 1300°C with average grain sizes of 300 nm and 2 μm at 100 kHz respectively. The dielectric tunability was found to be 4% and 14% respectively, under a dc electric field of 1.5 kV/mm.

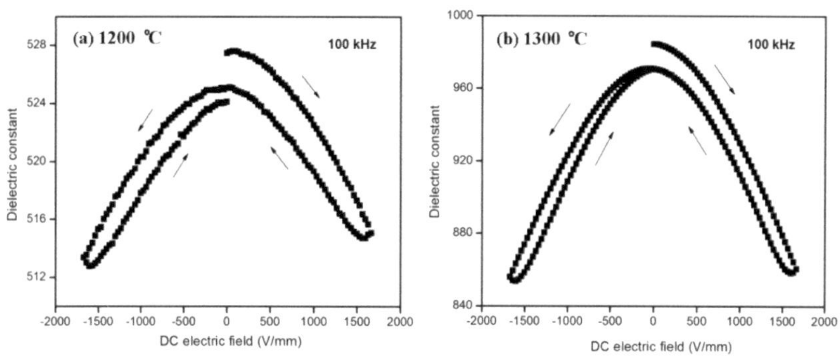

Figure 9. Room-temperature dielectric constant as a function of dc electric field for the ceramics sintered at 1200 and 1300°C at 100 kHz.

3.2. Microwave BST ceramic thick films

3.2.1. Sintering behavior

As shown in Fig. 1, the BST powder, being acted as filler in the BST ceramic slurry, plays an important role in the formation of dense and crack free films. It also strongly affects the surface morphology and resultant properties of the thick films. Therefore, well-crystallized nano-size or micron-size BST ceramic powders or their combinations, which had been pre-sintered at 1350°C for 7 h using conventional solid-state ceramic process, were dispersed into BST sol precursor to form uniformly suspended BST ceramic slurry. The BST ceramic slurry was spin-coated onto Pt substrates or Ag/alumina substrates at a spin rate of 3000 rpm for 20 s. The optimum thickness of a single deposition layer was about 1–2 μm. Thicker films were obtained by repeating the spin-coating process. For thick film deposition, the weight ratio of powder to precursor solution of the slurry was a critical parameter. The *wt%* BST powder in the ceramic slurry is defined as:

$$BST(wt\%) = \frac{p}{p+s} \qquad (1)$$

where $p$is the weight of BST powder added, $s$is the residue oxide weight of the BST sol solution after thermal annealing. On the basis of our results, the optimal BST (*wt%*) was between 70 and 80% in order to get the maximum thickness in a crack-free layer using one spin-coating process. In the experiment, the viscosity of the slurry was controlled by adjusting the amount of organic solvent. The as-deposited films were calcined at about 550°C to remove all organic solvent. Afterwards, the films were densified at a higher temperature of around 600–900°C, depending on the electrode used. The paste for screen printing was prepared by the following processing route: first, the ceramic slurry was calcined in order to remove solvents as well as organics brought in by the sol solution. The calcined powder was dispersed into an organic vehicle specifically designed for screen-printing. Green films were printed on Pd–Ag-electroded alumina substrates by screen-printing. The thickness of each single layer was about 5–10 μm depending on the loading percentage of the ceramic powder and the viscosity of the printing paste. In order to obtain thick films, the printing process was repeated. Each layer of the screen-printed wet film was allowed to settle at room temperature for 15 min and dried in an oven at 120°C for another 15 min. The final films were sintered at temperatures of around 1200°C for 2 h.

3.2.2. Microstructure and properties

Figure 10 shows the X-ray diffraction patterns of BST thick film deposited on different substrates annealed at different temperatures. It is quite evident that the crystal structures of the films were of cubic perovskite structure. Besides the diffraction peaks arisen from BST phase as well as the Pt substrate or silver electrodes and alumina substrates, no other impurity phases were observed.

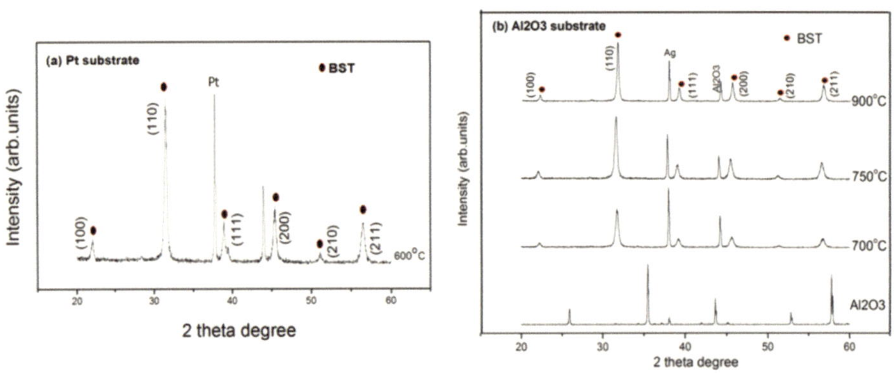

Figure 10. X-Ray diffraction patterns of BST thick films with various calcining temperatures deposited on either Pt or Ag/alumina substrates.

Figure 11 displays the SEM pictures of the surface (Figs. 11(a) and (c)) and the cross-section (Figs. 11(b) and (d)) of the BST thick films deposited on various substrates. It is noticed that the films were, in general, dense, uniform and crack-free. In Fig. 11 (b), the film, with only single deposited layer, had a film thickness over 1 μm. On the other hand, in Fig. 11(d), the film, with 10 deposited layers in totally, had a film thickness of about 10 μm. Thus, the average thickness for one coated layer was estimated to be about 1 μm, similar to that observed in Fig. 11(b). The sol-gel derived phase, shown as crystallites between the granular grains, was clearly seen in the microstructure. This sol-gel derived phase acted as a binder for the granular grains and filled the interstitials of the added powders. The agglomeration, which easily exists in conventional sol-gel derived ceramics and nanocomposite route, was not observed in Figs. 11. Therefore, we believe that agglomeration was eliminated by the hybrid process, demonstrating the advantage of this novel route.

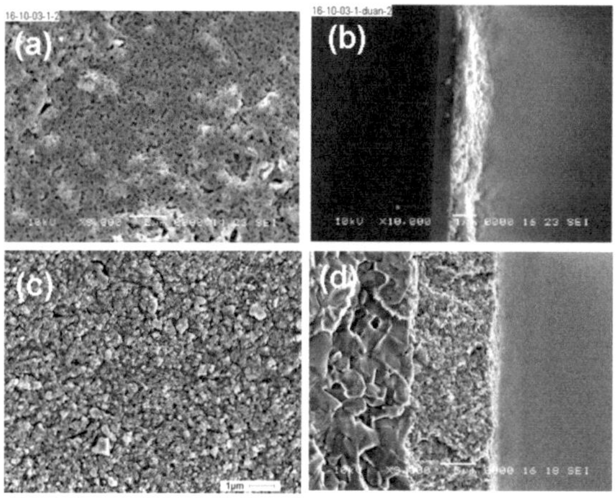

Figure 11. The surface and cross-sectional morphologies of thick films prepared from the slurry using the spin-coating method deposited on Pt substrates with one layer and Ag/alumina with 10 layers annealed at 600 and 700°C, respectively.

Figure 12 shows the temperature dependence relative dielectric constant ε_r and loss tangent (tan δ) for BST thick films deposited on (a) Pt substrate annealed at 600°C, and (b)–(d) Ag/alumina substrates annealed at 700, 750 and 900°C respectively, in the temperature range of -35°C to 120°C for four selected frequencies (1 kHz, 10 kHz, 100 kHz and 1 MHz). A relatively broad phase transition was clearly identified at about 0°C for all samples. With increasing annealing temperature, the dielectric constant increased gradually. For example, at 10 kHz at zero temperature, the dielectric constant of the thick films deposited on Pt substrate (Fig. 12(a)) was found to be about 150 with a loss tangent slightly below 0.02, for the thick films deposited on Ag/alumina substrates annealed at 900°C, the dielectric constant (Fig.12(d)) was about 370 with a loss tangent slightly below 0.01.

Figure 13 illustrates the room temperature tuning behavior of the BST thick films deposited on (a) Pt substrate annealed at 600°C, and (b)–(c) Ag/alumina substrates annealed at 750 and 900°C respectively at 100 kHz and under dc bias voltage. In Fig. 13(a), the room temperature dielectric constant at zero bias condition varied between 160 and 135, while the tunability of the dielectric constant of the thick film changed from 12% at 60 kV/cm with a loss tangent of about 0.015 to about 30% at 35 kV/cm with a loss tangent of 0.0067 (as shown in Fig. 13(b)). In Fig. 13(c) the tunability was 28% at 20 kV/cm with a loss tangent of 0.008. Usually, thick films with higher dielectric constants possess higher tunability of dielectric constant but poorer temperature stability [20]. General speaking, reducing the dielectric constant of the film will reduce its tunability but improve its temperature stability.

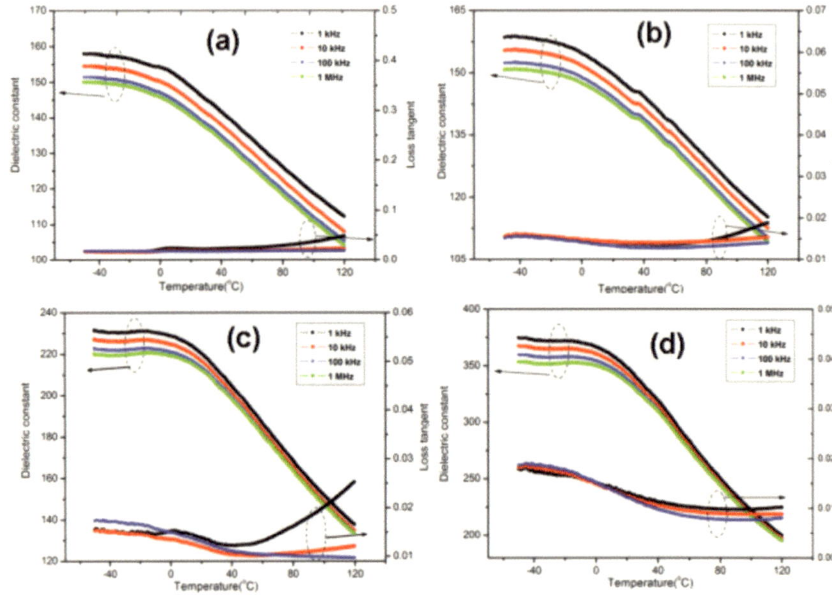

Figure 12. Temperature dependence dielectric constant and loss tangent for BST thick films deposited on (a) Pt substrate annealed at 600°C, (b) –(d) Ag/alumina substrates annealed at 700, 750 and 900°C, respectively.

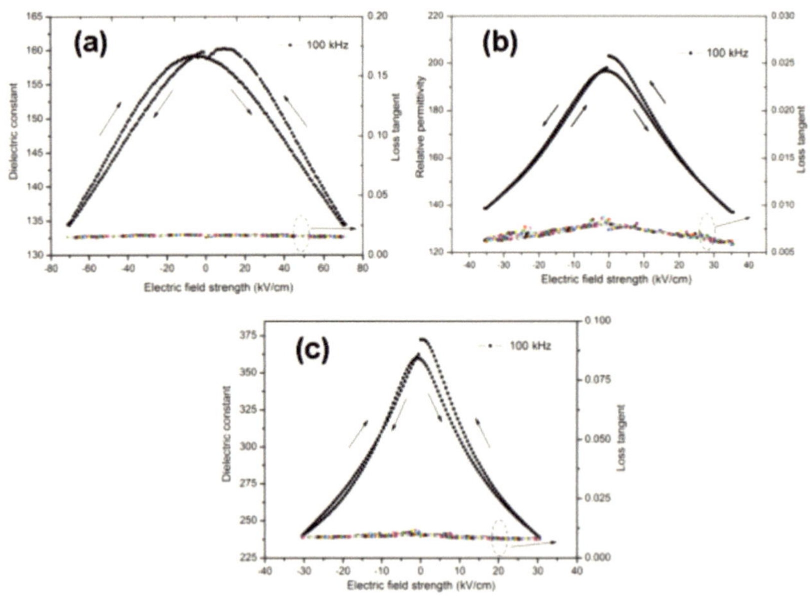

Figure 13. dc bias voltage dependency of the dielectric constant and loss tangent of thick BST films deposited on (a) Pt substrate annealed at 600°C, (b)–(c) Ag/alumina substrates annealed at 750 and 900°C respectively at 25°C.

The effects of the BST powder as filler in the sol precursor on the structural as well as dielectric properties of BST thick films prepared by screen-printing method were also investigated. One is the ceramic slurry was prepared by dispersing nano-sized BST powder in the BST sol precursor. Figure 14 shows the (a) XRD pattern, (b) surface and (c) cross-section micrographs of BST thick films deposited on Pd-Ag/ alumina substrate by screen-printing method. The films were sintered at 1200°C in air. As shown in Fig. 14(a), XRD pattern exhibited a pure BST perovskite structure in cubic phase with no observable intermediate as well as impurity phases, except the diffraction peaks arisen from Pd-Ag electrode and alumina substrate. As observed in Fig. 14(b) and (c), the film was dense, uniform and crack-free, similar to the surface morphologies of the films prepared by spin coating. For one single layer, the thickness was about 10 μm. This is much thicker than the thickness obtained in the spin-coated films. The temperature dependence dielectric constant and loss tangent, together with the tunability of dielectric constant of this thick film are shown in Figure 15. A broad and diffusive peak with the maximum dielectric constant of about 750 was observed. The loss tangent was below 0.01 at room temperature and the tunability was about 33% at 20 kV/cm. The film showed a larger dielectric constant and a slightly better tenability as compared to films prepared by spin coating.

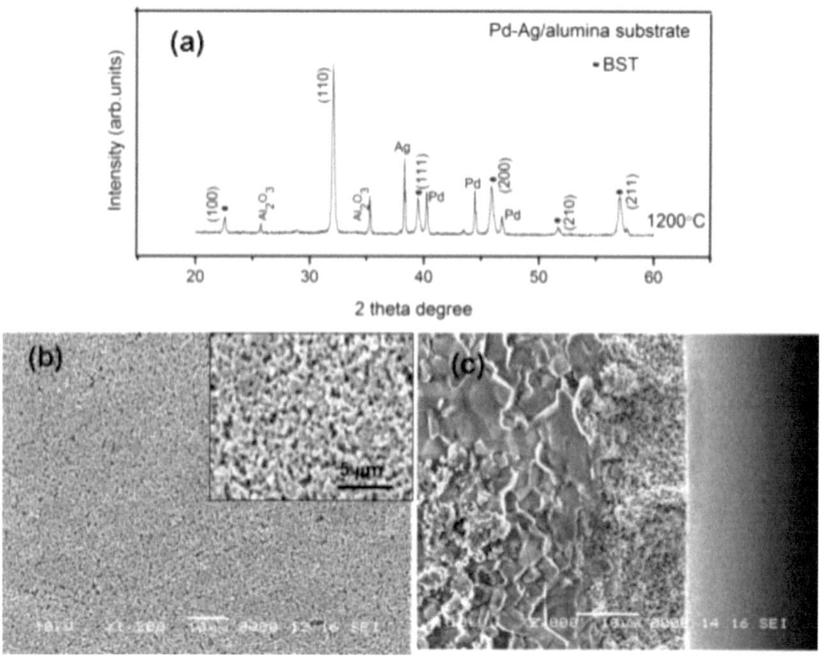

Figure 14. The BST thick films deposited on Pd-Ag/alumina substrate by screen-printing sintered at 1200°C: (a) the XRD pattern; (b) the surface, together with (c) the cross-section micrographs indicating the thickness of about 10 μm. Inset shows the magnified surface picture.

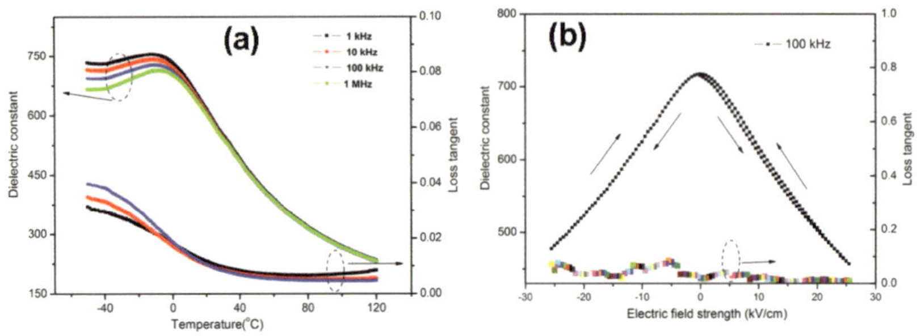

Figure 15. (a) Temperature dependence dielectric constant and loss tangent, and (b) the tuning behavior of dielectric constant of fine-grained BST ceramic at 10 kHz and room temperature under dc bias voltage.

While, Figure 16(a) shows the surface morphology of the BST thick film fabricated from ceramic slurry, which was prepared by combining nano-sized and micron-sized BST powders as the filler in the BST sol precursor matrix. Inset of Fig. 16(a) shows the magnified surface profile. The film was screen-printed on Pd-Ag/alumina substrate and annealed at 1200°C. It is interesting to find that inhomogeneous concrete-like structure, in which big grains in the diameter around ten or tens of micrometers (stones in concrete) and medium grains in the range of micrometer or sub-micrometers (sands in concrete) were adhered by gel derived substance. The dielectric constant was found to be about 1800 with a loss tangent slightly below 0.02 at 10 kHz at room temperature. These values were higher than those values shown in Fig. 15, but close to those values obtained in bulk ceramics formed by conventional solid-state technology. However, the detailed characteristics of the novel concrete-like structures still need to be further studied in future work.

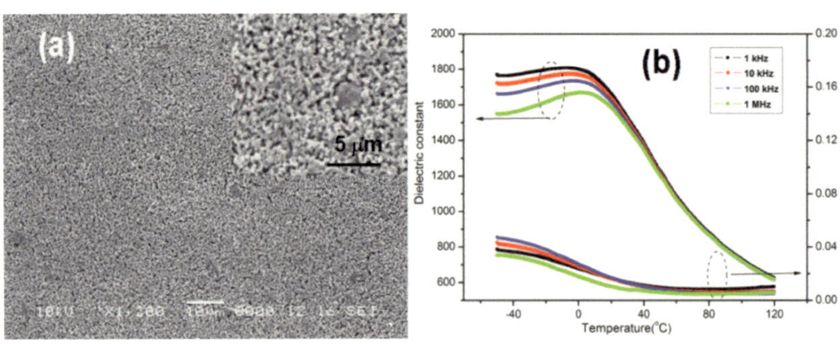

Figure 16. (a) Surface micrograph and (b) temperature dependence dielectric constant as well as loss tangent of the BST thick film which was prepared by combining nano- and micron-sized BST powders as filler and screen-printing on Pd-Ag/alumina substrate. The film was annealed at 1200°C. Inset shows the magnified surface picture.

http://dx.doi.org/10.5772/53255

3.3. Ferroelectric fine-grained glass-ceramic composites

Apart from BST ceramics composites, this modified hybrid process was employed to produce fine-grained Pb(Zr,Ti)O_3-based glass-ceramic composites. In general, fine ferroelectric glass-ceramics with high-crystalline phase content are desirable for better dielectric properties; however, the content is quite limited by using conventional melting/quenching glass-ceramics technique [26], because of the high volatility of the lead and the high melting temperature of the zirconium. Besides, zirconium is simply not soluble in many types of glass. On the other hand, in ferroelectric glass-ceramics derived from conventional sol–gel process, aggregation of ultrafine powders and formation of secondary phases are still difficult to be overcome [23]. Therefore, we applied the hybrid processing to fabricate the fine-grained Pb(Zr,Ti)O_3-based ferroelectric glass-ceramic composites at relatively low sintering temperatures of below 900°C. A dense, homogeneous, and fine-grained ferroelectric glass-ceramic composite were achieved successfully.

3.3.1. Sintering behavior

Nano-sized $Pb(Zr_{0.53}Ti_{0.47})O_3$ (PZT) powder was prepared via a polymer-assisted sol–gel method sintered at 800°C for 2h [17]. Alternatively, PbO-B_2O_3-SiO_2 (PBS) ternary phase with molar ratio of 2:2:1 was selected and used to form the PBS sol precursor. Then, PBS sol precursor was poured into glass vessel and kept at room temperature for three days to form a gel. The translucent gel was then dried at 120°C for 10 h. The dried gel was dissolved in ethanol together with the nano-sized PZT powder. The mixture was then mixed thoroughly to form $Pb(Zr_{0.53}Ti_{0.47})O_3$–PbO-$B_2O_3$-$SiO_2$ (PZT–PBS) powder-solution suspension/slurry (gel solution) by conventional ball milling for 2 h. The mass ratio of the PBS dried gel powder ($W_{gel}(wt\%)$) to the nano-sized PZT powder was in the range of 3–30 wt%. The final slurry was dried at 120°C, calcined at 450°C for 2 h, and granulated to obtain PZT–PBS ceramic powders. The powders were uniaxially pressed into disk pellets with a diameter of 13 mm and a thickness of about 0.7–1 mm at a pressure of 4 MPa in a stainless steel die. All the pressed pellets were finally sintered at 900°C for 4 h in air atmosphere, except for 30 wt% sample which was sintered at 850°C.

3.3.2. Microstructure and properties

The XRD patterns of (a) PBS gel powder, (b) nano-sized PZT powder, (c)-(f) PZT–PBS glass-ceramics composites with various glass-gel contents are shown in Fig. 17. For PBS-glass system, no sharp peaks were detected indicating the amorphous nature of the PBS dried gel powder. For nano-sized PZT powder calcined at 800°C, the XRD pattern exhibited a pure perovskite PZT structure. In Figs. 17(c)–(f), the XRD patterns indicated the presence of highly pure and crystalline pervoskite PZT structure in the PZT–PBS glass-ceramics. No intermediate or interfacial as well as impurity phases were observed in all the samples, even for the ceramic with 30 wt% PBS gel content. These results indicated the success in synthesizing PZT–PBS ceramic composites using the modified hybrid processing route. As expected, with increasing wt% of glass gel content, the peak intensity was slightly diminished.

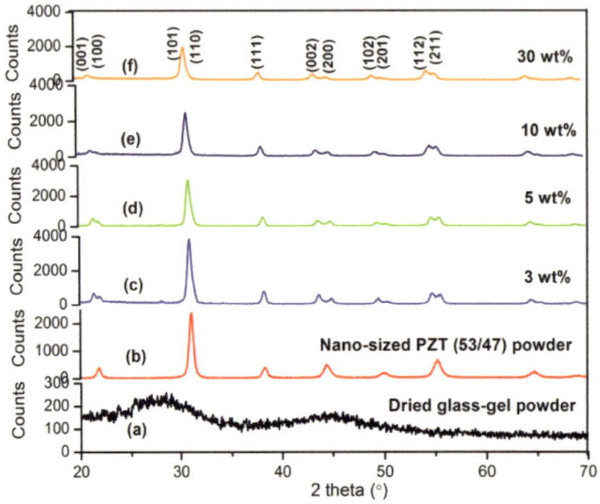

Figure 17. XRD patterns of PBS gel powder (a) dried at 120°C; (b) nano-sized PZT powder calcined at 800°C, together with (c)-(f) the glass-ceramic composites with different glass-gel additions.

The surface and cross-sectional morphologies of a representative sample of 5 wt% glass-gel content are shown in Figure 18. In Fig. 18(a), a dense, homogeneous and well-developed microstructure was obtained at low-firing temperature with a relative density of about 94% theoretical density. Moreover, it is also observed that the grain size of the PZT phases was about 1–2 μm. In Fig. 18(b), the cross-sectional SEM micrograph revealed a mixture of inter- and trans-granular fracture in the glass phase. The EDS analysis showed that the composition of the crystal grain (mark 2) contained only Pb, Ti, Zr and O elements; while for the grain boundary (mark 1), besides Pb, Ti, Zr and O elements, some Si element was detected. The minor amount of Si composition (in *Italics*) shown in the inset of Fig. 18(d) was negligible due to the error of the EDS detector (approximately ±5 at.%). On the basis of our results, we demonstrate that the modified hybrid process proposed in this study (i.e., gel solution in place of the sol precursor) can concisely control the amount of the crystalline phase content employed in the matrix. This new processing technique provides an easy, flexible, and reproducible feature, and should be extensively used in screen-printing, tape-casting or even in traditional ceramics and composites process to reduce the processing temperature.

Figure 19 shows the dielectric constant as a function of temperature for a typical 5 wt% sample at four different frequencies (1 kHz, 10kHz, 100 kHz and 1 MHz). The inset shows the corresponding temperature dependence loss tangents. The room-temperature dielectric constant was about 920 with a loss tangent of about 0.02. As the temperature increased, both the dielectric constant and the loss tangent increased. At about 400°C, the dielectric constant decreased with increasing temperature for different frequencies, except 1 kHz i.e. the dielectric constant of the sample at about 400°C did not drop with increasing temperature at 1 kHz. Thus, the dielectric spectra exhibited a characteristic dispersion at 400°C, namely, the Curie temperature. This indicates that the glass-ceramic composite had a strong tendency of diffusion phase transition. This diffused peak may be owing to the fine structure and the glass-ceramics containing $PbO\text{-}B_2O_3\text{-}SiO_2$ continuous matrix [27]. The apparent increase in

http://dx.doi.org/10.5772/53255

dielectric constant at elevated temperature was probably due to increase loss in this temperature range. This trend is consistent with the fact that PZT crystals, surrounded by a glassy matrix, have a higher electrical conductivity at higher temperature. In amorphous materials, the ease of polarizability, and hence dielectric constant, increases with temperature because of their relatively weak bonding structure. Correspondingly, as shown in inset of Fig. 19, the loss tangent increased markedly as the measuring temperature was increased above the Curie temperature.

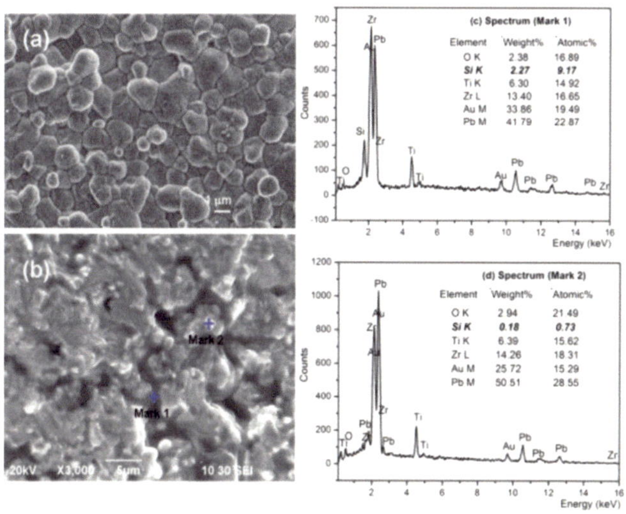

Figure 18. Surface and cross-sectional micrographs of a representative sample with 5 wt% glass-gel content, together with EDS analyses at two spots.

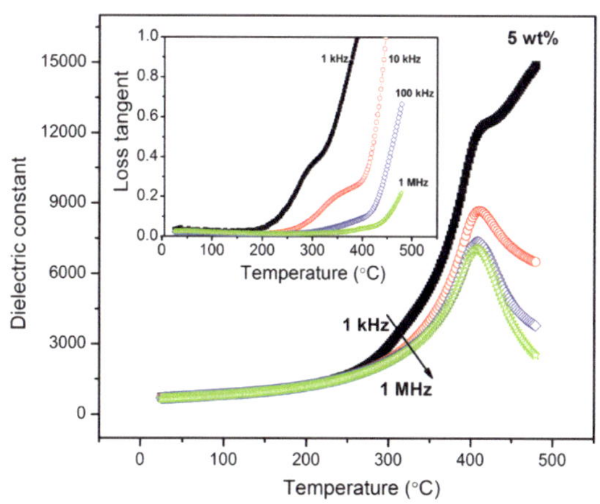

Figure 19. Temperature dependence dielectric constant for a typical 5 wt% sample at 1 kHz–1 MHz. Inset shows the loss tangent as a function of temperature.

Figure 20 shows the room temperature ferroelectric hysteresis loop ($P-E$) for a typical sample with 5 wt% glass-gel content at 100 Hz. The loop showed reasonable ferroelectricity of the embedded PZT (53/47) crystallites, demonstrating a high volume fraction of PZT phase existed in the sample. The measured P_s, P_r and E_c was 21.9 μC/cm², 10.8 μC/cm² and 2.19 kV/mm, respectively. These results demonstrate that these fine-grained glass-ceramics have a high application potential for micro-electronics.

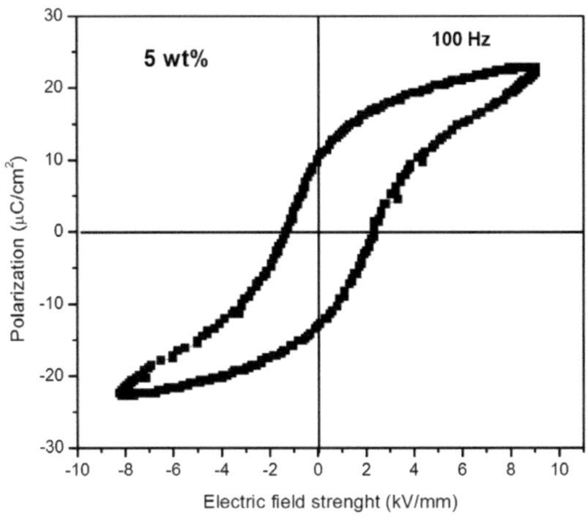

Figure 20. Ferroelectric hysteresis loop of a representative sample with 5 wt% glass-gel content at 100 Hz at room temperature.

3.4. Multiferroic magnetoelectric fine-grained ceramic composites

Following the general tendency of the microelectronic industry towards miniaturization and integration, it is of paramount important to find materials with best properties in small volume and possess more functions in the same structure. Multiferroic materials, which combine two or more primary ferroic orders such as ferromagnetic and ferroelectric orders [28], have attracted considerable research activity in recent years not only because they have the properties of both parental compounds, but also because they show multifunctionality caused by the coupling between the orders [19]. In here, the modified hybrid process as introduced above was employed to prepare the fine-grained multiferroic $Pb(Zr_{0.53}Ti_{0.47})O_3$-$(Ni_{0.5}Zn_{0.5})Fe_2O_4$ (PZT-NZFO) ceramic composites at a low sintering temperature of 900°C.

3.4.1. Sintering behavior

NZFO powder was prepared by a conventional solid-state process. Fe_2O_3, NiO, and ZnO powders were first mixed using conventional ball milling. After calcinating at 1000°C for 2 h, submicron-sized NZFO powder with an average size of 100-500 nm was obtained. Then, the calcined nano-sized PZT powder coated with glassy phase as described in section 3.3.1

was mixed with the NZFO powder by ball milling for 2 h. The resulting ceramic slurry was dried at 150°C for 10 h. The granulated powder was uniaxially pressed into disk (13 mm diameter, 0.50.7 mm thickness) at a pressure of 4 MPa in a stainless steel die. All the specimens were finally sintered in air at 900°C for 4 h. In the present work, wt% ratios of NZFO powder and glass coated nano-sized PZT particulates of 10/90, 15/85, 25/75, 50/50 and 60/40 were adopted (denoted as 10 wt%, 15 wt%, 25 wt%, 50 wt% and 60 wt% composites).

3.4.2. *Microstructure and properties*

Figure 21 shows the XRD patterns of PZT-NZFO composites at various mixing ratios. In Fig. 21(a), the diffraction pattern of NZFO powder calcined at 1000°C demonstrated a pure and crystalline Ni-Zn ferrite with spinel structure. For nano-sized PZT powder coated by glass and calcined at 450°C [Fig. 21(b)], all the peaks exhibited the presence of pure PZT phase with a perovskite structure. For the composites with 10, 15, 25, 50 and 60 wt% of ferrite contents [Figs. 21(c)-(g)], all the peaks were identified distinctly as either from ferroelectric (PZT) or ferrimagnetic (Ni-Zn ferrite) phases. Thereby, the preparation of a composite containing both ferroelectric and ferrimagnetic phases was successful.

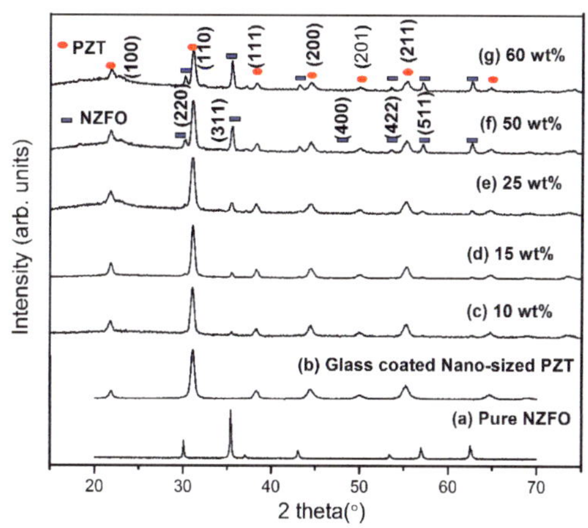

Figure 21. XRD patterns of the PZT-NZFO composite ceramics with various amount of NZFO phase contents sintered at 900°C for 4 h.

Figure 22 shows the surface morphologies of three selected composites with different ferrite contents: (a) 15 wt%, (b) 25 wt%, and (c) 50 wt%. General speaking, the surface of the composites were dense and homogeneous. Well-developed fine grains were observed with a good mixture of the two phases in the ceramics at low firing temperature. The EDX results suggested that grains of average size of 1- 2 μm were the PZT phase and those of size below 500 nm were the NZFO phase. It was clearly seen that the grain size of NZFO phase (about 100–500 nm) remained unchanged in all the samples. Here, well-crystallized NZFO powders were used as template (i.e. 3-0 connectivity pattern) for the in-situ preparation of the composites, the subsequent sintering for promoting the PZT grain growth practically did not change the crystallization of the NZFO component. Besides, the grain size of PZT phase in the ceramics was well-controlled and possessed a narrow size distribution with no formation of large grain. This means that excessive grain growth could be restrained by the modified hybrid process. For various compositions of PZT-NZFO ceramics sintered at same temperature, the proportion of the two phases determined from the relative peak intensity ratios of the parent phases roughly agreed with the nominal ones [Fig. 21(c)-(g)].

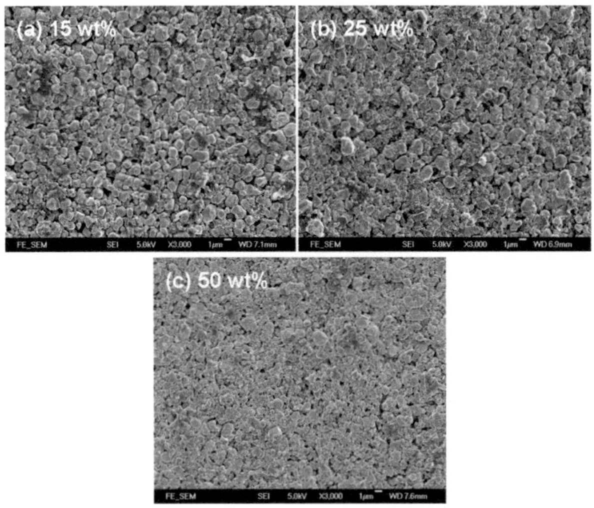

Figure 22. SEM surface micrographs of the composite ceramic with various amounts of NZFO content.

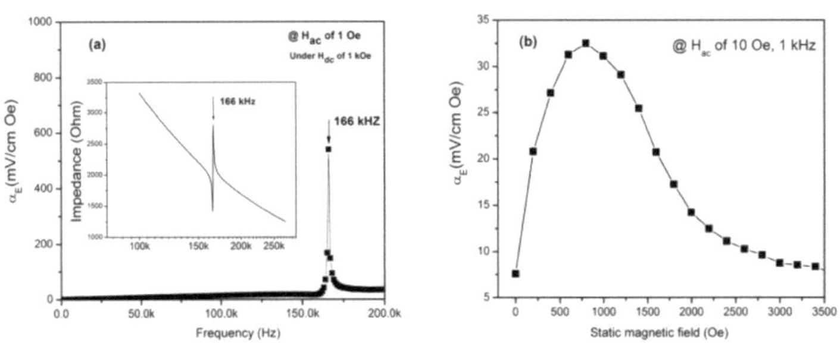

Figure 23. (a) Frequency dependence α_E under H_{dc}=1 kOe and H_{ac}=1 Oe, (b) the dc bias magnetic field dependence α_E under an ac magnetic input signal at 1 kHz and 10 Oe amplitude. The inset in (a) shows the impedance as a function of frequency in a poled 15 wt% composite.

Magnetoelectric (ME) coupling effect in the 15 wt% composite was studied by measuring the voltage response of the sample exposed to alternating and constant (bias) magnetic field as shown in Figure 23. In Fig. 23(a), a dramatic gain of up to 0.537 $Vcm^{-1}Oe^{-1}$ in the vicinity of electromechanical resonance was found. In Fig. 23(b), with an increase in dc bias magnetic field, α_E increased to a maximum value of about 33 $mVcm^{-1}Oe^{-1}$ at 1 kOe and then dropped with increasing dc bias field. It remained nearly constant beyond 3 kOe. We attribute the high α_E to high quality of fine-grained microstructure obtained by the modified hybrid processing at low firing temperature which avoids undesired phases and interphase diffusion of the constitutional atoms caused by the high-temperature sintering in conventional solid-state process. Furthermore, this low firing temperature lessens the thermal expansion mismatch due to the different sintering behavior between the constituent ferroelectric and ferrimagnetic phases. It also reduces the formation of microcracks and pores in the composites. Therefore, the ME response of the composite ceramics can be enhanced by transferring elastic strains fully when the components are in good contact. In this current work, the enhanced ME coupling in the fine-grained composite might find promising applications as high-frequency magnetic field sensors, transducers and magnetic field tunable microwave signal processors.

4. Summary

In this chapter, fine electronic ceramics, such as tunable microwave bulk barium-strontium titanate $(Ba_{0.6}Sr_{0.4})TiO_3$ (BST) ceramics, thick BST films, BST-MgO bulk ceramics, and ferroelectric $Pb(Zr_{0.53}Ti_{0.47})$ (PZT)–based glass-ceramics, as well as multiferroic $Pb(Zr_{0.53}Ti_{0.47})$-$(Ni_{0.5}Zn_{0.5})Fe_2O_4$ (PZT-NZFO) ceramic composites, have been successfully prepared by either hybrid or modified hybrid process at low sintering temperatures. For bulk BST ceramics, a well-densified and homogeneous microstructure was obtained at a sintering temperature of 1200°C, which was 200°C lower than that of the conventional solid-state process. The BST thick films were deposited on various substrates including Pt, Ag/alumina and Pd-Ag/alumina by spin-coating or screen-printing methods. High quality thick films were fabricated at annealing temperatures ranging between 600~1200°C. In addition, dense, homogeneous and well-sintered BST-MgO ceramics with grain size of 300 nm–2 μm were produced at sintering temperature < 1300°C, 100–200°C lower than that of the conventional solid-state ceramic process. At the same time, PZT based ferroelectric glass-ceramics were investigated. Dense and well-developed glass ceramic composites with a grain size below 2 μm were produced at a relatively low sintering temperature of 900°C by a modified hybrid process, and the apparent ferroelectric hysteresis loop demonstrated the high volume fraction of ferroelectric phase in the glass-ceramic composite, which overcome the drawback of both conventional melting/quenching glass-ceramic technology and sol-gel route. Finally, the modified hybrid process was introduced to prepare multiferroic magnetoelectric PZT-NZFO ceramic composite at a low sintering temperature of 900°C. Dense, well-developed and homogeneous microstructures were obtained with a well-controlled grain size and good mixing of the two phases in the ceramics. On the basis of our studies, we demonstrate that either the hybrid or the modified hybrid ceramic processing technology is a very promising new technology for fabrication of ceramics as well as composites. We consider that this approach should also be useful for almost all the ceramic materials and devices, realizing the dream of retaining a fine or ultra-fine grain size in fully sintered products at relatively low firing temperature by a simple, flexible and reproducible processing.

Acknowledgement

This work was supported by the Hong Kong Polytechnic University under Grant No. GYJ72.

Author details

Hongfang Zhang[1*], Chee-leung Mak[2], Helen Lai-Wa Chan[2] and Xi Yao[3]

*Address all correspondence to: zhf057@gmail.com

1 Department of Physics , Suzhou University of Science and Technology, Suzhou, China

2 Department of Applied Physics, The Hong Kong Polytechnic University, Hung Hom, Kowloon, Hong Kong

3 Functional Materials Research Laboratory, Tongji University, Shanghai, China

References

[1] Bell AJ. Ferroelectrics: The Role of Ceramic Science and Engineering. Journal of The European Ceramic Society 2008; 28 (7) 1307-1317.

[2] Setter N. and Waser R. Electroceramic Materials. Acta Materialia 2000; 48 (1) 151-178.

[3] Petrov VM, Bichurin MI, Laletin VM, Paddubnaya MN and Srinivasan G. Modeling of Magnetoelectric Effects in Ferromagnetic/Piezoelectric Bulk Composites. In: Fiebig M., Eremenko VV, Chupis IE (eds) MEIPIC-5: Magnetoelectric Interaction Phenomena in Crystals: proceeding of the fifth International Meeting on Magnetoelectric Interaction Phenomena in Crystals, NATO Advanced Research Workshop (ARW), 21-24 September 2003, Sudak, Ukraine.

[4] Kokubo T and Tashiro M. Dielectric Properties of Fine-Grained $PbTiO_3$ Crystals Precipitated in a Glass. Journal of Non-Crystal Solids 1973; 13(2) 328-40.

[5] Beall GH, Flats B, Pinckney LR. US Patent No. 5,491,116, February 13, 1996.

[6] Cai H, Gui Z, Li L. Low-sintering Composite Multilayer Ceramic Capacitors with X7R Specification. Materials Science and Engineering: B 2001; 83(1) 137-141.

[7] Newnham RE, Bowen LJ, Klicker KA. Cross LE., Composite Piezoelectric Transducers. Materials & Design 1980; 2(2) 93-106.

[8] Yang WH, Yu SH, Sun R, Du RX. Nano- and Microsize effect of CCTO Fillers on the Dielectric Behavior of CCTO/PVDF Composites. Acta Materialia 2011; 59(14) 5593-5602.

[9] PFISTERER INTERNATIONAL AG. Germany: Composite Insulators for High Voltage Applications. http://www.pfisterer.com

[10] Ruschau GR, Newnham RE, Runt J. Smith BE. 0–3 Ceramic/Polymer Composite Chemical Sensors. Sensors and Actuators 1989; 20 (3) 269-275.

[11] Dislich H. Glassy and Crystalline Systems from Gels: Chemical Basis and Technical Application. Journal of Non-Crystal Solid 1984; 63(1-2) 237-241.

[12] Muralt P. Ferroelectric Thin Films for Micro-sensors and Actuators: a review. Journal of Micromechanics and Microengineering 2000; 10(2) 136-146.

[13] Park GT, Choi JJ, Park CS, Lee JW, and Kim HE. Piezoelectric and Ferroelectric Properties of 1-μm-thick Lead Zirconate Titanate Film Fabricated by a Double-Spin-Coating Process. Appled Physic Letter 2004; 1794354 (85) 2322-2325.

[14] Bruchhaus R, Pitzer D, Primig R, Schreiter M, Wersing W. Sputtering of PZT Thin Films for Surface Micromachined IR-detector Arrays. Integrated Ferroelectrics 1999; 25(1-4) 1-11.

[15] Scott JF. High-dielectric Constant Thin Films for Dynamic Random Access Memories (DRAM). Annual Review of Materials Research 1998; 28: 79-100.

[16] [16]Mukhortov VM, Masychev SI, Golovko YI, Chub AV, and Mukhortov VM. Application of Nanodimensional Barium-Strontium Titanate Films in Tunable Microwave Devices. Technical Physics 2006; 51(10) 1359-1361.

[17] Zhang HF, Mak CL. Preparation and Characteristics of Fine-grained Ferroelectric Glass-ceramic Composites via a Modified Hybrid Route at Low Temperature Sintering. Journal of Electroceram 2011; 27:126-133.

[18] Zhang HF, Or SW, and Helen Chan WL, Yang F. Formation and Characterization of Three-ply Structured Multiferroic $Sm_{0.88}Nd_{0.12}Fe_{1.93}$–$Pb(Zr_{0.53}Ti_{0.47})O_3$ Ceramic Composites via a Solid Solution Process. Journal of European Ceramic Society 2011; 31(9) 1753-1761.

[19] Zhang HF, Or SW, and Helen Chan WL. Multiferroic Properties of $Ni_{0.5}Zn_{0.5}Fe_2O_4$–$Pb(Zr_{0.53}Ti_{0.47})TiO_3$ Ceramic Composites. Journal of Applied Physics 2008; 104: 104109.

[20] Zhang HF, Yao X, Zhang LY. Microstructure and Dielectric Properties of Barium Strontium Titanate Thick Films and Ceramics with a Concrete-like structure. Journal of American Ceramic Society 2007; 90 (8) 2333-2340.

[21] Zhang HF, Or SW, and Helen Chan WL. Synthesis of Fine-crystalline $Ba_{0.6}Sr_{0.4}TiO_3$–MgO Ceramics by Novel Hybrid Processing Route. Journal of Physics and Chemistry of Solids 2009; 70(8) 1218-1222.

[22] Zhang HF, Zhang LY, Yao X. Fabrication and Electrical Properties of Barium Strontium Titanate Thick Films by Modified Sol–Gel Method. Journal of Electroceramics 2008; 21(1-4) 503-507.

[23] Dislich H, Hinz P. History and Principles of The Sol-gel Process, and Some New Multicomponent Oxide Coatings. Journal of Non-Crystalline Solids 1982; 48 (1)11-16.

[24] Chang W, Sengupta L. MgO-mixed Ba0.6Sr0.4TiO3 Bulk Ceramics and Thin Films for Tunable Microwave Applications. Journal of Applied Physics 2002; 92(7) 3941.

[25] Sengupta L, Ngo E, Stowell S, O'Day M, Lancto R. US Patent No 5,427,988, 1995.

[26] Yao K, Zhang LY, Yao X, Zhu WG. Controlled Crystallization in Lead Zirconate Titanate Glass-Ceramics Prepared by the Sol–Gel Process. Journal of American Ceramic Society 1998; 81 (6) 1571-76.

[27] Zhai JW, Yao X & Zhang LY. The High Frequency Properties and Crystallization of $PbTiO_3$ Glass-Ceramics by Sol-Gel Process. Journal of Electroceramics 2000; 5(3) 211-216.

[28] Schmid, H. Multi-ferroic magnetoelectrics. Ferroelectrics 1994; 162 (1) 317-338.

Section 2

Sintering Of Metals

Chapter 7

Sintering of Ternary Copper Alloys (Powder Metalurgy) – Electrical and Mechanical Properties Effects

W. A. Monteiro, J. A. G. Carrió, M. A. Carvalhal,
A. K. Okazaki, C. R. da Silveira and M. V. S. Martins

Additional information is available at the end of the chapter

http://dx.doi.org/10.5772/53258

1. Introduction

In order to stay successful in an environment of competing shaping technologies, the manufacturers must also in the future improve every single processing step, which involves in decrease of the compaction press downtime caused by tool-setting interactions or tool readjustments.

The foreword of Paul Beiss (RWTH Aachen) in the book "Modeling of Powder Die Compaction", edited by Peter R. Brewin, Olivier Coube, Pierre Doremus and James H. Tweed (2008, Springer-Verlag London Limited) shows us that the "*Die compaction of powders that develop green strength on compaction is the absolutely dominating forming technology for powdered materials. Areas of application are structural parts, hardmetal and ceramic indexable inserts, pharmaceutical tablets, electrical contacts, filters, hard magnets, soft magnetic composites, friction materials and many others. In particular, multi cross-sectional net-shape geometries have been gaining importance continuously, because the ability to deliver complex shapes with higher and higher productivity has contributed to competitive advantages over alternative forming techniques. Since the raw material usage is better than 90% in die compaction, even in areas that could be served by competing manufacturing technologies, die compaction of powders is often the most economical solution. The industries applying this technique have seen tremendous innovation especially in shape capability, reproducibility and productivity over the last 15 years that has resulted in high added value, astonishing growth rates and increased employment also in high labor cost countries*".

To increase the strength, ductility and formability keeping good electric conductivity of these specific copper alloys, there have been used special thermal treatments, as well as variations in the chemical composition. In this work were synthesized ternary copper alloys

INTECH
open science | open minds

© 2013 The Author(s). Licensee InTech. This chapter is distributed under the terms of the Creative Commons Attribution License http://creativecommons.org/licenses/by/3.0), which permits unrestricted use, distribution, and reproduction in any medium, provided the original work is properly cited.

with the aim of obtaining a steadily alloys production by sintering (powder metallurgy) in laboratory scale and optimizing the electrical and mechanical properties.

The alloy elements are added to copper to improve its mechanical resistance, ductility and thermal stability, without causing considerable costs on its form, electric and thermal conductivity and its resistance to the corrosion, typical characteristic aspects of pure copper. It is well known that the mechanical resistance in metallic alloys depends on the size, type, form and regular distribution of precipitations, which is also fundamental to obtain an electrical conductivity similar to that of the copper matrix.

Samples of ternary Cu-Ni based alloys (precursors of high purity; powder form; weight %; alloy element: Al; Sn; Cr), initially compressed, sintered and sometimes homogenized were characterized by optical metallography (microstructure), mechanical strength (hardness Vickers), electrical properties (electrical conductivity) and structural characterization (X-rays diffraction) for the study of the influence of sintering (powder metallurgy).

The major contribution of these sintered metallic alloys (compression, homogenization and sintering of metallic powder) is the evolution of their electrical conductivity, mechanical properties (hardness tests) and microstructure changes after appropriate thermal treatments.

In all analyzed samples the conductivity showed a strong dependence of sintering temperature and also of special thermal treatment. Specially, some compositions of Cu-Ni-Cr alloy perform very interesting physical properties, like mechanical resistance of 1000 MPa and an average electrical conductivity of more than 75% IACS.

The copper has different applications in the modern society due to its excellent thermal conductivity, electric properties, resistance to the corrosion, resistance to the fatigue and good mechanical properties. Connectors, contact switches, heaters, valves, piping, pots for absorption of solar energy, radiators for automobiles, current driver, electronic driver, contact sheets, elements of thermostats are common applications. Copper can be used with high purity or with addition of alloy elements (Ni, Sn, Be, Pt, Cr, Nb, Pb, Al) that increase the principal properties.

Products based on copper alloys such as porous material filters, electric friction equipment, contacts and structural parts can be manufactured through the process of powder metallurgy, which have the advantages of making fine grained homogeneous structures, forming complicated shapes with close dimensional tolerances and the ability to produce parts with a superior surface finishing. These advantages reduce or eliminate costly machining processes and allow less scrap loss, compared to other forming methods [1-13].

Diverse types of products based on copper alloys can be manufactured through the process of powder metallurgy: porous, material filters, electric friction equipment, contacts and structural parts [14-18]. The aim of this work is to obtain metallic alloys with high mechanical strength and high electric conductivity after adequate optimization of sintering and thermal treatments (powder metallurgy) followed by structural, microstructural, electrical and mechanical characterization of ternary Cu-Ni based alloys.

The alloy elements are added to copper with the purpose to improve its resistance, ductility and thermal stability, without causing considerable damages on its form, electric and thermal conductivity and resistance to the corrosion, typical characteristic aspects of pure copper. The choice of these current alloys is related to the studies carried through previously in ternary alloys to similar copper nickel base the chosen ones [1-7; 13-16].

The process of making powders, compacting them into useful shapes and then sintering them is costly, but the finished parts have some specific advantages over wrought or cast parts. The main advantages are: the possibility to make fine grained homogenous structures; the ability to form complicated shapes with close dimensional tolerances. Costly machining processes are thus reduced or eliminated and consequently there is less scrap loss compared to other forming methods [13-16].

The mechanical resistance in metallic alloys depends on the size, type, form and a regular distribution of precipitations, which is also fundamental to obtain an electrical conductivity similar to that of the copper matrix. To increase the mechanical resistance, ductility and formability keeping good electric conductivity of these alloys, there have been used special thermal treatments as well as variations in the chemical composition.

The main production of metallic materials is acquired by casting [1-8]. The contribution to the production of metallic parts by powder metallurgy is increased of consistent outline, supported for inestimable advantages. The list of benefits in the industrial processes of sintering is great and, not rare, surprises the coordinators of production of different industries that had not yet tried the technology and attend the explanation of specialists in the area [15-18].

It is therefore most economical to use powder metallurgy for the high volume production of small, intricately shaped, and/or very precise parts such as gears and links. In addition, the process offers the potential to produce a wide variety of alloys with different material properties such as high temperature toughness and hardness.

This work attempts to obtain systematic stages of the sintering and homogenization of the ternary copper nickel alloys utilizing powder metallurgy. Being an alternative process to the conventional processes, the powder metallurgy also allows, in some cases, the structural manufacture of parts and components in economically and more advantageous conditions.

Varied types of products of copper based alloys can be manufactured through the process of powder metallurgy for electric friction, contacts and structural parts. The alloy elements are added to copper with intention to improve the resistance, the ductility and the thermal stability, without causing considerable damages on its shape, electric and thermal conductivity, and also resistance to the corrosion [3-5].

Next to casting, mechanical forming and machining, powder metallurgy (PM) technology is an important method of manufacturing metal parts. Undesirable characteristics of ingot based metals can be greatly reduced, and desired properties of metals, which would normally not alloy easily, can be achieved by combining different metal powders or mixtures of metal and non-metal powders [14-18].

Costly machining processes are thus reduced or eliminated and consequently there is less scrap loss compared to other forming methods. It is therefore most economical to use powder metallurgy for the high volume production of small, intricately shaped, and/or very precise parts such as gears and links. In addition, the process offers the potential to produce a wide variety of alloys with different material properties such as high temperature toughness and hardness [17, 18].

Powder production and mixing is a highly specialized and complex process which produces custom made powder mixes designed to satisfy the needs of a specific application. A good powder mix not only has the ability to produce the required properties of a specific alloy, but also needs to facilitate handling, compacting and sintering.

For instance, the easy flow of powder and its capability to mix evenly with other powders is important for an even powder distribution before pressing, and ensures uniform properties of the finished part. Thinking in the theory for creep of dispersion-strengthened metals developed by Rösler and Arzt [19] to predict the optimum particle size for a given service temperature and to illustrate the need for a high interfacial energy. Resistance to coarsening leads to a requirement for low diffusivity and solubility of particle constituent elements in the matrix.

Based on the needs for a low difference in the coefficients of thermal expansion to minimize thermal-mechanical fatigue damage and low diffusivity and solubility of the constituent elements, several candidate ceramic phases are compared using a weighted property index scheme. The results of this quantitative comparison suggest that CeO_2, MgO, CaO and possibly Y_2O_3 may be good candidates for the dispersed phase in a copper matrix [18 - 20].

2. Experimental conditions

Powder production and mixing is a highly specialized and complex process which produces custom made powder mixes designed to satisfy the needs of a specific application. A good powder mix not only has the ability to produce the required properties of a specific alloy, but also needs to facilitate handling, compacting and sintering. Experimentally, for instance, the easy flow of powder and its capability to mix evenly with other powders is important for an even powder distribution before pressing, and ensures uniform properties of the finished part. In research and failure analysis, metallography is a major tool used to develop new products and improve manufacturing processes. In addition to chemical analysis, quality control also includes physical methods for checking density, dimensional changes, flow rate etc.

As the density of the compacted and sintered part influences its key properties of strength, ductility and hardness, a specific porosity is critical. For process control, metallography is used to check porosity, non-metallic inclusions and cross-contamination.

http://dx.doi.org/10.5772/53258

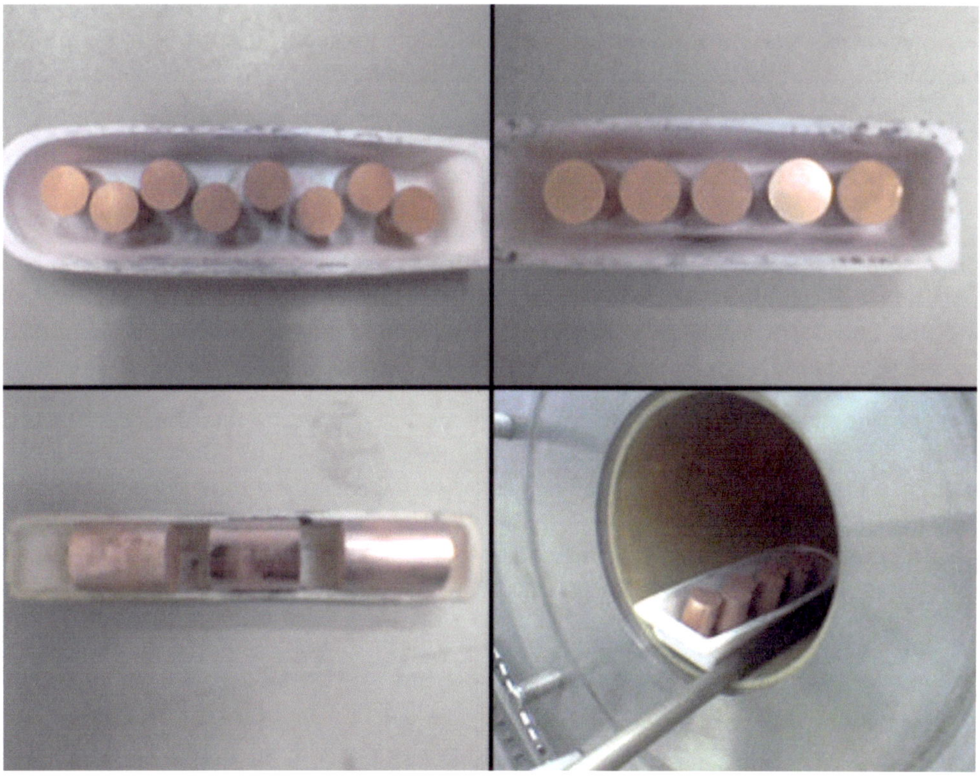

Figure 1. Ternary Cu-Ni alloys distributed in alumina boat for sintering and homogenizing heat treatment in vacuum furnace.

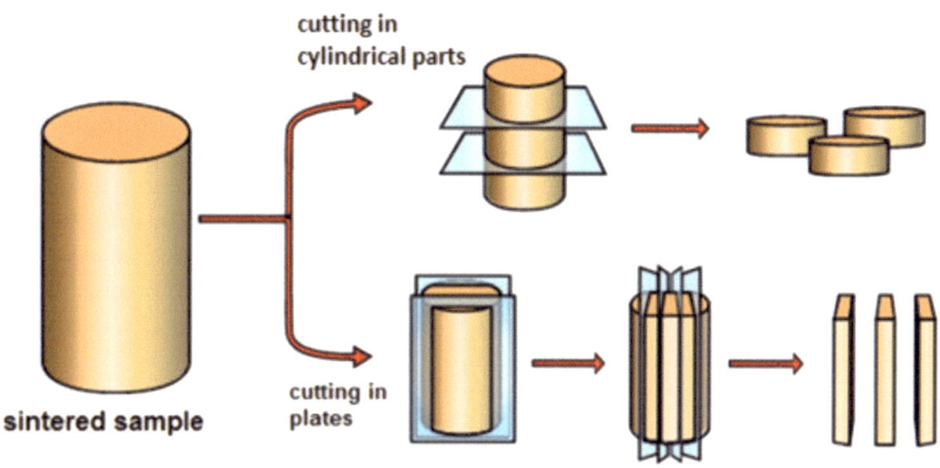

Figure 2. Chart for cutting planes of the cylinder samples on ternary Cu-Ni alloys

For the production of components the mixed powders are first compacted under high pressure in a suitable system. At this stage the part has the geometrical features of the finished component, but not its strength and is called the "green" part. The bonding occurs through diffusion between adjacent particles. In order to develop the mechanical and physical properties of the material, metallurgical bonding has to take place through sintering at high temperature in a sintering furnace. The bonding occurs through diffusion between adjacent particles (Fig.1).

Cold mounting of the sintered and homogenization samples was done by optical and hardness studies. The compacts were grinded with 400, 600, 800, 1000 and 1200 SiC papers followed by fine wet wheel polishing (diamond or alumina pastes). Vickers hardness of the polished specimens was measured on a hardness tester (HXD 1000TM – Pantec, load of 100 g).

To avoid oxidation, which would impair the inter-particle bonding, the sintering process is conducted in a protective atmosphere or convenient vacuum. The bonding increases the density, and pressed and sintered powder metal parts generally contain some residual porosity depending of the initial conditions. The as-pressed compacts were conventionally sintered in a vacuum Carbolite furnace that had a stabilized thermal gradient (hot zone) of about 150 mm.

In research and failure analysis, metallography is a major tool used to develop new products and improve manufacturing processes. The procedure for cutting the samples for metallographic and mechanical tests is represented in the scheme in Figure 2.

In addition to chemical analysis, quality control also includes physical methods for checking density, dimensional changes, flow rate etc. Acidic $FeCl_3$ was used as the etchant and the microstructures of selected etched samples were observed in an optical microscope. In the case of Cu-Ni-Cr alloys the apparent density of the samples was measured before and after the thermal treatments.

Special samples for electrical conductivity studies were characterized using an Agilent 4338B Milliohmmeter, through resistance measurements, that were performed repeatedly in all samples to avoid eventual errors.

X - rays powder diffraction data of various samples of the three alloys were collected with a conventional diffractometer with fixed monochromator. From samples of the Cu-Ni-Cr alloys the x rays data were collected using synchrotron radiation with a Hubert diffractometer at the beam line XRD2 of the Brazilian Synchrotron LNLS (Campinas, São Paulo State, Brazil), using energies of 8keV and 10keV and counting time of 2 seconds. The structures of the samples were refined by the Rietveld Method using GSAS [23]. The peak profiles were simulated using function number 4 of GSAS.

2.1. Cu-Ni-Sn alloys

The as-pressed compacts were conventionally sintered in a vacuum Carbolite furnace that had a hot zone of about 150 mm. At the utilized composition the Cu-Ni-Sn alloys can be consolidated by solid state sintering. The most important conditions are presented in Table 1.

Condition	Premixed		
Compaction pressure	20MPa		
Chemical alloy composition (wt %)	Cu-0.5%Ni-0.5%Sn; Cu-1.0%Ni-0.5%Sn; Cu-1.0%Ni-1.0%Sn; Cu-3.0%Ni-3.0%Sn; Cu-5.0%Ni-5.0%Sn		
Sample dimensions	ϕ = 10.2mm and h = 14.8mm (cylinder)		
Sample weight	6.5 g		
Sintering temperature and sample conditions	Sintering temperature (K)	Condition for premixed	Vacuum pressure
	823 – 1073	Solid state sintering	10^{-4} mBar
Sintering time	$1.8x10^3$ to $5.4x10^3$ s		
Homogenization time	3.6 to 28.8 $x10^3$ s		

Table 1. Sintering parameters of Cu-Ni-Sn alloys

2.2. Cu-Ni-Al alloys

The Cu-Ni-Al alloys can be consolidated by solid state sintering. The most important conditions are presented in Table 2.

Condition	Premixed		
Compaction pressure	20MPa		
Chemical alloy composition (wt %)	Cu-0.5%Ni-0.5%Al; Cu-1.0%Ni-0.5%Al; Cu-1.0%Ni-1.0%Al; Cu-3.0%Ni-3.0%Al; Cu-4.0%Ni-4.0%Al		
Sample dimensions	ϕ = 10.2mm and h = 14.8mm (cylinder)		
Sample weight	6.5 g		
Sintering temperature and sample conditions	Sintering temperature (K)	Condition for premixed	Vacuum pressure
	823 – 1073	Solid state sintering	10^{-4} mBar
Sintering time (s)	$1.8x10^3$ to $5.4x10^3$		
Homogenization time (s)	3.6 to 28.8 $x10^3$		

Table 2. Sintering parameters of Cu-Ni-Al alloys

2.3. Cu-Ni-Cr alloys

High purity powders of copper, nickel and chromium were mixed for a suitable time and then compacted under 20 MPa in a cold uniaxial pressing. Afterwards, the specimens were sintered in temperatures varying from 973 K up to 1073 K in a high vacuum Carbolite furnace that had a hot zone of about 150mm under vacuum. At last, the samples were homogenized at 823 K under vacuum, for special times. For process control, metallography was used to check porosity, non-metallic inclusions and cross-contamination. The most important preparation conditions of the samples are presented in Table 3.

Condition	Premixed		
Compaction pressure (MPa)	20		
Chemical alloy composition (wt %)	Cu-0.5%Ni-0.5%Cr; Cu-1.0%Ni-0.5%Cr; Cu-1.0%Ni-1.0%Cr; Cu-1.5%Ni-0.5%Cr		
Sample dimensions (mm)	ϕ = 10.2 and h = 12.4 (cylinder)		
Sample weight (g)	6.4		
Sintering temperature and sample conditions	Sintering temperature (K)	Condition for premixed	Vacuum pressure (mBar)
	823 – 1073	Solid state sintering	10^{-4}
Sintering time (s)	$1.2x10^3$ to $5.4x10^3$		
Homogenization time (s)	21.6 to $172.8x10^3$		

Table 3. Sintering parameters of Cu-Ni-Cr alloys

3. Results and discussion

3.1. Cu-Ni-Sn alloys

The important data with copper-nickel-tin samples is shown in Table 4 concerning mixing, compacting, sintering, homogenizing treatments and also values of hardness and electrical conductivity.

Cu-Ni-Sn Alloys	Sintering		Homogenizing		Mechanical Resistance	Electrical Conductivity
	T(K)	t(s)	T(K)	t(s)	(MPa)	(% IACS)
Cu-1.0%Ni-0.5%Sn	823	5400	--	--	(520 ± 9)	(35 ± 4)
Cu-1.0%Ni-0.5%Sn	823	5400	773	21600	(410 ± 8)	(40 ± 4)
Cu-1.0%Ni-1.0%Sn	823	5400	773	32400	(540 ± 9)	(37 ± 4)
Cu-5.0%Ni-5.0%Sn	823	5400	773	21600	(620 ± 9)	(34 ± 4)

Table 4. Mechanical and electrical properties of the copper-nickel-tin alloys obtained by powder metallurgy

The mechanical resistance in metallic alloys depends on the precipitation distribution to obtain similar electrical conductivity of the copper (matrix). To increase the strength, ductility and formability keeping good electric conductivity of these alloys, there have been used special thermal treatments as well as variations in the chemical composition. At the present time the mechanical strength (620MPa) and electrical conductivity (40%IACS) media values indicate a fine appliance for these alloys utilizing powder metallurgy as a substitute of conventional metallurgy processing.

The Rietveld refinements using x rays powder diffraction data indicate that the utilized amounts of nickel and tin did not distorted the copper matrix structure and the refined profile parameters did not showed any detectable effects in the microstructure.

Concerning the microstructural aspects, figures 3 to 5, show optical micrographs of some Cu-Ni-Sn alloys. Fine grains presences but with inadequate porosity and second phases show that a new homogenization treatments will be necessary to overcome this situation and also investigations with scanning and transmission electron microscopy to identify the presence of second phase on these alloys.

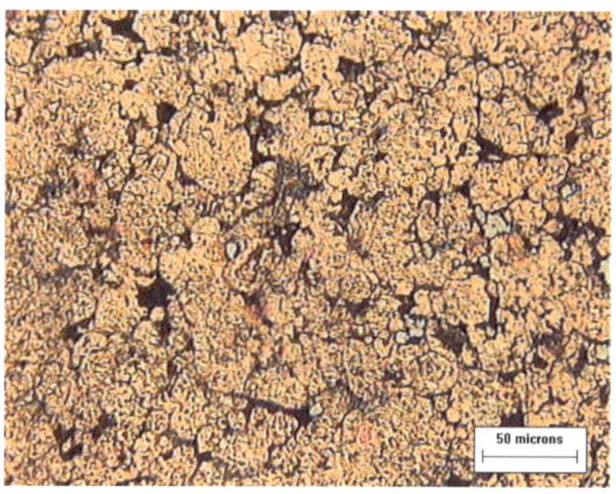

Figure 3. Optical micrograph of the ternary alloy Cu-1.0%Ni-0.5%Sn, cold compact (20MPa) and sintered at 823K for 1200s

Figure 4. Optical micrograph of the ternary alloy Cu-1.0%Ni-0.5%Sn, cold compact (20MPa) and sintered at 823K for 1200s.

Figure 5. Optical micrograph of the ternary alloy Cu-1.0%Ni-1.0%Sn, cold compact (20MPa) and sintered at 823K for 5400s.

3.2. Cu-Ni-Al alloys

The Table 5 resumes some data with the copper-nickel-aluminum alloys samples concerning mixing, compacting, sintering, homogenizing treatments and also values of hardness and electrical conductivity obtain the best condition for electrical and mechanical application with powder metallurgy processing.

Cu-Ni-Al Alloys	Sintering		Homogenizing		Mechanical Resistance	Electrical Conductivity
	T(K)	t(s)	T(K)	t(s)	(MPa)	(% IACS)
Cu-1.0%Ni-0.5%Al	1053	5400	--	--	(420 ± 90	(30 ± 3)
Cu-1.0%Ni-0.5%Al	1053	5400	773	21600	(280 ± 7)	(35 ± 4)
Cu-1.0%Ni-1.0%Al	1053	5400	--	--	(240 ± 5)	(29 ± 3)
Cu-1.0%Ni-1.0%Al	1053	5400	773	32400	(370 ± 9)	(30 ± 3)
Cu-5.0%Ni-5.0%Al	1053	5400	773	21600	(400 ± 8)	(28 ± 3)

Table 5. Some mechanical and electrical properties of the copper-nickel-aluminum alloys obtained by powder metallurgy

The mechanical resistance in metallic alloys depends on the precipitation distribution to obtain similar electrical conductivity of the copper (matrix). At the present time the mechanical strength (400MPa) and electrical conductivity (35%IACS) values indicate a good application for these alloys utilizing powder metallurgy instead conventional metallurgy.

Concerning the microstructural aspects, figures 6 to 10 show optical micrographs of some Cu-Ni-Al alloys. Fine grained presences but with inadequate porosity and second phases until now show that new homogenization treatments will be necessary to overcome this situation and also investigations with scanning and transmission electron microscopy to identify the presence of second phase on these alloys.

The Rietveld refinements using X rays powder diffraction data indicate that the utilized amounts of nickel and aluminum do not distorted the copper matrix structure significantly. Furthermore, no special broadening of peak profiles was detected, which indicates that crystallite sizes are not affected.

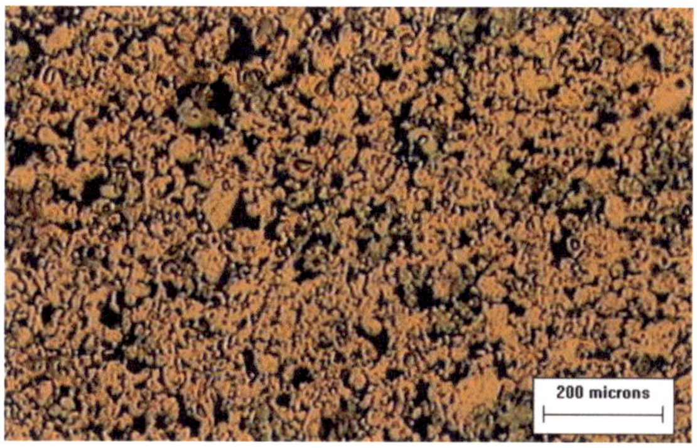

Figure 6. Optical micrograph of the ternary alloy Cu-1.0%Ni-1.0%Al, cold compact (20MPa) and sintered at 923K for 1200s

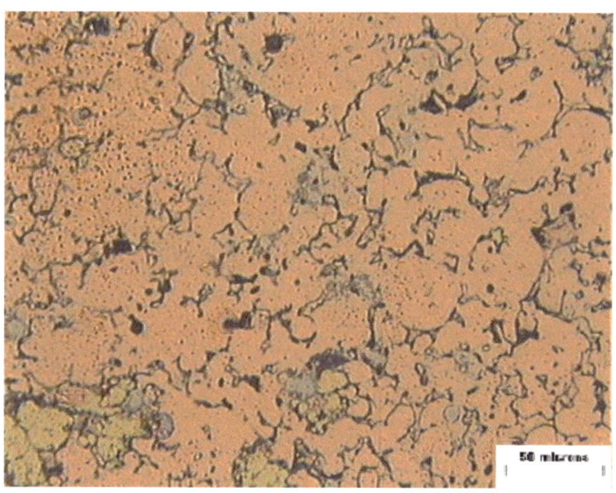

Figure 7. Optical micrograph of the ternary alloy Cu-1.0%Ni-0.5%Al, cold compact (20MPa) and sintered at 1053K for 5400s.

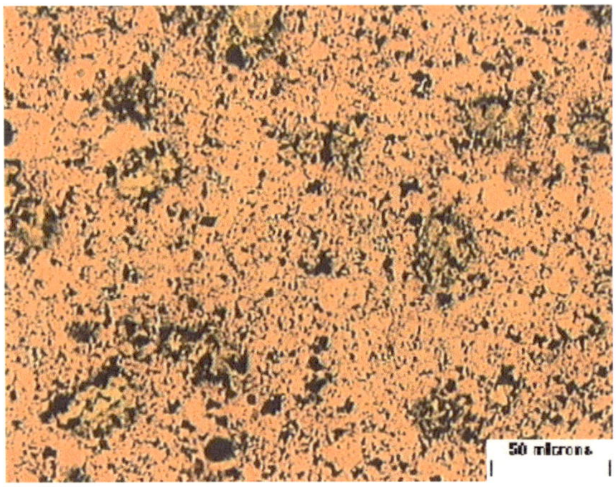

Figure 8. Optical micrograph of the ternary alloy Cu-1.0%Ni-0.5%Al, cold compact (20MPa) and sintered at 1053K for 5400s).

Figure 9. Optical micrograph of the ternary alloy Cu-5.0%Ni-5.0%Al, cold compact and sintered at 1053K for 5400s and also homogenized at 773K during 21600s.

http://dx.doi.org/10.5772/53258

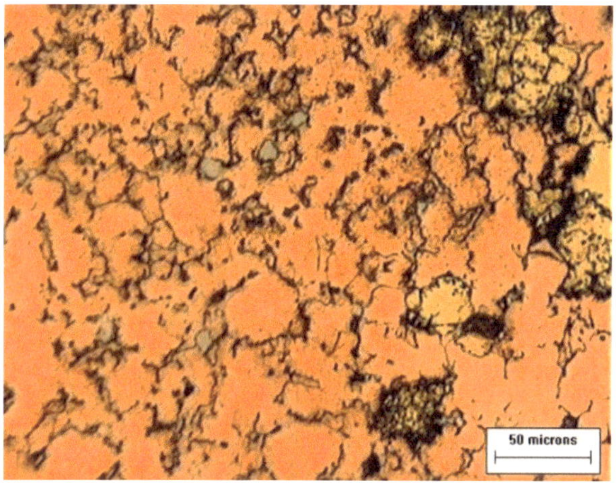

Figure 10. Optical micrograph of the ternary alloy Cu-1.0%Ni-1.0%Al, cold compact and sintered at 1053K for 5400s and also homogenized at 773K during 21600s.

3.3. Cu-Ni-Cr alloys

Due to the results for the compositions Cu-1.0%Ni-0.5%Cr and Cu-1.0%Ni-1.0%Cr, special attention was given to these samples. The mechanical strength of 1000MPa and the electrical conductivity of 79%IACS for the Cu-1.0%Ni-0.5%Cr homogenized sample indicate a good application for this alloy, using powder instead conventional metallurgy. This sample also presented an increase in its apparent density of approximately 5% after the thermal treatments. Using the synchrotron X - rays data with energy 8 keV, a Rietveld refinement was performed for only for copper with Fm-3m space group, because the amounts of nickel, chromium and copper oxide were almost insignificant. The discordance factors were R_{wp} = 13.96%, R_p = 7.47%, R_{Bragg} = 0.4% and χ^2 = 25.12%. Using the Gaussian broadening formula from GSAS, a crystallite size of approximately 99 nm was calculated for this sample.

For the sample with composition Cu-1.0%Ni-1.0%Cr the increase in apparent density was less than 2% after the thermal treatments. This sample showed also a significant but lower electrical conductivity of $2.6.10^{7}\ ^{-1}.m^{-1}$, i.e. 44.40%IACS. The Rietveld refinement using synchrotron radiation with energy 8 keV was performed until discordance factors of R_{wp} = 2.95%, Rp = 2.18%, R_{Bragg} = 10.48% and χ^2 = 2.89%. The amounts of nickel and chromium were not detectable for this sample and a crystallite size of approximately 124 nm was calculated using the Gaussian broadening formula.

Although both values are of the order of 100 nm, the lower size could be due to the more effective homogenization process of the first composition (Cu-1.0%Ni-0.5%Cr), which also exhibit the higher electrical conductivity and mechanical strength. Further thermal treatments shall be applied to continue the study of these alloys and to obtain the best conditions for electrical and mechanical applications using powder metallurgy processing.

The results of mechanical and electrical characterization in dependence of treatment of the samples are resumed in Table 6.

Cu-Ni-Cr Alloys	Sintering		Homogenizing		Mechanical Resistance	Electrical Conductivity
	T (K)	t (10^3 s)	T (K)	t (10^3 s)	(MPa)	(% IACS)
Cu-0.5%Ni-0.5%Cr	1053	5.4	773	32.4	(330 ± 5)	(32 ± 3)
Cu-0.5%Ni-0.5%Cr	1073	5.4	--	--	(420 ± 5)	(37 ± 3)
Cu-1.0%Ni-0.5%Cr	1053	5.4	773	21.6	(1000 ± 9)	(79 ± 5)
Cu-1.0%Ni-0.5%Cr	1053	5.4	--	--	(460 ± 5)	(33 ± 3)
Cu-1.0%Ni-0.5%Cr	1073	5.4	--	--	(430 ± 5)	(27 ± 3)
Cu-1.0%Ni-1.0%Cr	1073	5.4	--	--	(370 ± 4)	(37 ± 3)
Cu-1.0%Ni-1.0%Cr	1073	5.4	773	172.8	(400 ± 5)	(44 ± 4)
Cu-1.5%Ni-0.5%Cr	1053	5.4	773	32.4	(370 ± 5)	(35 ± 3)

Table 6. Mechanical and electrical properties of the copper-nickel-chromium alloys obtained by powder metallurgy

Optical micrographs of some of the samples are presented in Figures 11 to 18 to show their microstructural aspects. Fine grained presences but with inadequate porosity and second phases indicate that further treatments will be necessary to overcome this situation and also investigations with scanning and transmission electron microscopy to identify the presence of second phase on these alloys.

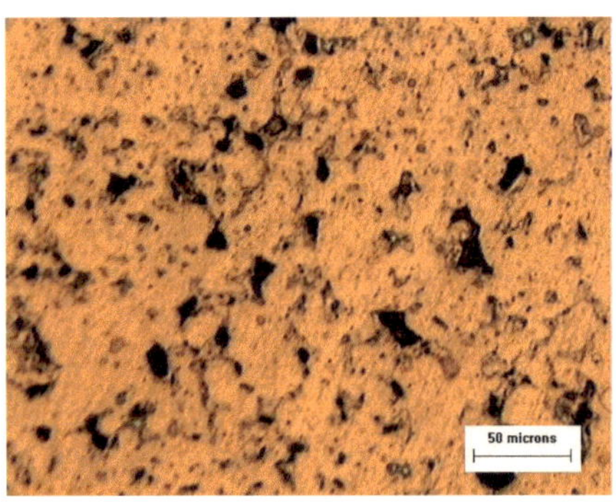

Figure 11. Optical micrograph of the ternary alloy Cu-0.5%Ni-0.5%Cr, cold compact (20MPa) and sintered at 1053K for $5.4x10^3$s and also homogenized at 773K for $32.4x10^3$s.

http://dx.doi.org/10.5772/53258

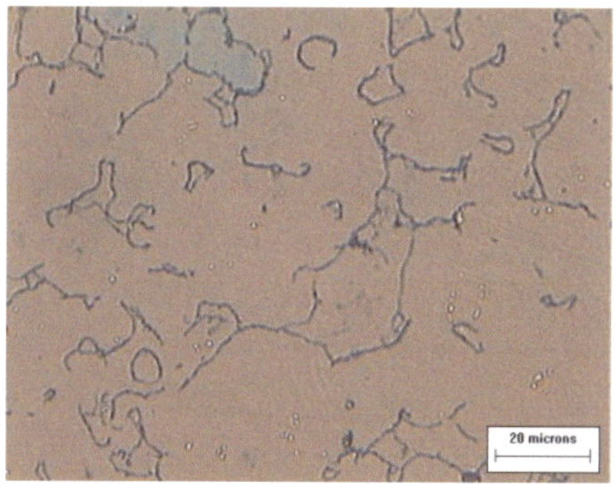

Figure 12. Optical micrograph of the ternary alloy Cu-1.0%Ni-0.5%Cr, cold compact (20MPa) and sintered at 1053K for $5.4x10^3$s and also homogenized at 773K for $21.6x10^3$s.

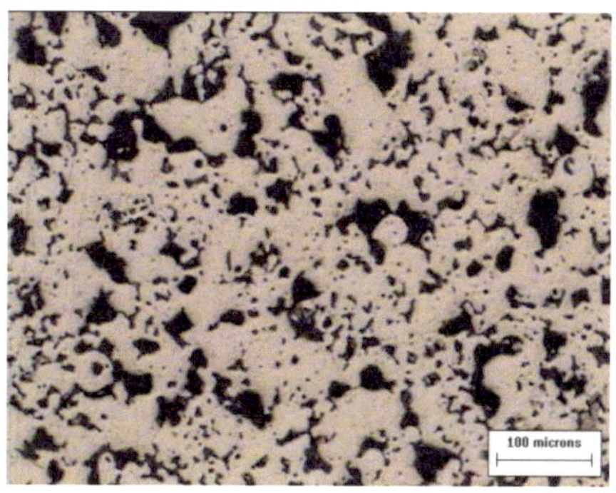

Figure 13. Optical micrograph of the ternary alloy Cu-1.0%Ni-0.5%Cr, cold compact (20MPa) and sintered at 1053K for $5.4x10^3$s and also homogenized at 773K for $21.6x10^3$s.

Figure 14. Optical micrograph of the ternary alloy Cu-1.0%Ni-0.5%Cr, cold compact (20MPa) and sintered at 1073K for $5.4x10^3$s and homogenized at 773K for $172.8x10^3$s.

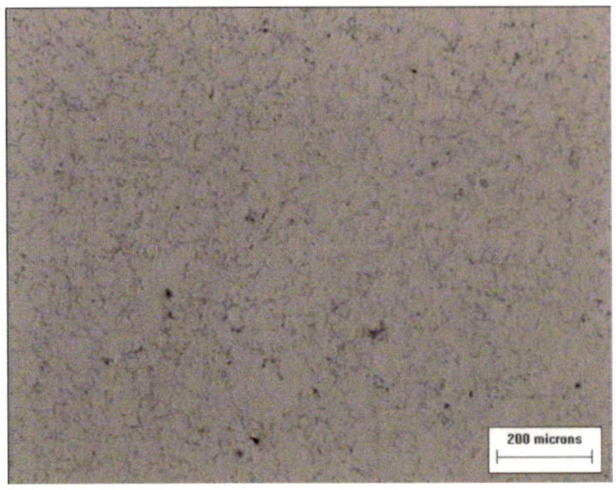

Figure 15. Optical micrograph of the ternary alloy Cu-0.5%Ni-0.5%Cr, cold compact (20MPa) and sintered at 1073K for $5.4x10^3$s and homogenized at 773K for $172.8x10^3$s.

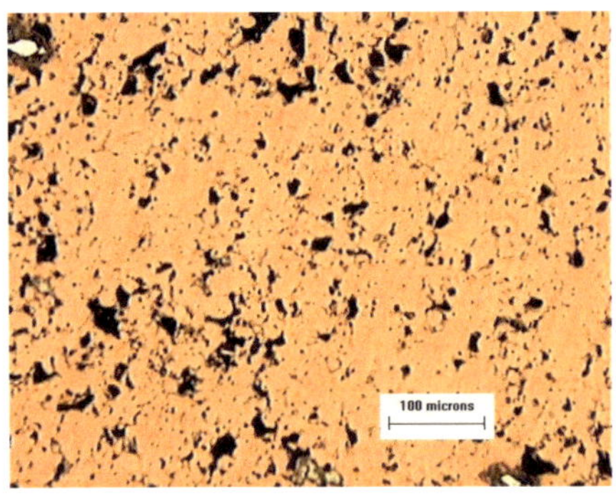

Figure 16. Optical micrograph of the ternary alloy Cu-0.5%Ni-0.5%Cr, cold compact (20MPa) and sintered at 1053K for $5.4x10^3$s and homogenized at 773K for $172.8x10^3$s.

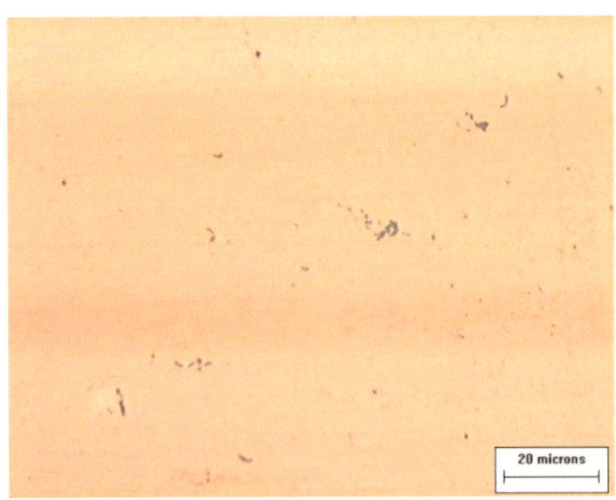

Figure 17. Optical micrograph of the ternary alloy Cu-1.0%Ni-0.5%Cr, cold compact (20MPa) and sintered at 1073K for $5.4x10^3$s and homogenized at 773K for $172.8x10^3$s.

Figure 18. Optical micrograph of the ternary alloy Cu-1.0%Ni-1.0%Cr, cold compact (20MPa) and sintered at 1073K for $5.4x10^3$s and homogenized at 773K for $172.8x10^3$s.

4. Conclusions

The practical powder metallurgy processing steps on the copper-nickel-base alloys corroborate a first-rate mechanical strength and electrical conductivity values (Cu-Ni-Sn: 620Mpa, 40% IACS; Cu-Ni-Al: 400MPa, 35% IACS; Cu-Ni-Cr: 1000MPa; 79% IACS) that indicate a good quality application for these alloys utilizing powder metallurgy instead conventional metallurgy processing. Furthermore, the results for these alloys open the possibility to search and make fine grained homogeneous structures with interesting physical properties, also the capacity to produce parts with a superior surface finishing in different dimensions.

Acknowledgements

The authors would like to thank to UPM (PIBIC and Mackpesquisa), IPEN, CNPq, LNLS and CAPES (Brazil) for financial support.

Author details

W. A. Monteiro[1,2], J. A. G. Carrió[1], M. A. Carvalhal[1], A. K. Okazaki[1], C. R. da Silveira[1] and M. V. S. Martins[1]

1 School of Engineering, Presbyterian Mackenzie University, São Paulo, Brazil

2 Materials Science and Technology Center, Energetic and Nuclear Researches Institute, São Paulo, Brazil

References

[1] Crane J. and Winter J., Copper: Properties and alloying, Encyclopedia of Materials Science and Engineering, vol 2, Ed. MB Bewer, Pergamon Press and the MIT Press 1986, p. 848 – 855.

[2] Taubenblat P. W., Copper Selection of high conductivity alloys, Encyclopedia of Materials Science and Engineering, vol 2, Ed. MB Bewer, Pergamon Press and the MIT Press 1986, p. 863-866.

[3] ASM Specialty Handbook: Copper and Copper Alloys, Metal, chapter 1, section 1, ASM International, 2001.

[4] Brooks C. R., "Heat Treatment, Structure and Properties of Nonferrous Alloys", ed. ASM, Metals Park, 1988, chapter 8.

[5] Butts A., "Copper, the Science and Technology of the Metal, its Alloys and Compounds", ed. Reinhold Publishing Corporation, New York, 1954, 3th printing 1960.

[6] Gosh G., Kiyake J. and Fine M. E., JOM, March (1997), p. 56-60

[7] Rotem A., Shechtman D. and Rosen A., Metall. Trans. A.1988, vol. 19A, p. 2279-2285.

[8] Steiner D., Beddoe R., Gerold V., Kostorz G. and Schmelczer R., Scripta Metall., 1983, vol. 17, p. 733-36

[9] Guha A.," Development of a high-strength, high-conductivity Cu-Ni-Be Alloy, high conductivity Copper and Aluminum Alloys, eds. E. Ling and P. Taubenblat W., TMS – AIME Publ. 1984, p. 133-145.

[10] Sakai Y., Inoue K., and Maeda H., Acta Metall. Mater. 43 (1995), p. 1517.

[11] Monteiro W. A., Proceedings of IX CBECIMAT, Águas de São Pedro- SP, Brazil dez. 90, p. 546-549.

[12] Monteiro W. A., Silveira M. A. G. e Júlio Jr. O., Metalurgia & Materiais, v. 51, may 95, p. 440-444.

[13] Monteiro W. A., Cosandey F. and Bandaru P., The Effect Of Thermomechanical Treatment On The Microstructure Of A Cu-Ni-Be Alloy, Proceedings of THERMEC'97, Wollongong, Australia, July 1997.

[14] Stadtler W. A., Production of Metallurgy Parts ASTM, Powder Metallurgy, Ohio, p. 449-463, 1989.

[15] Ghadiri M., Farhadpour, F.A., Clift R., Seville, J. P. K. Particle characterization size and morphology, The Institute of Metals Series on Powder Metallurgy - An Overview. London, p. 56-75, 1991.

[16] Powder metallurgy - An overview. 1. Ed. The Institute of Metals Series on Powder Metallurgy, London, Great Britain, 1991.

[17] Kaysser W. A. Solid State Sintering. The Institute of Metals Series on Powder Metallurgy - an overview. London, p 45-53, 1991.

[18] Thümmler F. & R. Oberacker, Introduction to Powder Metallurgy, The Institute of Materials, 1993, ISBN 0-901716-26-X)

[19] Rösler J., Arzt E., A new model-based creep equation for dispersion strengthened materials, Acta Metallurgica et Materialia, Volume 38, Issue 4, April 1990, p. 671-683

[20] J Groza. R. and Gibeling J. C., Principles of particle selection for dispersion-strengthened copper, Materials Science and Engineering: A, Volume 171, Issues 1-2, 1 November 1993, Pages 115-125

[21] Okazaki A. K., M Martins. V. S., Carvalhal M. A., Monteiro W. A., Carrió J. A. G., Mechanical and Structural Characterization of Cu-Ni-Ag/Y_2O_3 Composites Obtained by Powder Metallurgy, Proceedings of the Eighth International Latin American Conference on Powder Technology, November 06 to 09, Florianopolis, SC, Brazil, p 397 – 402, 2011.

[22] Carrió, J. A. G., Monteiro W. A., Rodrigues V. A., Terence M. C., Masson T. J., De Miranda L. F., Structural analysis of influence of dopants in the electrical conductivity of CuNi alloys, European Powder Diffraction Conference – EPDIC 11, 19-22 September 2008, Warsaw, Poland.

[23] Larson, A. C.; Von Dreele, R. B. General Structure Analysis System (GSAS), Los Alamos National Laboratory Report LAUR 86-748, 2004 (software).

Chapter 8

Lead and Zinc Sintering

Baojun Zhao

Additional information is available at the end of the chapter

http://dx.doi.org/10.5772/54618

1. Introduction

1.1. Background of lead and zinc production

Galena (PbS) and sphalerite (ZnS) are the major ore minerals for Pb and Zn respectively. Pyrometallurgical processing of Pb and Zn metals usually includes four stages: ore dressing, smelting, drossing and refining. Sinter – blast furnace route is one of the most important smelting technologies which was adopted from iron and steel industries. Lead-rich sinter is treated in lead blast furnace to produce lead metal. Sometimes it is difficult to separate lead and zinc minerals, Imperial Smelting Process is used to treat mixed zinc-lead ores for production of lead and zinc metals simultaneously. A typical flowsheet of lead-zinc smelter is shown in Figure 1 [1].

Imperial Smelting Furnace (ISF) is also called zinc blast furnace which was developed for mixed zinc-lead ores. Charges to the furnace are lump sinter, hot briquettes, and coke. Slag and lead are continuously tapped from the bottom of the furnace into a forehearth where separation occurs. The slag containing zinc overflows to the fuming plant to recover zinc. Liquid lead containing copper and precious metals is transferred by ladle to the decopperizing plant. Zinc vapor generated in the furnace enters the lead splash condenser where it is quenched and absorbed by fine lead droplets splashed by the rotors. The hot lead containing zinc is pumped from the condenser to a brick-lined cooling launder fitted with immersion boiler panels. On cooling the recirculating lead stream, crude zinc is released and separated by specific gravity from lead. The lead is then returned to the splash condenser.

The smelting of lead and zinc includes oxidisation (sintering) and reduction stages. The main chemical reactions involved are

Sintering: $2PbS + 3O_2 = 2PbO + 2SO_2$

$2ZnS + 3O_2 = 2ZnO + 2SO_2$

INTECH
open science | open minds

© 2013 The Author(s). Licensee InTech. This chapter is distributed under the terms of the Creative Commons Attribution License http://creativecommons.org/licenses/by/3.0), which permits unrestricted use, distribution, and reproduction in any medium, provided the original work is properly cited.

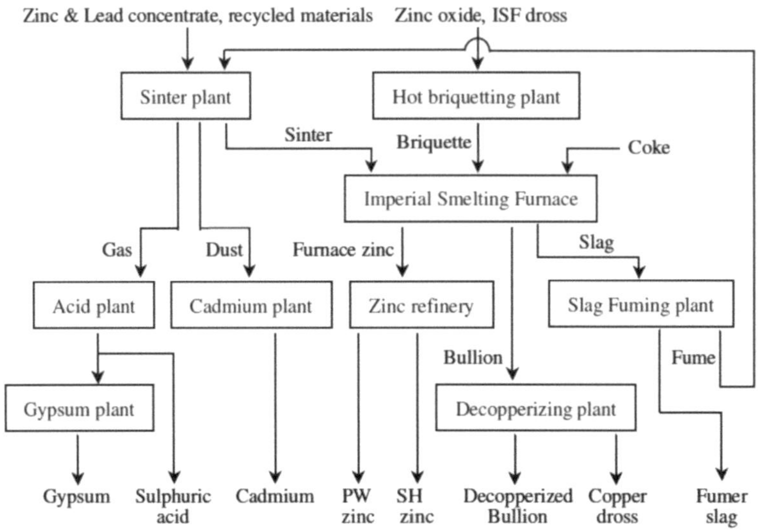

Figure 1. A typical flowsheet of lead-zinc smelter [1]

Reduction: $2PbO + C = 2Pb + CO_2$

$PbO + CO = Pb + CO_2$

$2ZnO + C = 2Zn + CO_2$

$ZnO + CO = Zn + CO_2$

1.2. Why sintering

Lead and zin metals are produced in blast furnace. In a blast furnace, ore, coke and flux are continuously supplied through the top of the furnace, while air (sometimes with oxygen enrichment) is blown into the lower section of the furnace, so that the chemical reactions take place throughout the furnace as the material moves downward. The downward flow of the ore and flux in contact with an upflow of hot, carbon monoxide-rich combustion gases is a countercurrent exchange process. Fine particles cannot be used as the feed of a blast furnace because they will be blown out by the high pressure gas flow. Zinc and lead sulphides occur as small particles in the rock. There are only a few percentage of zinc and/or lead present in commercial ores and they have to be concentrated by mineral dressing before smelting. The fine particles produced during the dressing cannot be fed into a blast furnace directly. The suitable feed for a blast furnace should be strong lump made of oxides which can be obtained by sintering. The main goals of sintering are 1) produce hard oxide lump; 2) remove sulphur; 3) produce strong SO_2 for acid plant.

1.3. Sintering process

Sinter is the primary feedstock for blast furnace to produce lead and zinc metals and it is obtained by the oxidation sintering of fine particulate zinc/lead sulphide concentrates in

http://dx.doi.org/10.5772/54618

updraught sinter machines. A typical flowsheet of sinter process is shown in Figure 2. The product is a partially fused porous agglomerate, which is crushed and large lump material is fed to the blast furnace. Fine sinter material is recycled or returned to the sinter feed to act as a support for fresh feed. The return sinter and fresh feed (sulphide concentrate, fluxes, fume, sludge) are agglomerated to form granulated spherical feed with 5-10 mm diameter to sintering machine.

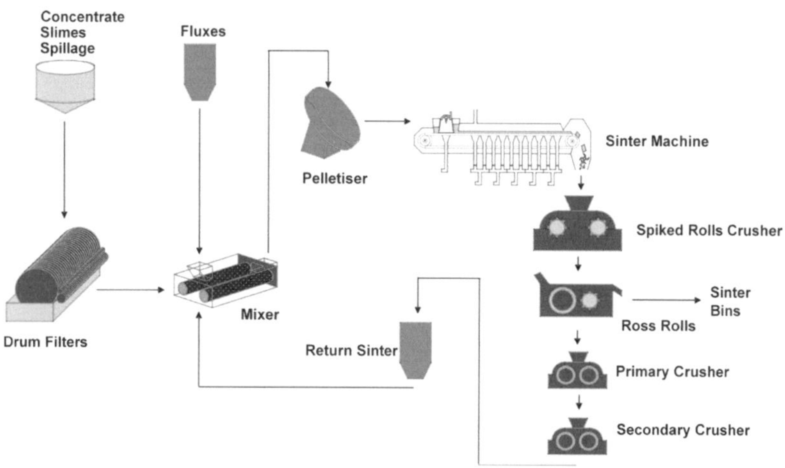

Figure 2. A typical flowsheet of sinter process

Sample	Composition (wt%)								
	Zn	Pb	Fe	CaO	SiO_2	Al_2O_3	MgO	Cu	S
concentrate	33-42	15-19	4-9	1-5	2-6	0.7-2	0.2-1	0.5-1.3	19-26
sinter	37-44	15-22	7-12	3-6	3-5	0.6-2	0.2-1	0.6-1.2	0.3-1.6

Table 1. Composition ranges of concentrate and sinter for ISF

Sample	Composition (wt%)								
	Zn	Pb	Fe	CaO	SiO_2	Al_2O_3	MgO	Cu	S
concentrate	4-7	38-46	6-8	2-11	5-13	0.7-2	0.2-0.6	0.6-0.7	9-12
sinter	4-9	36-52	8-14	6-12	7-11	0.8-2.4	0.3-2	0.3-1.2	1.1-3.4

Table 2. Composition ranges of concentrate and sinter for lead blast furnace

2. Sinter quality and conditions affecting sinter-making

2.1. Sinter quality

To obtain high productivity in all packed bed processes, it is essential to maintain a high bed voidage. In lead and zinc blast furnace this high bed voidage is achieved through careful control of sinter characteristics, ensuring in particular that the sinter does not deform under the process conditions experienced in the furnace. The quality of sinter is associated with its properties such as macrostructure and microstructure, and represented quantitatively by softening temperature.

2.1.1. Macrostructure

The macrostructures of the sinter varied from partially-fused feed granules to well-fused material. Most sinters were well-fused and physical strength of these sinters is high at room temperature. An example of a well-fused lead sinter is presented in Figure 3a. Presence of original feed granules (see Figure 3b) in some sinters indicates that during sintering these materials did not attain high temperature. Similarly, there are also well-fused part (Figure 4a) and partially-fused part (Figure 4b) in ISF sinter. The partially-fused or even non-fused parts of the sinter do not have required physical strength and they usually contain high level of sulphur. These sinter particles will be returned to the sinter machine as "core" of the feeds.

(a) (b)

Figure 3. Typical macrostructures of lead sinter

2.1.2. Microstructure

Pieces of sinter lump with approximately 20mm diameter were selected from different parts of the as-received sinters. These samples were mounted, polished and carbon-coated for examination. Microstructural analysis was performed using a Phillip XL30 scanning electron microscopy (SEM). The typical microstructures of "as-received" lead sinter samples are presented in Figure 5.

In the lead sinter samples melilite [2(Ca,Pb)O (Zn,Fe,Mg)O 2(SiO_2,Al_2O_3)], spinel [(Zn,Fe^{2+},Mg)O (Fe^{3+},Al)$_2$$O_3$], lead oxide (PbO), calcium silicate sulphate [$Ca_5(SiO_4)_2SO_4$] and

http://dx.doi.org/10.5772/54618

(a) (b)

Figure 4. Typical macrostructures of ISF sinter

glass are found to be common phases. In addition, some other phases such as undissolved blast furnace slag, di-calcium silicate [2(Ca,Pb)O SiO_4], Pb-Zn silicate (larsenite, $PbZnSiO_4$), Pb-Ca silicate ($Pb_3Ca_2Si_3O_{11}$), lead sulphide (PbS), copper sulphide (CuS), Ca sulphates (CaO $CaSO_4$ and 2CaO $CaSO_4$), silica (SiO_2), Pb metal and Cu metal are also observed in lead sinters. Examples of the phases are shown in Figures 5a to 5d.

It is found that there are two extremes of melilite structures, high aspect ratio (length to thickness) and low aspect ratio melilite. At high volume fractions the high aspect ratio melilite forms a framework structure which is expected to support the sinter lump at high temperature in the blast furnace. Low aspect ratio melilite are always present as isolated crystals surrounded by matrix material which consists of the relict of the melt from which this phase is formed.

Inspection of a large number of lead sinter microstructures shows that structures of the lead sinters are not uniform; even within a 20mm diameter sinter granule the structure of the lead sinter can vary significantly. It can be seen from Figure 5a that framework melilite structure is present and the proportions of the low melting point matrix phases are very low in this area. However, significant proportions of lead metal, remaining blast furnace slag and sulphides are also observed in the same sample. Various structures such as framework melilite, blocky melilite, remaining blast furnace slag and lead metal can always be observed in each sample. This indicates that it is difficult to characterize lead sinter quality simply on the basis of microstructure.

Typical microstructures of the ISF sinter from different lumps are presented in Figure 6. Zincite [(Zn,Fe)O] and spinel [$(Zn,Fe^{2+},Mg)O$ $(Fe^{3+},Al)_2O_3$] were found to be the major phases present in all ISF sinters. The matrix including glass, lead oxide, larsenite ($PbZnSiO_4$), di-calcium silicate [2(Ca,Pb)O SiO_4] is also present in all samples. In addition, melilite [2(Ca,Pb)O (Zn,Fe,Mg)O 2(SiO_2,Al_2O_3)] and lead-zinc sulfides are also common phases observed in the ISF sinter. The microstructure of the ISF sinter is generally determined by the shape and the arrangement of the zincite phase. There are usual two typical microstructures in the sinter lump: framework zincite and blocky zincite. In the region of the framework zincite the zincite has relatively high aspect ratio and forms continuous interlocking refractory structure (Figure 6a). In the region of the blocky zincite the zincite has relatively lower aspect ratio and exists

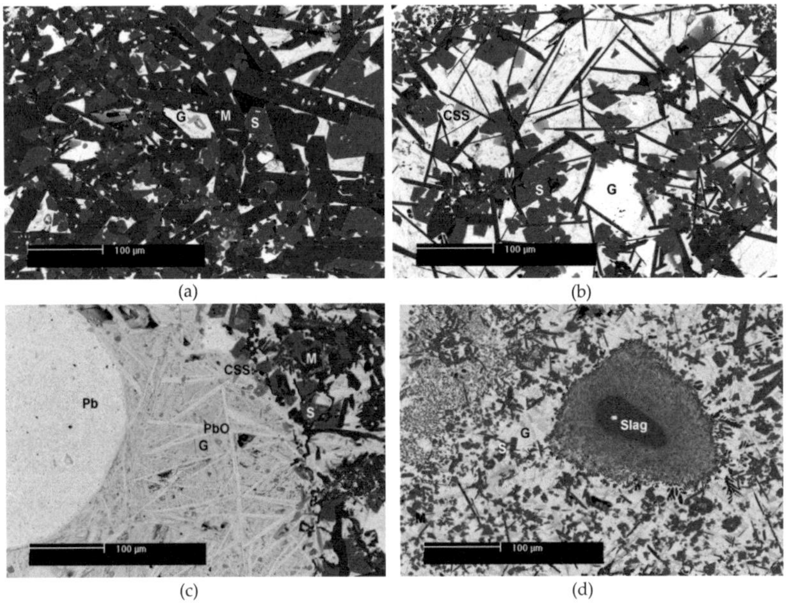

Figure 5. Typical microstructures of lead blast furnace sinter, CSS=$Ca_5(SiO_4)_2SO_4$; G=glass; M=melilite; Pb=lead metal; S=spinel [13]

as isolated phase (Figure 6b). In some of the sinters the framework zincite is dominated and in other sinters the blocky zincite is dominated.

In general, the microstructure of the ISF sinter is more uniform than that of the lead sinter. The microstructures shown in Figure 5 can be observed within a single lump of the lead sinter. However, the microstructures shown in Figure 6 are not usually to be observed within a single lump of the ISF sinter.

2.1.3. Softening temperature

The extent of the deformation is quantitatively expressed by the softening of the sinter on controlled heating rate. Softening temperature of lead sinter lump was measured in air using a vertical tube furnace shown schematically in Figure 7 [2]. The furnace was preheated to 650 °C for lead sinters and 800 °C for ISF sinters. A sinter sample with approximately 30 mm height was placed on an alumina plate and slowly raised to the hot zone of the furnace. The displacement probe (8mm OD alumina sheath with R-type thermocouple inside) was gently lowered onto the top of the sample. An aluminum plate is attached to the alumina sheath so that the total mass of the probe is approximately 235 g for lead sinters and 400 g for ISF sinters. The displacement meter was positioned on the top of the aluminum plate, which is fixed to the probe. The displacement meter was set to about half of its shaft's travel to allow for expansion and softening of the sample. The displacement meter gauge was set to zero. The furnace temperature was increased with a uniform heating rate of 400 degrees per hour, which was controlled by a programmable temperature controller (Philips KS 40). Initially the sinter

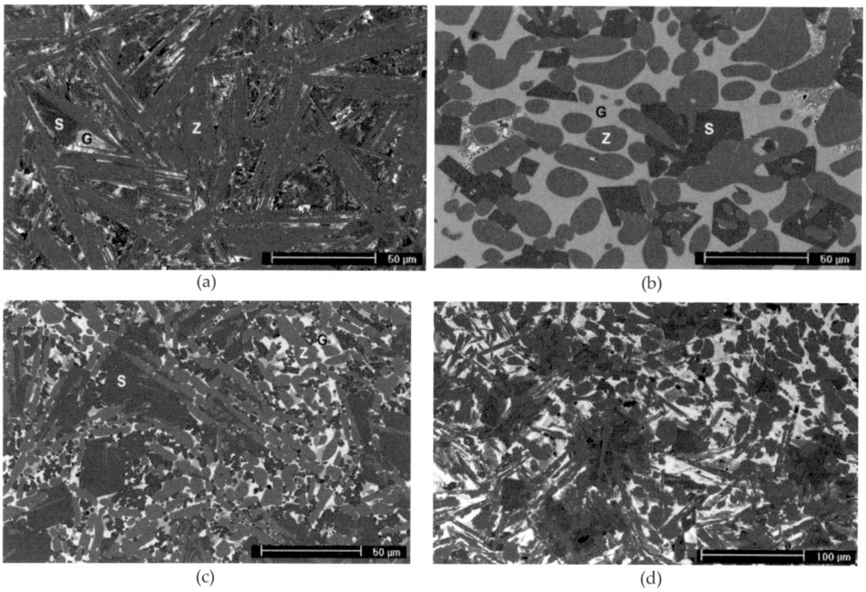

Figure 6. Typical microstructures of ISF sinter, G=glass; S=spinel; Z=ZnO [3]

was observed to expand slightly on heating, and then the displacement meter starts to go down as result of sinter softening. The measurement was stopped when the reading of the displacement meter reaches maximum (10 mm) corresponding to 30% reduction of the sinter height or the furnace temperature reached 1200 °C. A computer system with a data logger was used to simultaneously record the displacement and temperature readings throughout the tests through the thermocouple and the displacement meter.

From the data collected a softening curve can be drawn by plotting the displacement versus temperature. Typical examples of the results of the sinter softening test carried out on the samples obtained from the same batch are presented in Figure 8 for lead sinter and Figure 9 for zinc sinter. Several measurements (up to 7) were conducted for each sinter. The difference in the softening-behavior within the same sinter reflects the inherent variation of the sinter samples.

Ideally a sinter is softening when certain proportion of liquid is formed and softening curve should be smooth. However, the softening curves shown in Figures 8 and 9 do not have the same behavior. In addition to the idea softening curve some of lead sinters could show several steps softening (Figure 8). This may be caused by a few reasons: 1) sinter is porous material. If the displacement meter falls into the pore a false contraction can be shown; 2) lead sinter is not uniform in composition. Some low melting temperature materials such as lead metal, sulphide and sulphate can melt locally at relative lower temperature. Sinter lump itself does not softening but softening occurs locally; 3) lead sinter varies in structure and composition from area to area.

Initial softening has been observed to occur at as low as 700 °C in some lead sinter samples. Final softening temperatures, the temperatures at which complete collapse of the sample under load takes place, were found to be in the range of 800-1100 °C for the lead sinters. Several separate measurements were conducted for each sinter material, and it is clear that there is some inherent variation in the behavior of the samples from the same sinter. It can be seen from Figure 8 that sudden partial collapse of the sinters occurs in some of the samples as the probe penetrates the sample. Examination of these samples has shown that samples having small variations in softening temperature have more uniform microstructure. Samples having a large variation in softening temperature, and which exhibited frequent and random collapse with temperature, have uneven microstructures. These local differences in the sinter structure in the form of changes in the phases present, and the proportions of phases, are the result of local compositional variations. These variations in turn lead to differences in the softening and melting temperatures, and the apparent viscosities of the materials.

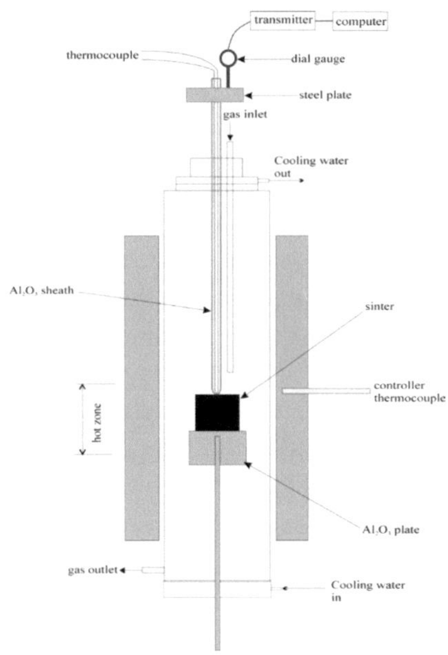

Figure 7. Apparatus used for softening temperature test [2]

ISF sinters have relative uniform composition and structure within a batch of sample. The softening curves for a given ISF sinter do not variation as great as the lead sinter shown in Figure 8. The softening curves shown in Figure 9 are from different sinter samples and they can be described as three typical types. Type 1 (G3 in Figure 9) shows a softening curve in which the sinter softening starts at approximately 950 ºC and the displacement reaches 5 mm at 1140 ºC. Type 2 (M3 in Figure 9) shows a softening curve in which the sinter softens very slow. Although the softening started at 1050 ºC, this was not great, and the sample still retained its strength at temperature. Because the softening test stopped at 1200 ºC it was difficult to obtain a temperature in which the displacement drops dramatically. Type 3 is the common softening curve observed in most of the ISF sinters and their softening behaviors are between

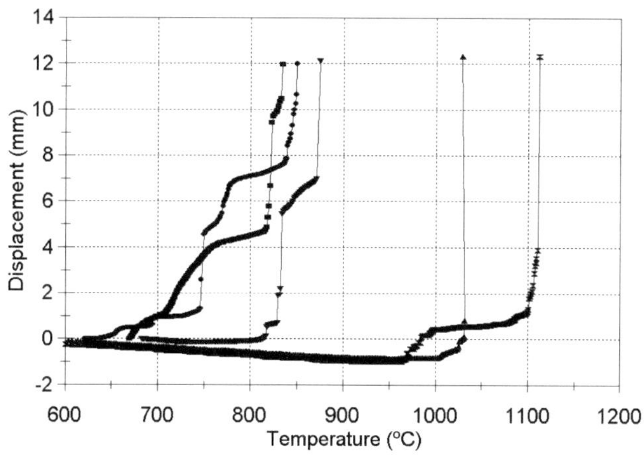

(heating rate 400 °C/hour, load 235g and 8 mm diameter sheath)

Figure 8. Softening Curves of a Typical Lead Sinter in Air [13]

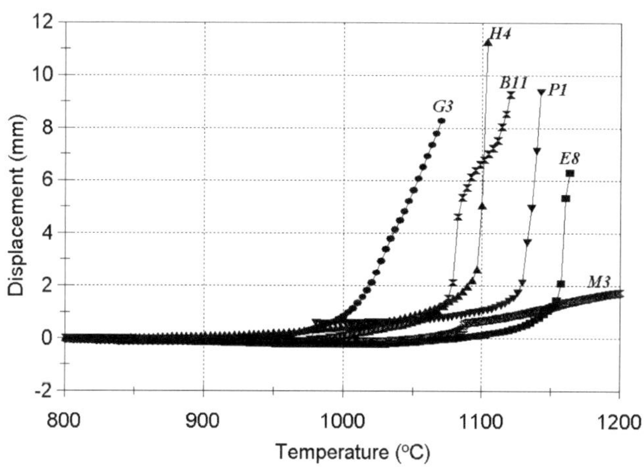

(heating rate 400 °C/hour, load 400g and 8 mm diameter sheath)

Figure 9. Typical softening curves of ISF sinters in air [2]

Type 1 and Type 2. Examination of the ISF sinters shows that the sample G3 has the microstructure shown in Figure 6b which contains the highest proportion of glass phase and low aspect ratio zincite. In contrast, the sample M3 has the microstructure shown in Figure 6a which contains the lowest proportion of glass phase and high aspect ratio zincite. Other samples such as B11, E8, H4 and P1 have the microstructures shown in Figures 6c and 6d. In brief, it can be explained that the softening in Type 1 mainly depends on the proportion of the liquid phase formed. The softening in Type 2 mainly depends on the zincite framework present in the sample. The softening in Type 3 depends on both of the proportion of the liquid phase and zincite framework.

2.2. Factors to determine the sinter quality

During the oxidative sintering of zinc sulphide concentrates in the sinter machine, and the reduction of sinter in the blast furnace t, the sinter is subjected to load from the overlying burden. Crucial to the operation and high productivity of both sintering and reduction processes is the ability to maintain bed high bed voidage, thus providing low resistance to gas flow.

Previous studies [4-5] have shown that the ISF sinter softening characteristics at temperature are related to bulk composition and microstructure of the sinter. It was found that sinters containing a high proportion of connected glass phase exhibited low softening temperatures. High softening temperature sinter is always associated with formation of microstructures containing high aspect ratio crystals of refractory zincite (ZnO), which at high proportions of this phase can give rise to a 3D interlocking network structure. Pure ZnO is known to have a melting temperature of 1975 ºC [6], far in excess of the silicate matrix. Although a number of microstructural types of the ISF sinter were identified in these earlier studies [4-5], to date no quantitative relationships between bulk composition and softening temperature are available to assist in the design of sinter materials.

Recent research by the authors [7] using synthetic sinter (PbO-ZnO-Fe_2O_3-CaO-SiO_2-Al_2O_3) materials has shown that the microstructures produced in air depend principally on the bulk composition and temperature. The aspect ratio of the zincite present in these sinters was found to increase with increasing peak bed temperature and increasing CaO/SiO_2 ratio. The relationships between high temperature softening behaviour and material microstructure in these synthetic materials is systematically investigated and compared to the results obtained from industrial sinters.

The synthetic sinters samples used for softening temperature tests were prepared from mixtures of pure oxide powders. The details of preparation and compositions of these synthetic materials are given in a previous publication by the authors [7]. The synthetic materials selected for the softening tests all contained fixed (ZnO+PbO) = 76.2% , PbO/ZnO = 0.40 by weight, having selected % Fe_2O_3 and CaO/SiO_2 ratios. The bulk compositions of the various materials, labeled as the ZM series, are summarized in Table 3.

	Bulk wt% Fe_2O_3		
CaO/SiO_2	10	14	18
0.35	ZM1	ZM6	ZM11
0.93	ZM3	ZM8	ZM13
1.5	ZM5	ZM10	ZM15

((ZnO+PbO) = 76.2%, PbO/ZnO = 0.40 by weight)

Table 3. Summary of synthetic materials used in softening tests in the present study

http://dx.doi.org/10.5772/54618

To prepare these materials water was added to each batch of approximately 50g of oxide powder mixture, the material was then shaped to a 30 mm diameter and 30 mm height cylinder. The sample was dried at 120 ºC and placed in a castable alumina plate. The sample was heated at 900 ºC for 120 minutes, and then at 1250 ºC for 30 minutes in air. After heat treatment the sample was cooled to room temperature in air in readiness for the softening temperature test. Typical microstructures of these materials are illustrated in Figure 10 [7]. The glass phase (light grey) and zincite (grey oval-shaped to plate-like) crystals are present in all samples; the angular (dark grey) phase is spinel which is only present in some of the samples.

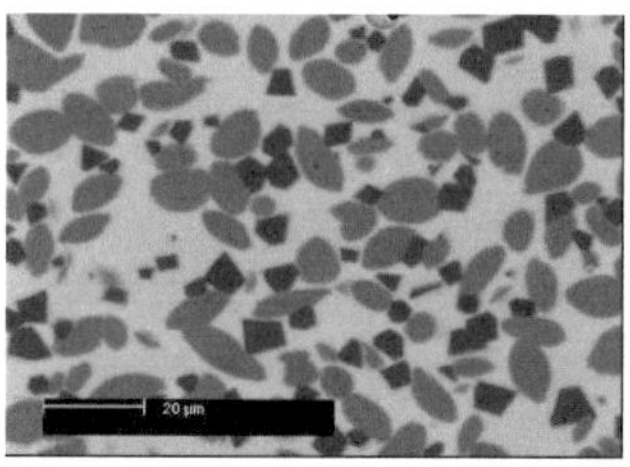
a. CaO/SiO_2=0.35, 10 wt% Fe_2O_3

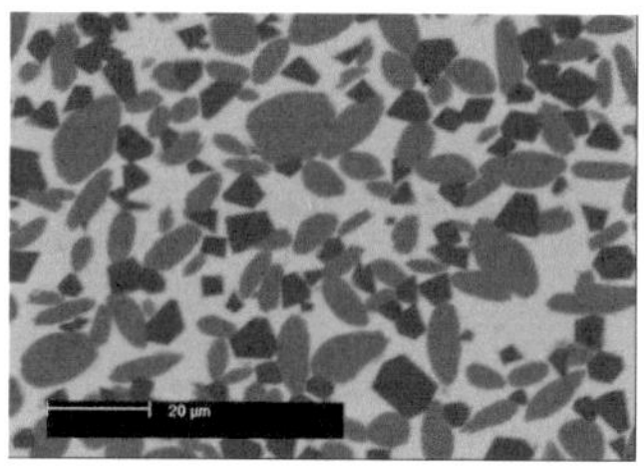
b. CaO/SiO_2=0.35, 14 wt% Fe_2O_3

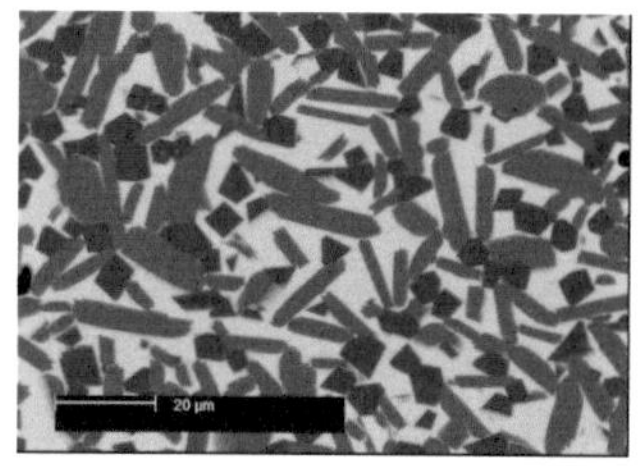
c. CaO/SiO_2=0.35, 18 wt% Fe_2O_3

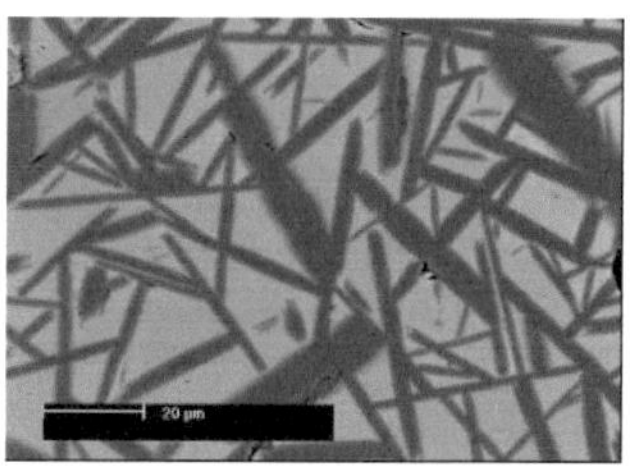
d. CaO/SiO_2=0.93, 10 wt% Fe_2O_3

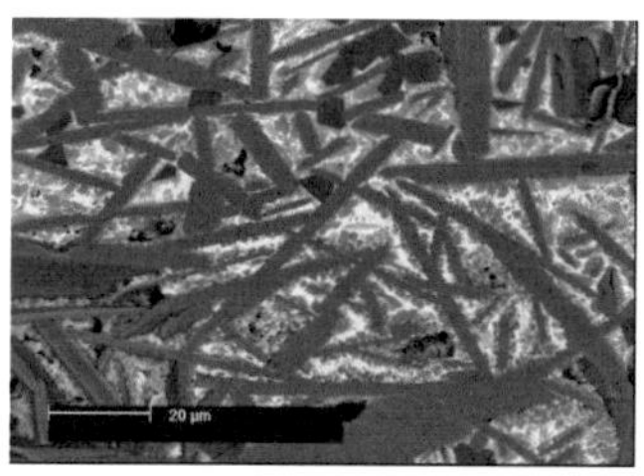
e. CaO/SiO_2=0.93, 14 wt% Fe_2O_3

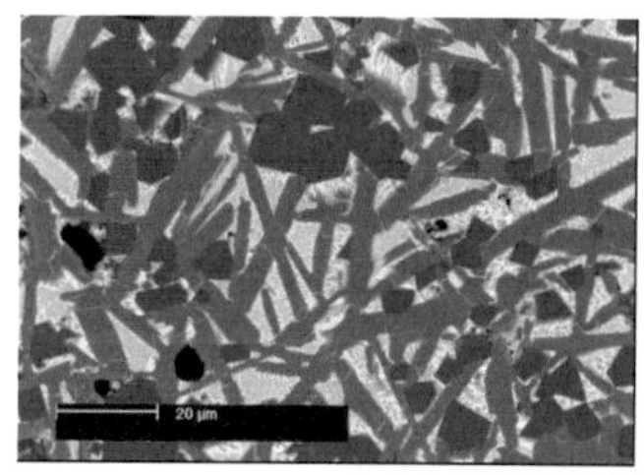
f. CaO/SiO_2=0.93, 18 wt% Fe_2O_3

g. CaO/SiO_2=1.5, 10 wt% Fe_2O_3

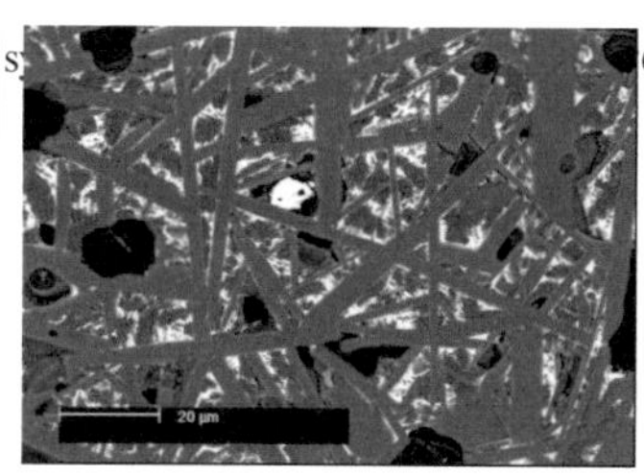
h. CaO/SiO_2=1.5, 14 wt% Fe_2O_3

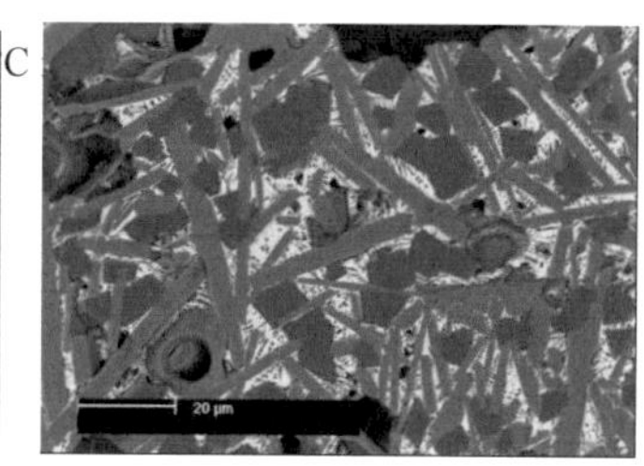
i. CaO/SiO_2=1.5, 18 wt% Fe_2O_3

Figure 10. Typical microstructures of synthetic sinters quenched from 1250 ºC in air, scale bar 20 µm [2]

Industrial sinter feed materials and final sinter product samples have also been obtained from operational ISF plants [7-8]. Representative samples were selected from the industrial sinter lumps for examination and testing.

2.2.1. Softening temperature measurement in air

The sinter softening test used was designed to reproduce conditions that may be experienced by individual sinter samples in sintering or packed bed reactors; the test involves applying a fixed load to 30x30mm cylinders of individual sinter samples and measurement of the displacement as function of time and temperature. The tests are intended to provide information on the comparative behaviours of the materials rather than to obtain absolute measures of mechanical properties.

The softening temperatures of individual sinter lumps selected from the bulk samples were measured using the apparatus shown schematically in Figure 7. A typical example of the softening curve of sample sinter is given in Figure 11. A positive displacement reflects a softening of the sinter. The sinters initially expand slightly with increasing temperature from room temperature, then softening and contraction of the sample occurs.

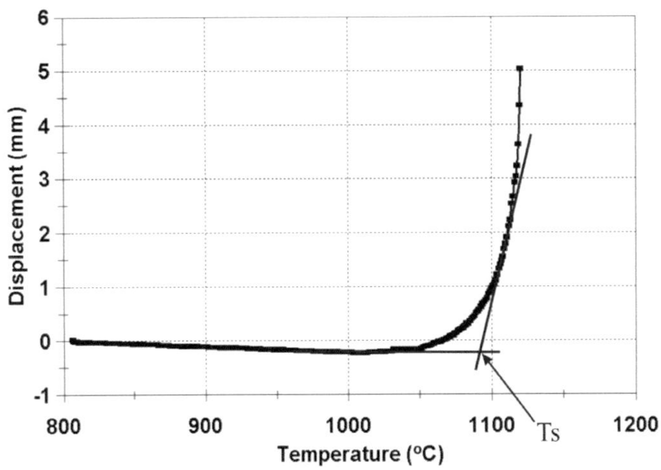

Figure 11. A typical softening curve of the synthetic ISF sinter [2]

The initial softening temperature T_s can be defined in a number of ways. In the present study it is defined by selecting an arbitrary value of displacement at a given temperature, from the intersection of the limiting lines describing expansion and contraction of the sample or the change in slope of the displacement vs temperature curve from –ve to +ve. In the case of synthetic and industrial sinters there is no sharp transition from one behaviour to the other, there is always some uncertainty in defining the softening temperature since the value is dependant on the shape of the displacement vs temperature curve.

2.2.2. Phase Assemblages at high temperature

Assuming chemical equilibrium is achieved during softening the phases present in the synthetic sinter, their proportions and compositions can be predicted as a function of temperature and bulk composition. Calculations of the phase assemblages for the samples used in the present study are shown in Figures 12 to 20. These predictions were made using FactSage databases and computer package [9]. The volume fractions were calculated using the densities of each of the component with the following assumptions:

1. Partial molar volumes of each component in the sinter are the same as for pure liquid component;
2. No thermal expansion is taken into account; and
3. Partial molar volume of liquid component is the same as for solid component.

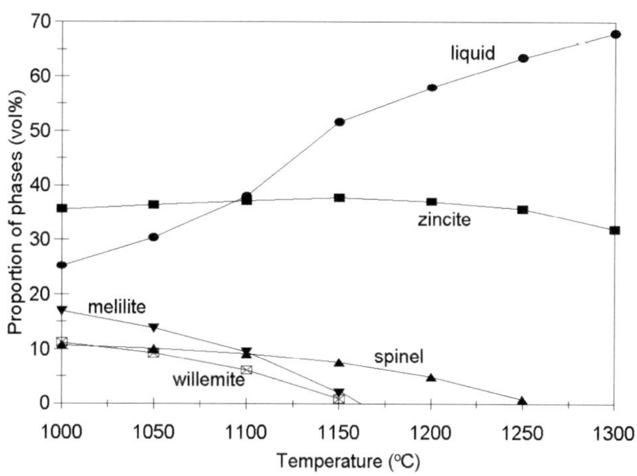

Figure 12. Proportions of phases for ZM1 calculated by FactSage in air [2]

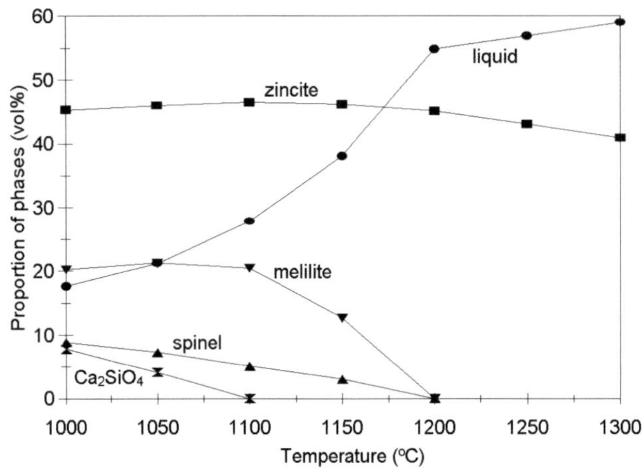

Figure 13. Proportions of phases for ZM3 calculated by FactSage in air [2]

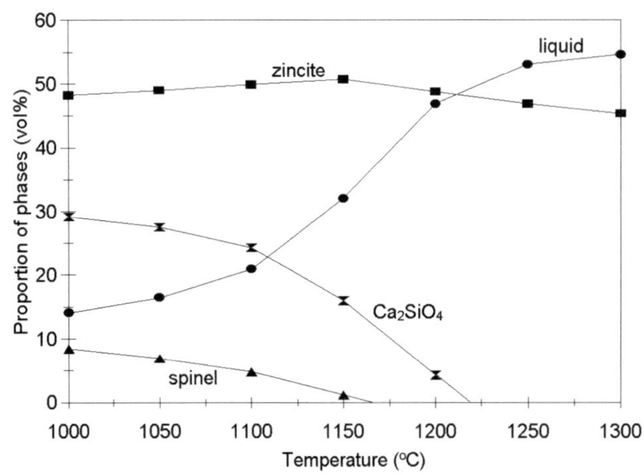

Figure 14. Proportions of phases for ZM5 calculated by FactSage in air [2]

http://dx.doi.org/10.5772/54618

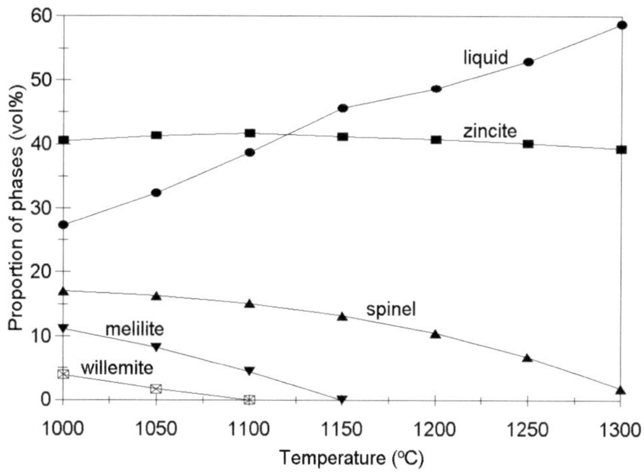

Figure 15. Proportions of phases for ZM6 calculated by FactSage in air [2]

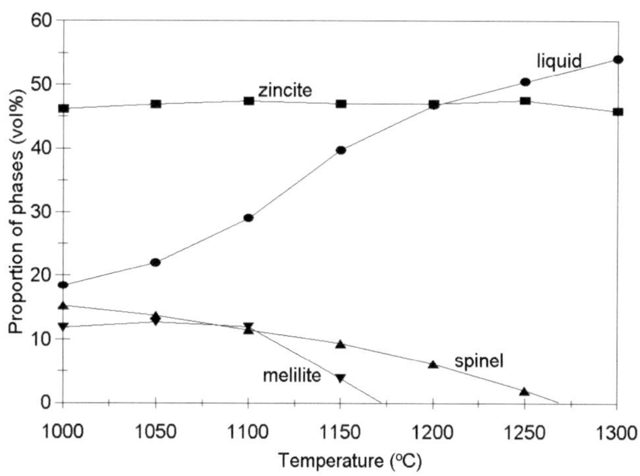

Figure 16. Proportions of phases for ZM8 calculated by FactSage in air [2]

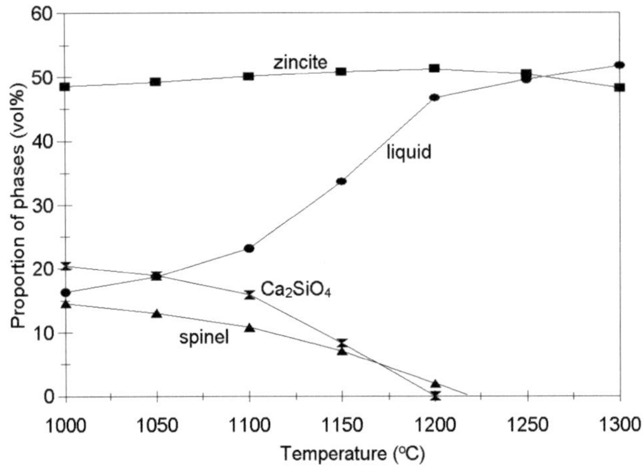

Figure 17. Proportions of phases for ZM10 calculated by FactSage in air [2]

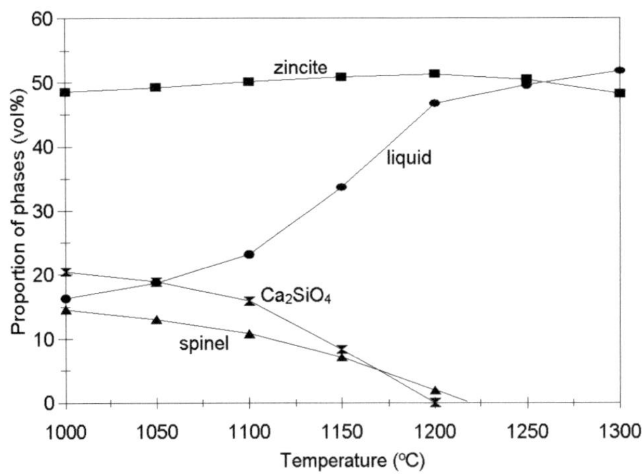

Figure 18. Proportions of phases for ZM11 calculated by FactSage in air [2]

http://dx.doi.org/10.5772/54618

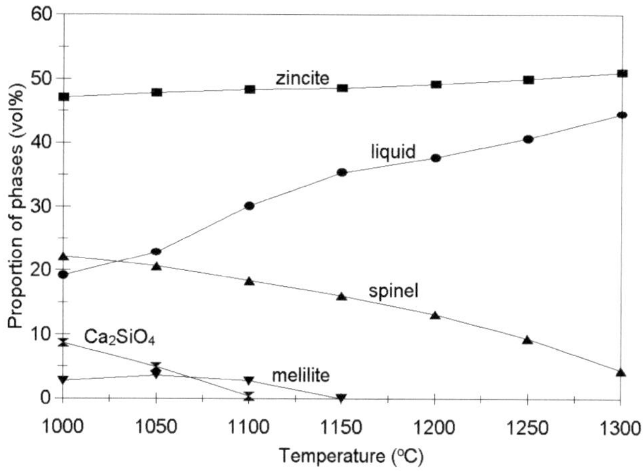

Figure 19. Proportions of phases for ZM13 calculated by FactSage in air [2]

It can be seen from Figures 12 to 19 that the three phases that are present in all samples are liquid, zincite solid solution (Zn,Fe)O, and spinel solid solution (Zn,Fe)O.Fe_2O_3. The volume of zincite phase remains almost constant over the temperature range 900-1200 °C; all samples apart from ZM1 contain 40-50 vol% zincite, ZM1 contains approximately 35 vol% zincite. All samples apart from ZM1 contain liquid phase as low as 900 °C, liquid in ZM1 appears above 950 °C. The proportion of liquid phase present increases with increasing temperature; the extent of the change in the proportion of liquid is dependent on the bulk composition of the samples. In general the % liquid at any temperature decreases with increasing CaO/SiO_2 ratio; the % spinel increases with increasing % Fe_2O_3.

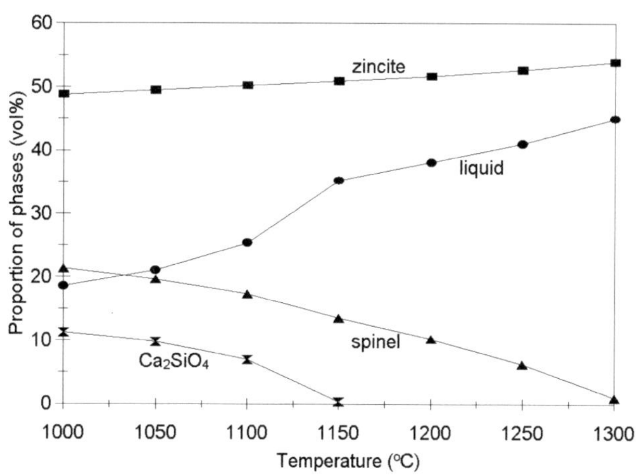

Figure 20. Proportions of phases for ZM15 calculated by FactSage in air [2]

2.2.3. Softening Temperatures of Synthetic Sinters in Air

The effects of CaO/SiO_2 ratio, "Fe_2O_3" concentration, and preparation temperature, on the softening characteristics of synthetic sinters in air are discussed in the following section.

Effect of CaO/SiO_2 ratio

Figures 21 and 22 show the effect of bulk CaO/SiO_2 ratio on the softening behaviour of the synthetic sinters containing 14 (ZM6, ZM8, ZM10) and 18 wt% "Fe_2O_3" (ZM11, ZM13, ZM15) in bulk sinter respectively. It can be seen that for both sets of data the initial softening temperatures, T_s, as given by the change in slope of the displacement vs temperature curves, increase with increasing CaO/SiO_2. For ZM6, ZM8 and ZM10 T_s is 1100, 1160 and 1180 °C respectively; at these temperatures the predicted % liquids are 38, 42 and 44 vol% respectively. For ZM11, ZM13 and ZM15 T_s is 1050, 1100 and 1160 °C respectively; at these temperatures the predicted % liquids are 30, 36 and 36 vol% respectively. Note in general that increasing CaO/SiO_2 ratio decreases the % liquid present in the sinters for given % Fe_2O_3.

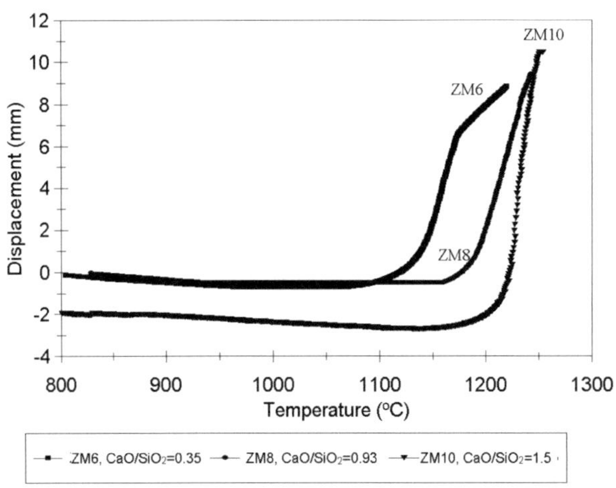

Figure 21. Effect of CaO/SiO_2 ratio on softening temperature in air at fixed 14 wt% Fe_2O_3 and PbO/ZnO=0.40 [2]

In addition the extent of softening, i.e. the positive displacement, at any temperature decreases with increasing CaO/SiO_2 ratio from 0.35 to 1.5.

Reference to the original sinter microstructures (Figure 10) shows an increase in the aspect ratio of the zincite crystals with increasing CaO/SiO_2 ratio at 1250 °C. For the samples containing 14 wt% Fe_2O_3 aspect ratios increase from 2.3 (ZM6) to 12.4 (ZM10); for 18 wt% Fe_2O_3 the aspect ratios increase from 3.6 (ZM11) to 13.3 (ZM15) with increase of CaO/SiO_2 ratio from 0.35 to 1.5.

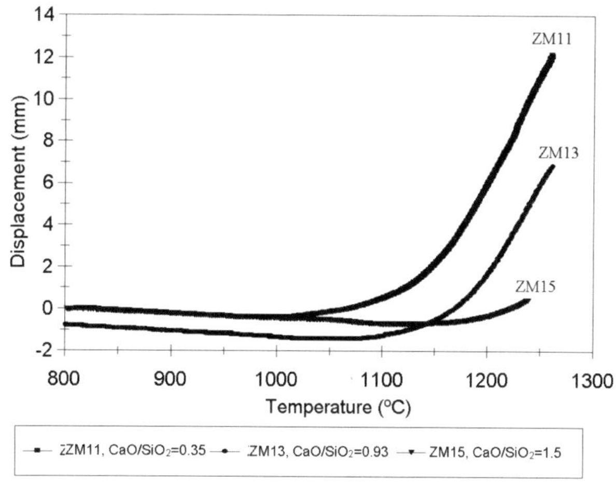

Figure 22. Effect of CaO/SiO_2 ratio on softening temperature in air at fixed 18 wt% Fe_2O_3 and PbO/ZnO=0.40 [2]

Effect of "Fe_2O_3"

It can be seen from Figure 23 samples (ZM1, ZM6 and ZM11) with CaO/SiO_2 = 0.35, having "Fe_2O_3" concentrations in bulk sinter of 10, 14 and 18 wt% respectively, have initial softening temperatures in the range 1100-1150 °C. The mean zincite crystal aspect ratios observed in these samples are in the range 2 – 4 [7], and the predicted equilibrium % liquids at 1100 °C are in the range 33-38 vol%.

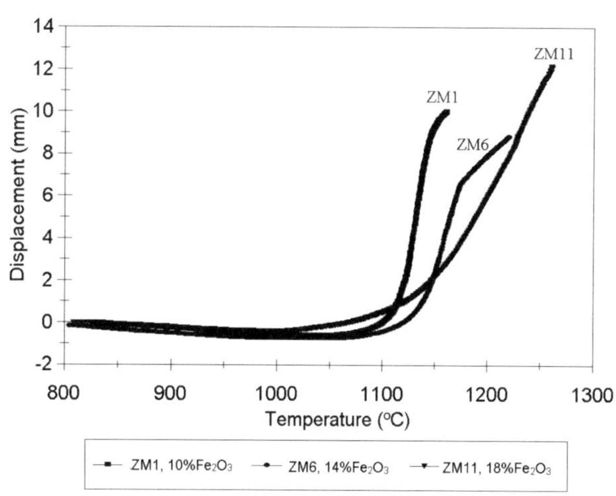

Figure 23. Effect of Fe_2O_3 concentration in sinter on softening temperature in air at fixed CaO/SiO_2 ratio of 0.35 and PbO/ZnO=0.40 [2]

It can be seen from Figure 24 samples (ZM3, ZM8, ZM13) with CaO/SiO_2 ratio of 0.93, the initial softening temperatures, appear to be the range 1150-1200 °C. The mean zincite crystal aspect ratios are in the range 9-11 [7], yet at 1150 °C the % liquid is in the range 35-40 vol%.

It can be seen from Figures 23 and 24 that at fixed CaO/SiO_2 ratio, increase of "Fe_2O_3" concentrations in bulk sinter from 10 to 18 wt% does not have significant effect on softening temperature.

Effect of preparation temperature

The effect of sinter preparation temperature on softening temperature in air is shown in Figure 25. It can be seen that for sinter ZM8 (14 wt% "Fe_2O_3", CaO/SiO_2 = 0.93 and PbO/ZnO = 0.40) the softening temperature of the sample prepared at 1300 °C in air is significantly higher than either of the samples prepared at 1200 to 1250 ºC. The softening temperatures of the samples prepared at 1200 and 1250 ºC are similar.

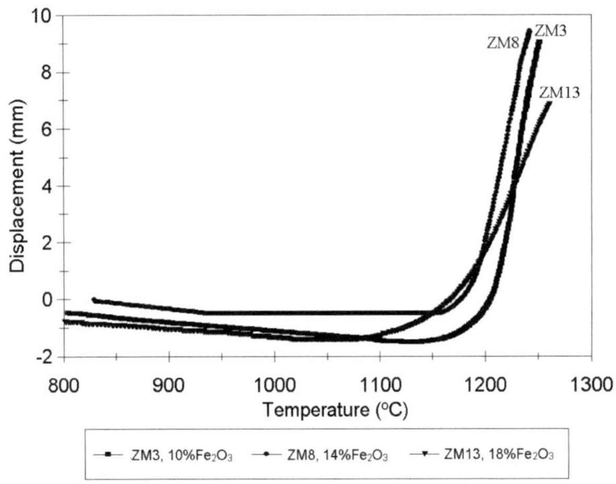

Figure 24. Effect of Fe_2O_3 concentration in sinter on softening temperature in air at fixed CaO/SiO_2 ratio of 0.93 and PbO/ZnO=0.40 [2]

Previous study [7] have shown increasing the preparation temperature for a given bulk composition results in an increase in % Fe_2O_3 dissolved in the zincite crystals and the corresponding increase in aspect ratio of the crystals. For a given volume fraction of zincite increased aspect ratio leads to increased framework formation.

All of the softening tests undertaken show consistent trends between microstructure and the extent of softening of the sample at a given temperature; increased zincite aspect ratio leads to increased resistance to deformation at temperature.

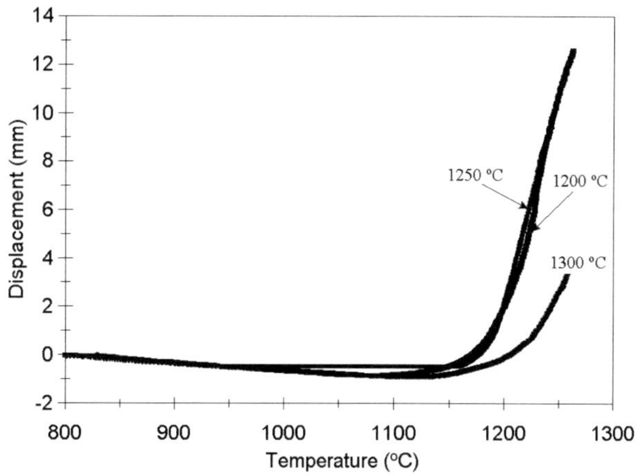

Figure 25. Effect of treatment temperature on softening temperature in air for sinter ZM8 (14 wt% Fe_2O_3, CaO/SiO_2=0.93 and PbO/ZnO=0.40) [2]

2.3. Conceptual sinter softening model

To assist in the understanding and analysis of the results obtained in the present and previous test work a conceptual sinter softening model is developed.

It is well established that the strength of a composite material is dependent on the physical properties of the phases present, the volume fractions of the phases and the macro- and micro-structure of the composite. The application of any stress, compressive or shear, to a liquid phase will result in viscous flow of this phase. Crystalline solid materials in contrast are rigid and behave elastically at low stress, permanent or plastic deformation, or fracture requires that a critical yield or fracture stress be exceeded [10].

The mechanical properties of a composite materials consisting of an isolated solid phase completely surrounded by a high volume fraction of second liquid or partially liquid phase matrix (see Figure 26) will largely be dependent on the properties of this matrix material. In this limiting condition at room temperature the matrix material is rigid, being in the form of crystalline or amorphous solid. If the matrix consists of a single crystalline solid phase, as the sample is heated this material becomes fully liquid at a given temperature. If the matrix material consists of a glass, on heating the glassy matrix reaches its glass transition temperature T_g, the glass then becomes liquid and the matrix material loses its compressive strength and begins to flow. With increasing temperature the viscosity of, for example, a high–PbO liquid silicate such as encountered in the samples studied in the present investigation, decreases and the deformation rate increases; since the material behaviour is determined by the properties of the liquid phase complete collapse of the sample under load occurs. In the limiting case for the formation of low viscosity liquid matrix at a given temperature the matrix phase loses its compressive strength and the structure will completely collapse, resulting in a step change in deformation at that temperature; the maximum compression of the sample C_{max} = 100%. This material behaviour is exemplified by sample ZM1, which consists of isolated zincite and spinel crystals surrounded at room temperature by a glassy matrix (Figure 10a)

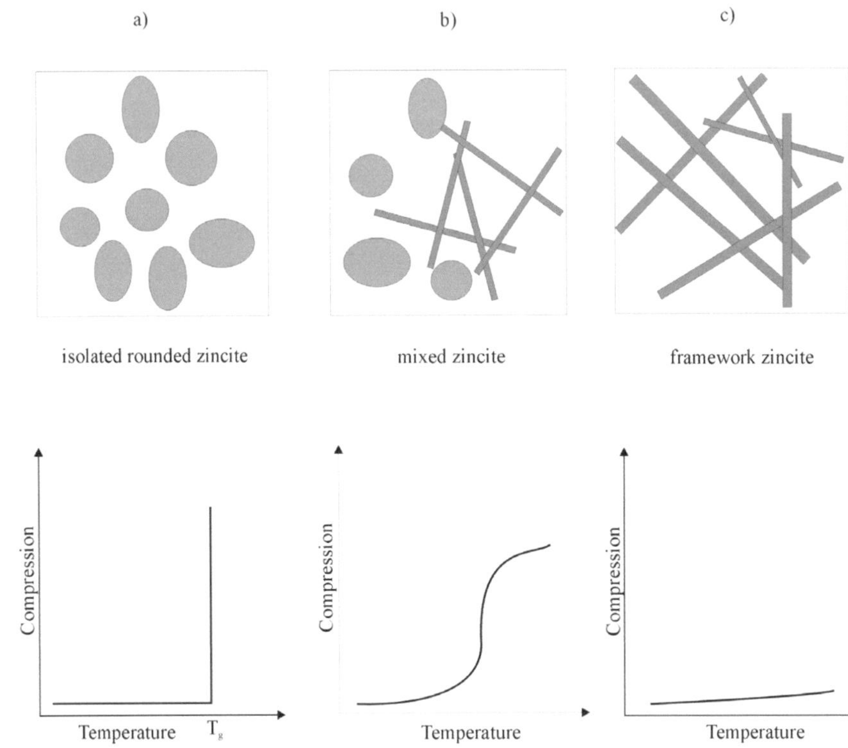

Figure 26. Idealised sinter softening model [2]

In contrast if the composite consists of a rigid 3-dimensional framework structure of a high melting temperature solid material the deformation characteristics of the composite material will not change when the matrix material becomes molten, since the material will be fully supported by the rigid 3D framework. In this case there is no compression of the material, i.e. $C_{max} = 0$, as the temperature is increased through and beyond the melting temperature of the matrix (see Figure 1d). On this basis the higher the proportion of framework zincite present in the material the lower the maximum compression of the sinter, C_{max}.

In the case of the synthetic sinters under study there are ranges of microstructure types, and melting occurs over a range of temperatures. Nevertheless this simplified analysis of sinter softening process indicates that two measures should be used to characterise sinter softening behaviour,

i. Ts, initial softening temperature microstructures, which is determined primarily by the strength of the matrix material, and

ii. The proportion of framework zincite present in the material, which determines C_{max}, the maximum deformation and the deformation rate of the sinter at any temperature.

The measured softening curves for typical synthetic ISF sinters are presented in Figures 21-25. Comparison with the calculated phases fractions from Figures 12 to 20 shows that the initial softening temperatures T_s in all cases corresponds to the temperature with 35-50 volume pct of liquid phase. Based on purely geometric considerations this would correspond to the minimum volume required to form a continuous liquid phase between the randomly distributed solid phases present in the individual samples.

The extent of formation of the framework zincite in the structures will depend on the volume fraction of zincite phase and the aspect ratio of these crystals. The thermodynamic analysis indicates that all synthetic sinter samples contain approximately 40-50 vol% zincite over the temperature range 900-1300 °C; this variation is insufficient on its own to explain the differences in softening behavior of the materials.

It has been well established [7] that the aspect ratio of the zincite can change significantly with bulk composition and thermal history. The softening tests carried out in the present study indicate displacement as a function of temperature for a given applied stress and heating rate (400 °C/h). The displacement vs temperature plots may also be interpreted as displacement vs time. The slopes of these softening curves reflect the strain rate at constant stress. It can be seen from the data shown in Figures 21 to 25 that, within experimental uncertainty, there are general trends in behavior with decreasing strain rate obtained with increasing zincite aspect ratio, increasing CaO/SiO_2 ratio, and increasing bulk iron concentration.

It appears that even though complete 3-D networks of zincite are not formed in all cases the increased aspect ratio of the zincite increases the resistance to deformation under load, i.e. increases the effective viscosities of these composite materials.

2.3.1. Implications for industrial practice

The results obtained from the present study are consistent with softening temperature measurements and microstructural analysis carried out on industrial ISF sinters [8]. Examples of the different softening behaviours observed for ISF sinters have been shown in Figure 9. It can be seen that initial softening of sample G3 starts at approximately 950 °C and collapses of the structure is complete over a relatively narrow range of temperatures. In contrast, although the softening of the sinter M3 starts at approximately 1050 °C but retains its strength at temperature until at least 1200 ºC. This contrasting behaviour is reflected in the very different microstructures observed in these samples [7]; G3 contains isolated zincite crystals of low aspect ratio (2-3), whereas M3 consists almost entirely of high aspect ratio zincite that forms an interlocking 3-D network of refractory material (aspect ratio >10). The CaO/SiO_2 ratios are 0.80 for G3 and 1.14 for M3 respectively.

Most ISF sinters have softening characterisics that are between these two extremes; the microstructure of these materials show this contain both isolated zincite and zincite having a range of aspect ratios. This is demonstrated in sinter B11 [7].

In developing an understanding of the structural factors influencing the strength of sinters these have to be considered at Macro-, Meso-, and Micro-structural levels. The product sinter consists of partially fused granules of oxidised zinc concentrates and flux materials. The porosity of the sinter depends on the degree of fusion of the granules. The porosity and density of the ISF sinter lumps were measured using an Australian Standard AS 4133.2.1.2-1993 "Rock porosity and density test-Determination of rock porosity and dry density-Saturation and buoyancy techniques". It was found that industrial ISF sinters have typical porosities in the range 25-40%. The sinter materials experience differences in thermal history during the sintering process due to differences gas flow path and combustion within the ignition, heating and sintering layers of the bed; these result in *macro-structural* differences over the range (10 – 100mm).

In the ISF sintering process approximately 65-75% by weight of the sinter is recycled. The product sinter is crushed, separated by size with lump sinter 100-200mm diameter sent to the ISF; the remainder is further crushed, sized and returned to the feed preparation. This recycled or "return" material is coated with fresh concentrate to form approximately spherical feed granules to the sinter process. These feed granules consist of individual return particles or and composites, agglomerates consisting of a number of coated return particles. *Meso-structural* differences in the range (1–10mm) are the result of gross differences in composition at granule level due to differences in thermal history arising from; local variations in the proportions of returns/fresh feed; single particle granule or composite granules, incomplete oxidation of sulphides, incomplete dissolution of fluxes e.g. CaO, or recycled materials, e.g. slag.

Microstructural differences in the range (10-100μm) reflect differences in crystal shape, phases formed, proportion of phases, and are typically the result of local compositional variation, cooling rate/crystallisation.

Trailing thermocouple tests on ISF sinter plants [11-12] have shown that although sulphur elimination is largely achieved because of the low peak bed temperatures, strong fusion of the granules does not occur in the bottom third of the updraft sinter bed. It has also been shown [7,11] that, far from helping to strengthen the structure, reheating the framework zincite structure at these low temperatures leads to the breakdown of the interlocking plates and reduction in zincite aspect ratio.

The top part of the sintering layer achieves the highest peak bed temperature, since the gas is preheated before it reaches the combustion zone. It is in this zone that is to be expected that most material is converted to high aspect ratio zincite.

Important factors that assist in achieving high peak bed temperatures include

- Fuel loading
- Feed ignition temperature
- Moisture content
- Feed granule size distribution

- Blowing rate and profile

Examination of the feed characteristics of all industrial sinters tested shows that significantly the sinter M3, which consistently contain high % framework structure, were produced from feed materials with small mean size (4mm) and narrow size distribution (σ_{size} = 5mm). Small granule diameter favours thinner fresh sulphide feed layer thickness on the granules because of the high surface area. Combustion of fresh feed is therefore more likely to take place rapidly. The smaller granule size also means that heat transfer to the return material at the core of the granules is rapid; this ensures that the return material is heated to a high peak bed temperature and not degraded by reheating to temperatures below the optimum value.

2.4. How to make high quality sinter

An extensive program of experimental and modeling studies has been carried out to improve the understanding of the factors affecting the quality of lead blast furnace sinter and lead smelting slag, and to develop methods used to characterise these materials, with a view to improving plant operations. [13] The program includes the collection of sinter samples and plant information from four different lead sinter plants, the characterization of these sinter samples and samples from subsequent plant trials, consultation with plant engineers and laboratory-based studies at the University of Queensland.

As part of the research program plant trials have been undertaken at Mount Isa lead sinter plant. In the Mount Isa Mines lead smelter of Xstrata Zinc, an updraught lead sinter machine is used to prepare feedstock for the lead blast furnace. [14] The aims of the process are to

a. produce self-fluxing lump material that maximises blast furnace productivity, and

b. remove excess sulphur from the charge.

The process combines fine particulate materials sourced from metal concentrates, recycled materials and fluxes, into lump material that is physically strong, both at ambient temperatures and within the blast furnace, and, chemically reactive in the blast furnace so that metal values can be recovered.

To incorporate all feed materials into the sinter and to form a uniform strong structure it is important to obtain high peak bed temperature during sintering process. [5,11,15] The sintering temperature has been shown to be related to a number of factors, including fuel content, composition, size distribution and moisture of the sinter feed.[[12,15-16] The gas temperature above the sinter bed is usually used as an indication of the sintering temperature, or a heat balance can be performed to predict the sintering temperature. Since the lead sinter machine is a closed and moving system it is very difficult to use conventional trailing thermocouples to measure the temperature profile within the sinter bed.

The aim of this study is to determine the temperature profile during sintering and the effect of thermal history on the sinter properties. This is achieved through the use of a wireless temperature probe developed at the University of Queensland, metallographic studies of product sinters and measurement of sinter softening characteristics.

2.4.1. *Sintering temperature measurement*

Lead sinter plant trials have been carried out at the Xstrata Zinc sinter plant (Mount Isa) by the author and Xstrata Zinc staff. The design of the hood at the Mount Isa sinter plant is such that there is a 10 cm gap between the ventilation hoods for the main feed hopper and the sinter machine. This space is sufficient for thermocouples to be inserted through the side wall of the sinter pallet after the main layer of sinter feed material was charged onto the sinter machine. The thermocouple once in place progresses through the length of the sinter bed and is discharged with the sinter lump. A wireless temperature probe was used to measure the temperature within the sinter bed. A K-type thermocouple was inserted into the sinter bed and connected to a radio transmitter, which is attached to the outside of the pallet and moves with the sinter bed as shown in Figure 27. A radio receiver remote from the sinter machine is connected to a computer. The temperature measured by the thermocouple inside the sinter bed is converted to the radio signal by the transmitter. The receiver outside then converts the radio signals to digital output, which is recorded on a computer.

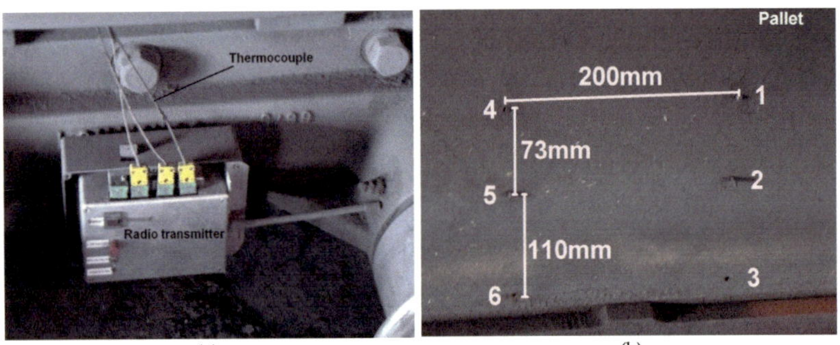

(a) (b)

Figure 27. The radio transmitter fixed on the pallet and positions of holes on side of the pallet at Mount Isa lead sinter machine [19]

Two sets trials have been conducted. The first set trials were conducted using a single channel wireless transmitter. An upgraded wireless transmitter with four channels was used in the second set trials.

The total initial sinter bed depth was 440mm. In the first set trials two holes of 5 mm diameter were drilled through the side of the sinter machine pallet. The bottom hole was made at ½ bed depth (220 mm) and the top hole was made at 2/3 bed depth (293 mm), height measured from grate surface.

In the second set trials six holes were drilled through the side of the sinter machine pallet as shown in Figure 27. The bottom holes (holes 3 and 6) are 110 mm above the sinter bed grate that are ¼ height of the initial sinter bed. The holes in the middle (2 and 5) are ½ height of the initial sinter bed and the holes in the top (1 and 4) are 2/3 height of the initial sinter bed.

The maximum working temperature of the transmitter is 50 ºC. The success of the trials indicates that the temperature on the outside of the pallet remains below 50 ºC throughout the cycle. The sinter machine motion had to be briefly interrupted for the insertion of the thermocouples. Due to the time limitation for stopping the sinter machine it was only possible to insert 2 or 3 thermocouples at a time. Usually two thermocouples were used in the trials. The speeds of the sinter machine were maintained in the range of 130-140 cm/min during the first set trials and 120–130 cm/min for the second set. In both sets of the trials the sinter samples and plant operation conditions corresponding to the temperature profiles were collected.

The temperature profiles recorded during the first set trials are presented in Figure 28. It can be seen that in trials 1.1 and 1.2 peak bed temperatures in the range 1050-1080 ºC were observed. In trial 1.3 a peak bed temperature of 870 ºC was observed. In trials 1.1 and 1.2 the temperature remained above 800 ºC until the end of the sinter bed. In trial 3 the sinter bed temperature returned to ambient temperature after a period of approximately 10 minutes. The temperature profiles measured in trials 1.1 and 1.2 are most commonly observed when the thermocouple is inside the sinter lump. These temperatures represent real sintering temperature. In the case of trial 1.3 the temperature dropped rapidly after the peak temperature. This indicates that the thermocouple was most likely outside or in the void between the sinter lumps. Since these data are all taken from the same pallet but at different times it is an indicator of feed property variation within the process.

Results of the measurements during the second set trials are presented in Figure 29. In the trial 2.1 two thermocouples were inserted at 1/4 and 2/3 bed heights respectively. It can be seen that the temperature in the bottom hole rapidly increased from 100 ºC to approximately 1120 ºC and remained at the peak temperature for 10 minutes. A peak temperature of 1190 ºC in the top hole was achieved, with most of the temperature rise not occurring until after 15 minutes of travel; this is consistent with the passage of the combustion front up through bed. The peak temperature in the upper is much higher (170 ºC) than that in the lower part; again this is expected due to the preheating of the gas from the sintered charge in the lower bed.

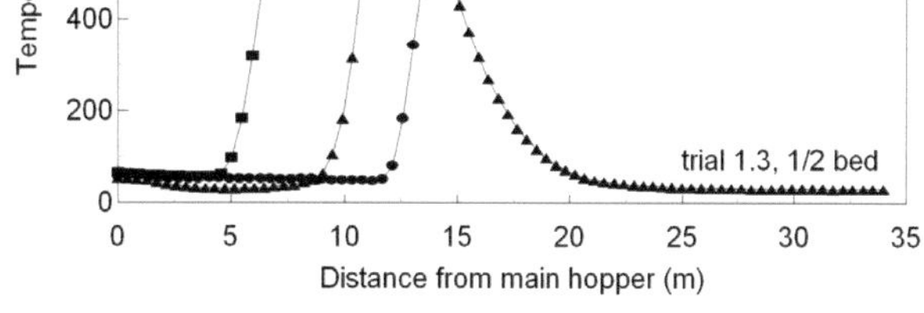

Figure 28. Bed temperature profiles against sinter pallet position at Mount Isa Mines measured during 1st set trials [19]

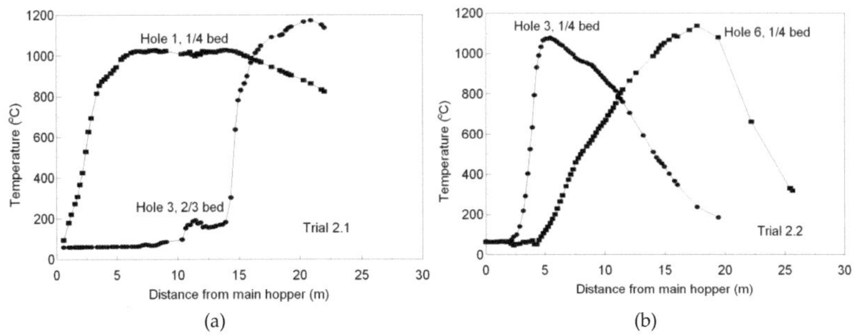

Figure 29. Temperature profiles measured during 2nd set trials [19]

In the trial 2.2 (Figure 29) two thermocouples were inserted into the pallet at the same bed height (1/4). It can be seen that the temperature in hole 3 increased rapidly to 1100 ºC and the temperature in hole 6 increased slowly to 1135 ºC. The horizontal distance between the holes are 200 mm but the peak temperatures were reached at different windbox positions. This variation of the temperature at different position of the sinter bed at the same height indicates possible differences in ignition conditions or variation in feed property along the bed.

2.4.2. Operating conditions and results of the analysis

During the plant trials, samples of the sinter machine feed and sinter lump were collected. The bulk compositions of these samples analysed by X-ray fluorescence (XRF) are given in Table 4. I1 and I5 are the sinter lump and feed collected during the first set trials and I8 and I10 are the sinter lump and feed collected during the second set trials respectively.

It can be seen from Table 4 that the fuel (sulphur) content in the sinter feeds I5 and I10 are close and the ratio of PbO/(CaO+SiO_2) in I5 (3.24) is much higher than that in I10 (2.21). The CaO/SiO_2 ratio in sinter I8 is higher than that in sinter I1. Ratio of raw feed to total feed is reported to be 0.165 in I5 and 0.175 in I10. CdO vapour pressure increases with increasing temperature; the concentration of CdO in sinter is usually used to be indication of the sintering temperature. It is seen in Table 1 that Cd contents in I8 and I10 are lower than those in I1 and I5. This confirms that the sintering temperature of I8 is higher than that of I1.

Sample	**Composition (wt%)**												
	Pb	Zn	Fe	CaO	SiO_2	Al_2O_3	MgO	S	Cu	Cd	As	CaO/SiO_2	PbO/(CaO+SiO_2)
I1–sinter	51.8	5.4	8.7	8.9	7.8	0.77	0.45	1.9	0.29	0.52	0.04	1.14	3.34
I5–feed	48.7	5.1	8.4	8.4	7.8	0.76	0.43	7.4	0.29	0.68	0.04	1.08	3.24
I8–sinter	45.1	6.6	10.2	10.9	8.4	0.97	0.41	1.6	0.67	0.19	0.1	1.30	2.52
I10–feed	41.4	5.9	8.4	11.1	9.1	0.87	0.36	7.1	0.56	0.31	0.05	1.22	2.21

Table 4. Bulk compositions of "as-received" sinter lump and sinter feed

The size analysis has been carried out for sinter machine feeds using mechanical screening. The size distributions of sinter machine feeds I5 and I10 are shown in Figure 30. At the time of 1st set trials the crusher for returned sinters was worn; this resulted in an uneven size distribution of the return sinter contained within I5. It can be seen from Figure 30 that in the sinter feed I5 the proportions of the feed smaller than 2mm and larger than 15mm are much higher than those in sinter feed I10. As a result of the bimodal size distribution the permeability of the sinter bed is low during the first set trials. This is reflected in the higher windbox air pressures of 1st set trials than that of 2nd set trials shown in Figure 31a.

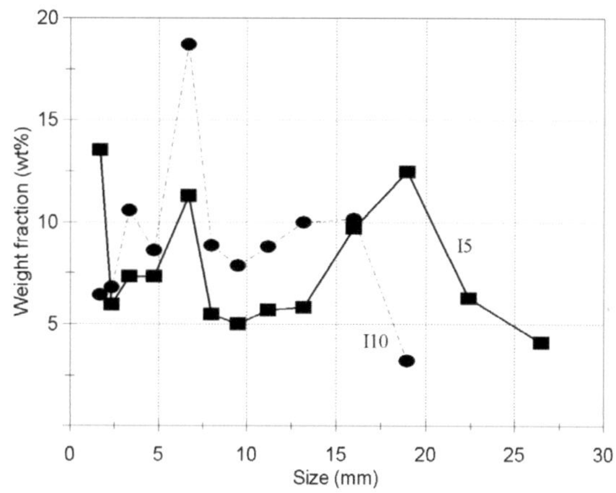

Figure 30. Size distributions of sinter feeds I5 from 1st set trials and I10 from 2nd set trials [19]

Figure 31b shows the hood temperatures measured during the two sets trials. Note that the hood thermocouple in windbox 6 was not working during the trials. It can be seen from Figure 31b that the peak hood temperature in 2nd set trials is 150 °C higher than that in 1st set trials.

Typical microstructures of sinter lump I1 and I8 are shown in Figure 32. It can be seen from the figure that plate-like melilite ($2CaO.ZnO.2SiO_2$) crystals, equiaxed crystals of spinel ($ZnO.Fe_2O_3$) and glassy lead silicate are all present in both sinters. However, the proportion of the plate-like melilite is much higher in I8 which forms an interconnected 3D network of refractory material that physically supports the structure during heating.

The softening characteristics of sinter lump have been investigated in air. Cubic sinter samples (30mm diameter) were used for softening temperature test. The tests were carried out by applying a fixed load of 235g to a 8mm OD closed-end alumina tube and a heating rate of 400 ºC per hour was used. A series of measurements have been carried out for each sinter sample. Figure 33 show the softening temperature curves for I1 and I8. It can be seen that the final softening temperatures of I1 are in the range of 920 to 1070 ºC. The wide range of softening temperatures is attributed to variations in sinter microstructure on a micro- and meso-scale, since the materials are composed of recycled (return) lump sinter coated and agglomerated with fused fresh feed. In contrast, the final softening temperatures of I8 are in the range of 1050 to 1130 ºC. The I8 sinter sample softens over a narrower temperature range and at higher temperature than I1. Sinter softening temperature is a direct indication of the sinter quality, in particular sinter strength and uniformity of properties. The results shown in Figure 33 indicate that the sinter lump produced during 2nd set trials is better than that produced during the 1st set trials.

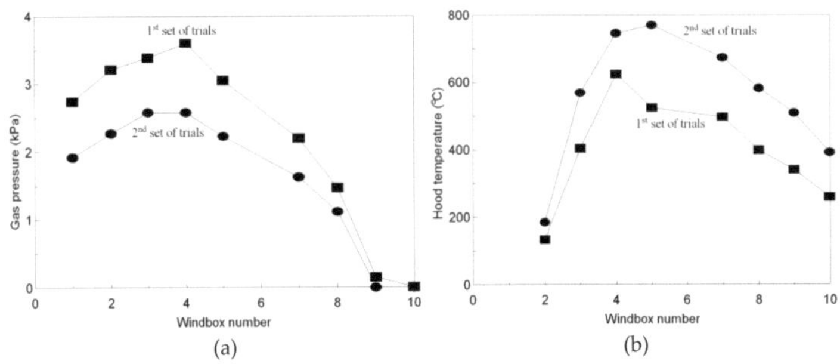

Figure 31. Gas pressures (a) and hood temperature (b) during the 1st and 2nd set trials [19]

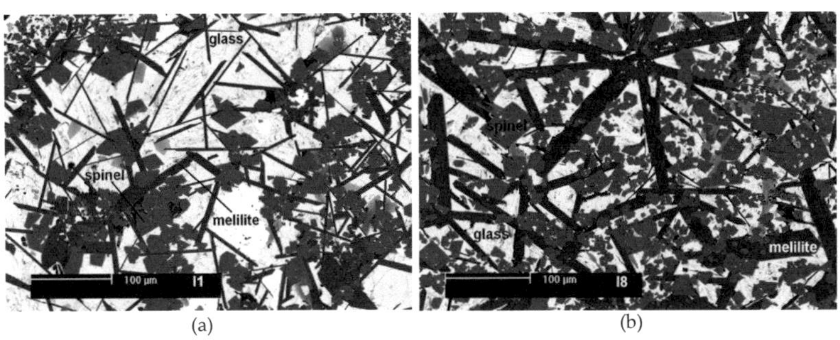

Figure 32. Typical microstructures of I1 and I8 collected during the trials [19]

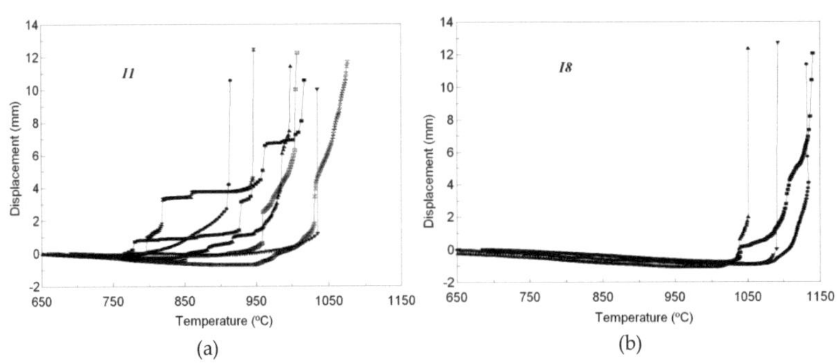

Figure 33. Softening curves of sinter lump I1 and I8 in air [19]

The predicted volume fractions of the major phases present in lead sinters I1 and I8 as a function of temperature are presented in Figure 34. It was calculated by FactSage and optimised thermodynamic database for the system ZnO-Fe_2O_3-FeO-PbO-CaO-SiO_2. [9,17-18] The weight fractions obtained from the calculations were converted to the volume fractions of the phases assuming: a) the partial molar volume of each component is constant and the same in the liquid and solid phases, and b) no thermal expansion takes place over the range of temperatures examined. It can be seen from Figure 34 that for both I1 and I8 the proportion of liquid phase increases with increasing temperature. The proportions of Ca_2SiO_4 and spinel decrease with increasing temperature. There is an optimum temperature range for formation of the melilite phase.

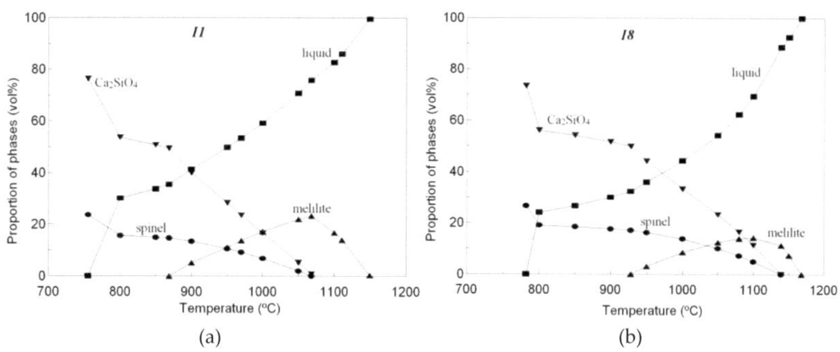

Figure 34. Phase volume fractions calculated by FactSage for as-received sinter I1and I8 [19]

2.4.3. Discussion of the correlations

The temperature profile data obtained during the plant trials can be used to correlate the attainment of peak bed temperature at a given depth in the sinter bed against the distance travelled along the sinter strand (Figure 35). The correlation assumes that the relative position of the thermocouples in the bed remained constant, i.e. the thermocouple at the ½ bed height position stays in that relative position despite the overall slumping of the sinter bed during reaction. It is known that the actual height of the sinter bed decreases as sintering proceeds; the final bed height can be 60-70% of the initial bed height.

The sinter strand is approximately 30 m long. The correlations indicate that in the case of the 1st set trials at ½ bed height the combustion front breakthrough approximately 12 m from the main hopper, allowing time for the top of the bed to cool slowly before exit from the sinter machine.

For the 2nd set trials peak bed temperatures between 1000 and 1160 ºC were achieved within the bed. The correlation between combustion front position and distance from the hopper indicates that breakthrough was only just achieved before exit from the sinter strand. The slow combustion rate achieved during the 2nd set trials compared to that for the 1st set trials.

It can be seen from the trials that there was a considerable variation in peak bed temperature and combustion front velocity, and hence significant variation in thermal history of sinter material within even a given charge of sinter material.

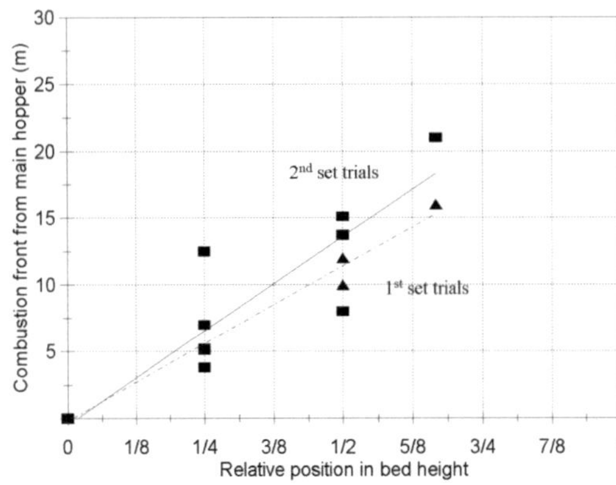

Figure 35. Correlation between combustion front and position in bed during lead sintering, initial bed height 440 mm [19]

Figure 36 shows the relationship between mean final softening temperature and peak bed temperature. It can be seen that in general the mean final softening temperature increases with increasing peak bed temperature. The average final softening temperature for the 2nd set trial sinters are 1079 ºC which is much higher than that of the 1st set trial sinters (1003 ºC).

Peak bed temperature is not monitored in everyday operation. Instead, temperature is commonly measured in the ventilation hood above the sinter strand. This is used in many plants as an indication of sinter machine performance. Relationship between peak bed temperature and peak hood temperature at Mount Isa lead sinter plant is shown in Figure 37. It can be seen that there is strong correlation between the peak bed temperature and the peak hood temperature; this indicates that the peak bed temperature in lead sinter plant can be related to the hood temperatures, the latter being available to the sinter plant operators as an on-line measurement.

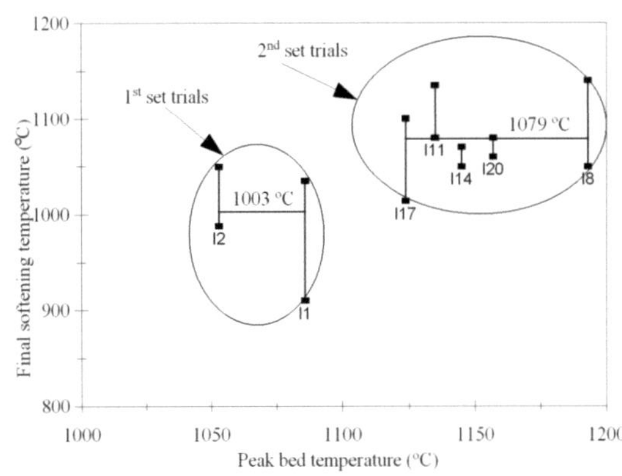

Figure 36. Relationship between final lead sinter softening temperatures and peak bed temperatures at Mount Isa lead sinter plant [19]

http://dx.doi.org/10.5772/54618

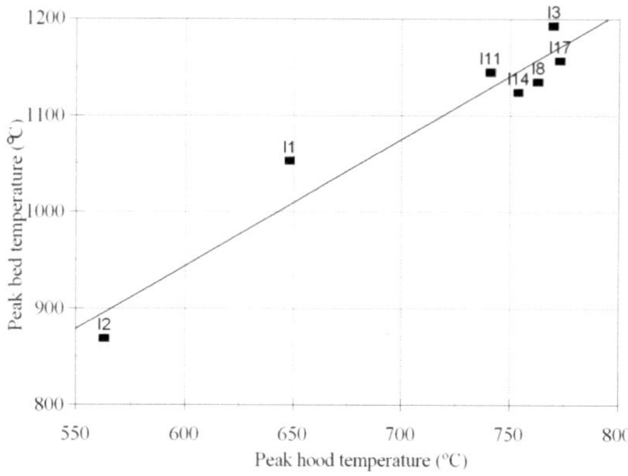

Figure 37. Relationship between peak bed temperatures and peak hood temperatures at Mount Isa lead sinter plant [19]

3. Conclusions

Softening temperature measurements, microstructural, compositional and thermodynamic analysis have been carried out on a range of synthetic and industrial ISF sinters. A simple conceptual has been proposed to assist in understanding the softening behaviour of the complex phase assemblages formed in synthetic and industrial sinters.

The "initial softening temperatures" of ISF sinters are shown to depend principally on the bulk chemical compositions of the materials. There is strong evidence to suggest that in the materials under investigation the initial softening temperatures are related to a critical volume fraction of liquid, equivalent to 30-40 vol% liquid.

The maximum compression and the rate of deformation of the sinters at temperature appear to be directly related to the aspect ratio of the zincite present in the materials; the higher the aspect ratio the lower the deformation rate of the materials. The principal factors determining the formation of plate-like zincite are:

- Peak sinter bed temperature
- Sinter composition

High the aspect ratio zincite is formed by obtaining high iron in solid solution in zincite, this condition is favoured by high iron in the liquid phase, high CaO/SiO_2 ratio and high sintering temperature.

A new wireless temperature probe with four channels has been successfully used on the lead sinter machine at Mount Isa Mines to measure the temperature profile within the sinter bed during operation. Strong correlations have been observed between peak bed temperature and peak hood temperature, and mean sinter softening temperature and peak bed temperature. For the feed material used in the trials, to obtain sinters with mean softening temperatures above 1050 ºC, peak bed temperatures above 1130 ºC are required. There appears to be a direct correlation between granule size distribution in the feed and peak bed temperature attainable. High peak bed temperatures are observed with narrow feed granule size distribution.

Acknowledgements

Part of the chapter was taken from the following papers, the co-authors of these papers are acknowledged for their contributions.

Zhao B., Burrows A., Jak E. and Hayes P.C., "Microstructures and Softening Behaviours of Lead Sinters and their Correlation to Sintering Temperatures in Mount Isa Lead Smelter", *Transactions of the Institutions of Mining and Metallurgy, Section C: Mineral Processing and Extractive Metallurgy,* Vol. 119(3), 2010, pp. 130-135.

Zhao B., Jak E. and Hayes P.C., "High Temperature Softening Behaviour or ISF Sinters in Air", *Canadian Met. Quart.*, Vol. 47(2), 2008, pp. 139–148.

Author details

Baojun Zhao*

Address all correspondence to: baojun@uq.edu.au

School of Chemical Engineering, The University of Queensland, Brisbane, Australia

References

[1] Noguchi, H. Zinc and Lead Smelting at Hachinohe Smelter. lead-zinc (2010). COM2010, Vancouver, Canada, 2010, , 615-620.

[2] Zhao, B, Jak, E, & Hayes, P. C. High Temperature Softening Behaviour or ISF Sinters in Air. Canadian Met. Quart., (2008). , 47, 139.

[3] Zhao, B, Jak, E, Hayes, P. C, Yang, G, Wang, J, Dong, Y, & Errington, B. Characteristics of ISASMELT Slag and Lead Blast Furnace Sinters. Lead & Zinc'05 Int. Symp., Kyoto, 17-19 Oct (2005). , 571-586.

[4] Holliday, R. J, & Shoobridge, P. H. *The Aus. I.M.M. Conference,* North Queensland, AusIMM, Melbourne, (1978). , 311-321.

[5] Holliday, R. J, Shoobridge, P. H, & Firkin, G. R. *Proc. Australia Japan Extractive Metallurgy Symp.*, Sydney, AusIMM, Melbourne, (1980). , 343-351.

[6] Hayashi, J, Morita, M, Kumagai, H, & Chiba, T. *Fuel*, (2003). , 82, 487-500.

[7] Zhao, B, Jak, E, & Hayes, P. C. *Canadian Met. Quart.*, (2006). , 45, 261-274.

[8] Zhao, B, Jak, E, Hayes, P. C, Holliday, R, & Lee, R. (2005). *Lead & Zinc'05 Int. Symp.*, Kyoto, 17-19 Oct 2005, , 1459-1474.

[9] Jak, E. Degterov SHayes., P.C. and Pelton A.D., *Proceedings of the Fifth International Symposium on Metallurgical Slags and Fluxes*, Iron and Steel Society, AIME, Sydney, publ. Iron and Steel Society, Warrendale, PA, (1997). , 621-628.

[10] Ralls, K. M, Courtners, T. H, & Wulff, J. *Introduction to Materials Science and Engineering*, John Wiley & Sons, (1976). , 400.

[11] Nitta, T, & Kikuta, K. *World Zinc'93*, edited by Matthew I.G., AusIMM, (1993). , 445-451.

[12] Bellot, J. P, Patisson, F, & Ablitzer, D. *Metall. Trans. B.*, (1993). , 24B, 27-38.

[13] Zhao, B, Jak, E, Hayes, P. C, Yang, G, Wang, J, Dong, Y, & Errington, B. Characteristics of Isasmelt Slag and Lead Blast Furnace Sinters', Proc. Pb-Zn'05, Kyoto, Japan, MMIJ, October (2005). , 2005, 571-586.

[14] Cameron, A. W, Eyers, S. A, & Nelson, N. L. The New Updraught Sinter Plant at Mount Isa Mines Limited', Proc. Aust. Inst. Min. Met., No. 226, Part (1968). , 2(1968), 131-149.

[15] Siemon, J. R, Kowalczyk, E, & Tuppurainen, J. The Estimation of Burn-through Point on a Lead Sinter Plant', The 3rd Conference on Control Engineering, Sydney, Australia, Institution of Engineers, Australia, May (1986). , 1986, 191-195.

[16] Siemon, J. R, Tuppurainen, J, & Hollis, R. G. Bed Temperature Prediction in a Lead Sinter Plant', Non-ferrous Smelting Symposium, Port Pirie, Australia, AusIMM, Septmber(1989). , 1989, 27-32.

[17] Jak, E, Zhao, B, Degterov, S, Pelton, A. D, & Hayes, P. C. Coupled Experimental and Thermodynamic Modelling Studies of the System PbO-ZnO-FeO-Fe_2O_3-CaO-SiO_2-Al_2O_3 for Lead and Zinc Smelting', Int'l Conf on Lead/Zinc, Calgary, Canada, CIM, August (1998). , 1998, 313-333.

[18] Jak, E, Degterov, S, Zhao, B, Pelton, A. D, & Hayes, P. C. Coupled Experimental and Thermodynamic Modelling Studies of Slags for Metallurgical Smelting and Coal Combustion Systems', Metall. Mater. Trans. B, (2000). B, , 621-630.

[19] Zhao B., Burrows A., Jak E. and Hayes P.C., "Microstructures and Softening Behaviours of Lead Sinters and their Correlation to Sintering Temperatures in Mount Isa Lead Smelter", *Transactions of the Institutions of Mining and Metallurgy, Section C: Mineral Processing and Extractive Metallurgy*, Vol. 119(3), 2010, pp. 130-135.

Chapter 9

Pulse Current Auxiliary Sintering

Zhang Chunping and Zhang Kaifeng

Additional information is available at the end of the chapter

http://dx.doi.org/10.5772/53469

1. Introduction

1.1. Origin and development

Pulse current auxiliary sintering (PCAS) originated from electric current auxiliary sintering technique (ECAS), whose history can be traced back to the year of 1922, when an American did a path-breaking work to heat oxide powders by an electric current and synthesized a dense material. Then in 1927, another man sintered metal powders by simultaneously applying an electric current and a uniaxial mechanical load [1]. Since then, the synthesis technique has been being improved and perfected by numerous researchers. Up to now, over 1800 papers have been reported on the subject of ECAS and the technique has come to a new level: the used electric current extended from constant direct current to alternating current or pulsed electric current; power source was firstly selected to produce electric current, then capacitor banks were used; loading conditions evolved from no loading to uniaxial mechanical loaded or even multi-axial loading, etc.. Since 1960s, the technique commences to be commercialized. So far, it has been used to synthesize a large family of materials, including ceramics, intermetallics, metal-ceramic and ceramic-ceramic composites with various powders, such as elemental powders, pre-alloyed powders, pre-synthesized powders, or mechanical milled powders.

As a method to synthesize powers by applying electric current and mechanical loading simultaneously, ECAS technique revealed its increasing importance in the coming-out of a large amount of relevant bibliographies. Fig.1 shows the statistics of the number of articles on the subject published since the year 1922. It can be known that since 1999, the number of publications have been increased exponentially. If countries are weighed by the number of relevant articles, Japan, China and Korea are listed in the top three, as given in Fig. 2., while the contribution of other countries to the field remains relatively insignificant.

INTECH
open science | open minds

© 2013 The Author(s). Licensee InTech. This chapter is distributed under the terms of the Creative Commons Attribution License http://creativecommons.org/licenses/by/3.0), which permits unrestricted use, distribution, and reproduction in any medium, provided the original work is properly cited.

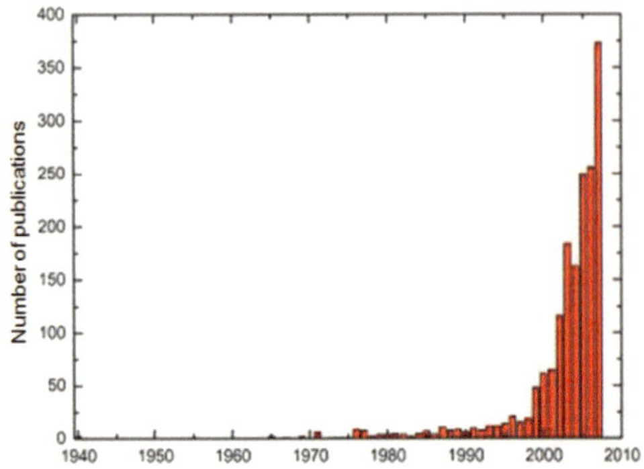

Figure 1. Number of publications related to ECAS processes by year [1]

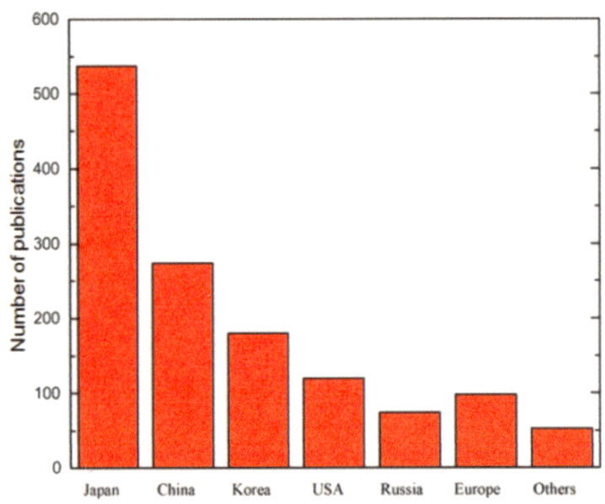

Figure 2. Publications related to ECAS numbered by country [1]

1.2. Classification

According to the standard classification in the sintering process, ECAS belongs to pressurized solid compaction [2], as drafted in Fig.3.

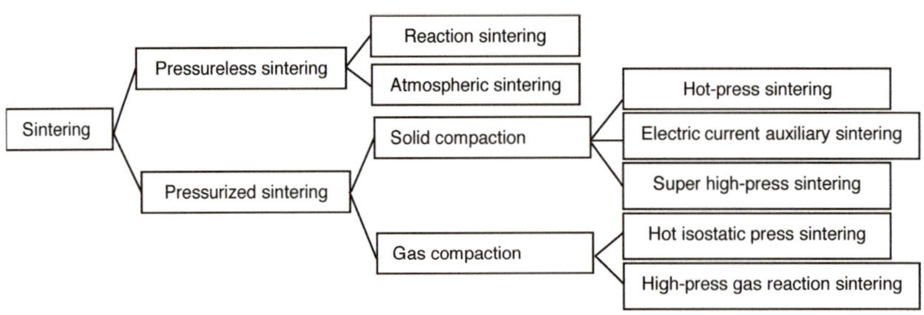

Figure 3. Standard classification of sintering process

When based on characteristics of the applied electric current, ECAS itself can be divided into two categories: Resistance Sintering (RS) and Electric Discharge Sintering (EDS) [3]. The former, RS, adopts direct current, alternating current or pulsed current with low voltage (dozens of volts) and high current (thousands of amperes) to sinter powders in action time ranging from 10^0s to 10^3s. In contrast, in EDs, powders were sintered by the electrical current with high voltage and high current from capacitor bank, in which abundant electrical energy is stored and can be launched instantaneously. But, most especially, the sintering process is often accompanied by electromagnetic phenomenon [4]. The distinction between the two categories has been listed in Table 1. Fig. 4 presents the statistics on the number of articles regarding the two categories of techniques [1]. As seen, studies on RS are far more numerous. The subject of this chapter, PCAS, is in the range of RS.

	Current origin	**Current waveform**	**Voltage /current**	**Electromagnetic phenomenon**	**Action duration (s)**
RS	power	constant direct current/ alternating pulsed current	low voltage high current	no	$10^0 \sim 10^3$
EDS	capacitor bank	large pulsed current	high voltage high current	yes	$10^{-5} \sim 10^{-3}$

Table 1. Differences between Resistance Sintering (RS) and Electric Discharge Sintering (EDS)

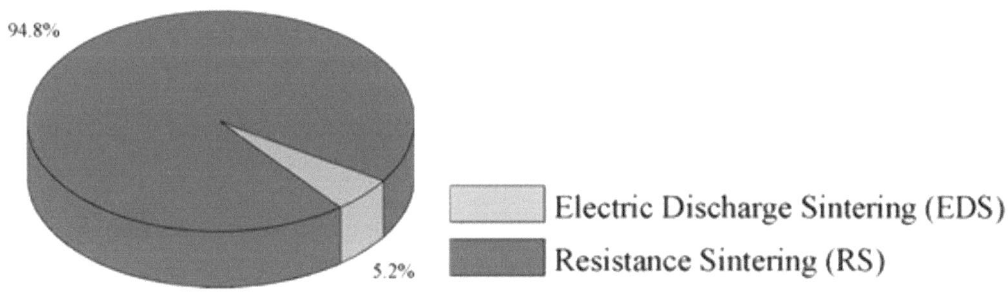

Figure 4. Relative amount of scientific reports regarding the two main ECAS processes

RS equipments used by researchers in different countries were purchased directly from commercial corporations, or developed by scientific institutions or researchers themselves for facilitate their own studies. Apart from the RS equipments developed by the two Japanese companies (Spark Plasma Sintering apparatus, well known as SPS), user-built machines accounts for about 40% among all the current RS machines, as introduced in Fig.5, where other companies refer to Sodick Co. Ltd [5]., Superior Graphite Co. Ltd.(USA) [6], Materials Modification Inc.(USA) [7], Eltek Co.(Korea) [8]etc.. All the user-built machines were named diversely to distinguish them from the commercial equipments. Fig. 6 shows a huge variety of RS designations. It is worth noting that almost all the names reflect the use of electric current [9]. For example, plasma activated sintering (PAS), pulse current hot pressing (PCHP), and so on. In the lab where the author of the chapter works, a similar apparatus was also developed and named ZLY-60 Pulse Current Auxiliary Sintering (PCAS) to distinguish it from other used machines. The photo of the machine has been given in Fig.7. In RS, 4 electric current waveforms have been mentioned [10-18], as shown in Fig. 8. Among them, the third one, pulsed direct current, was the most common current. In PCAS, such a current was adopted for a series of studies.

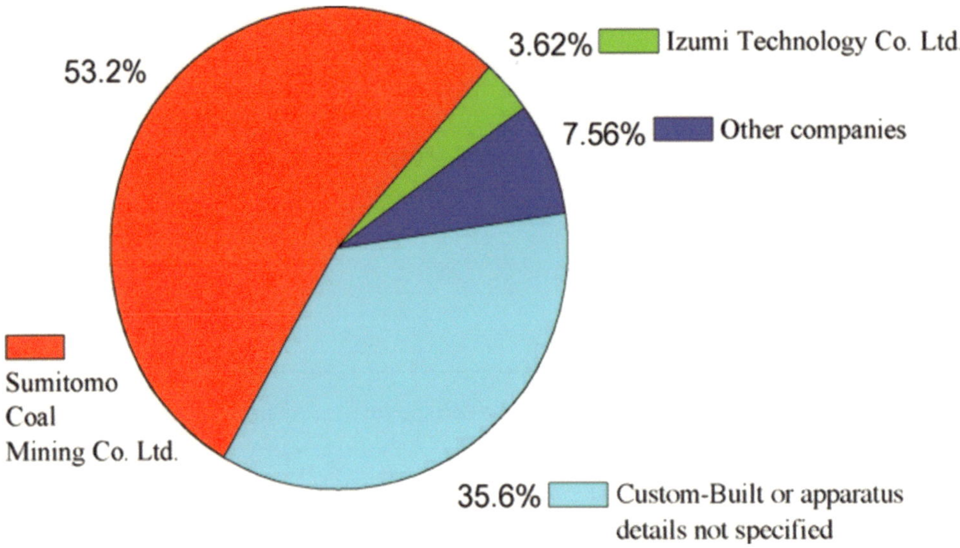

Figure 5. Relative distribution of apparatuses adopted in the RS processes as reported

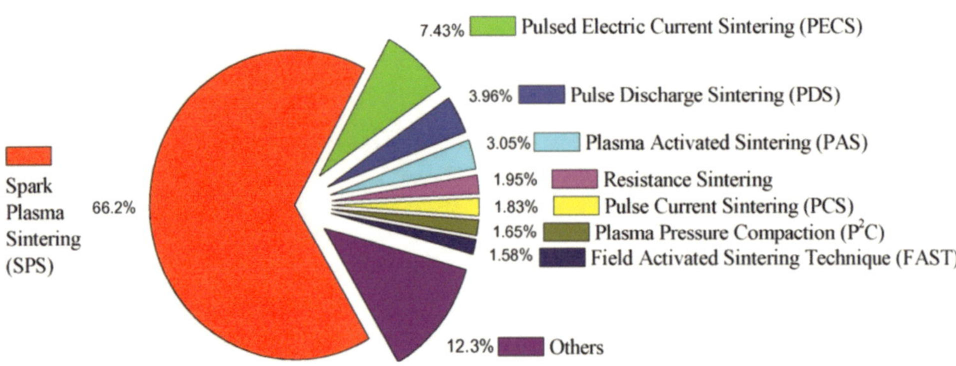

Figure 6. RS processes designation [1].

http://dx.doi.org/10.5772/53469

Figure 7. ZLY-60 pulse current auxiliary sintering apparatus

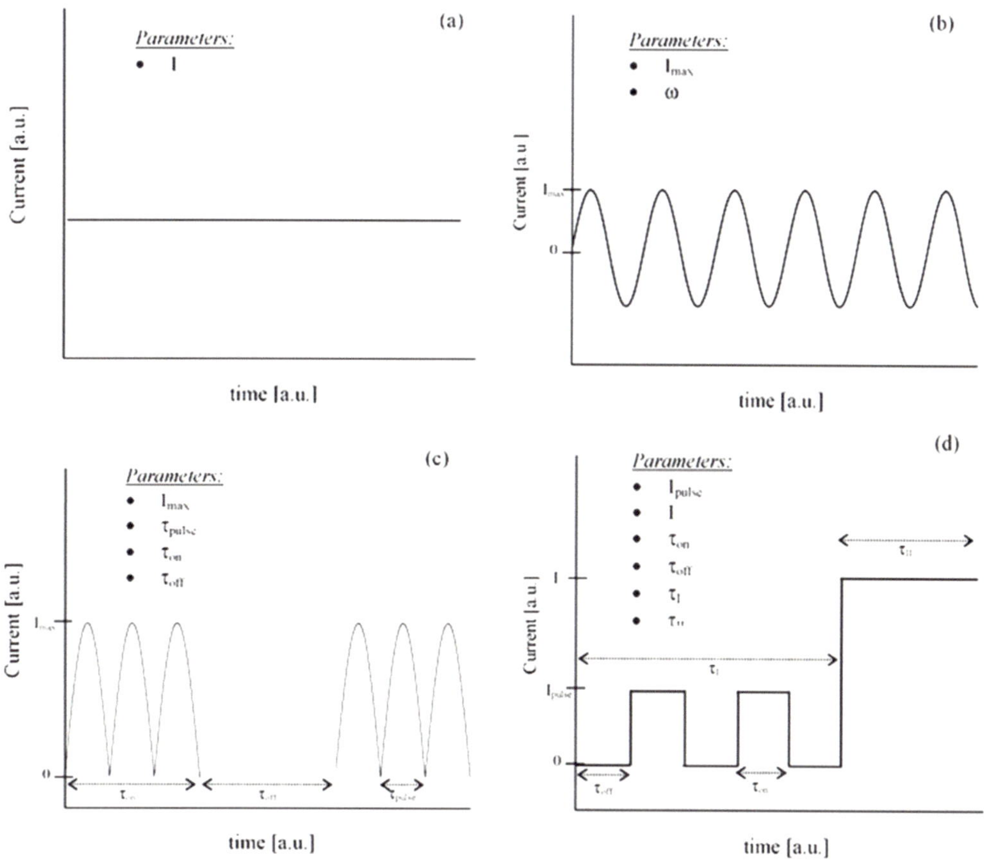

Figure 8. Typical electric current waveforms applied in the RS processes; (a) constant direct current (DC)(b) alternating current (AC) (c) pulsed direct current(d) pulsed direct current + direct current

1.3. Device structure and working fundamental mechanisms

PCAS equipment consists of three parts, as shown in Fig.9: (1) sintering mold and the device to provide an axial force, the force which can be adjusted in terms of synthesized material properties; (2) pulsed DC generator to generate electrical current to activate material powders; (3) electrical resistance heating part. When PCAS is in use, loose powders are firstly put into the die, then the furnace is vacuumed before the die is pressurized. In the sintering process, heat is provided by passing an electric current through the upper and lower punches, the powders and their container. Because of the multi-channel to transfer the current, the heat transfer rate is that high and the temperature can arrive at 1000~2000°C in a few seconds or minutes. What's more, heating rate and sintering temperature are both in control by adjusting the magnitude of pulsed current. Coupled with the load regulating system (the applied load can be quite low, say, 20~30MPa, or quite high, 500~1000MPa), PCAS can be utilized to synthesize metals, ceramics, and composite materials.

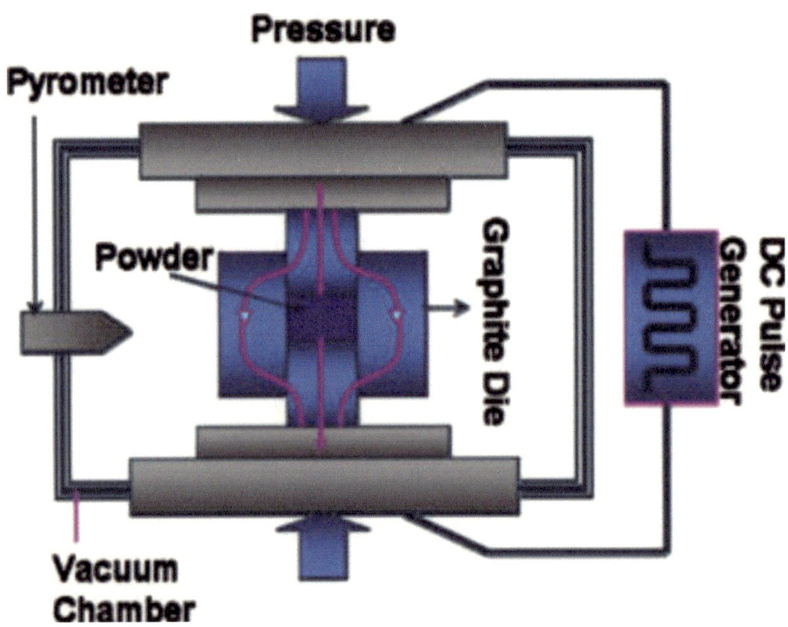

Figure 9. Schematic of the PCAS process

As an advanced technology for materials synthesis and processing, RS presents a lot of potential applications. But its sintering mechanism is still in dispute. It is generally acknowledged that in the sintering procedure, when an electric current runs through, the consequent Joule heat effect and the plastic deformation caused from temperature-rising and loading help the sintering process, but other elements play a much greater role. For example, when pulse DC voltage is on, spark discharge/plasma generates between adjacent particles [19,20], and individual ones are spontaneously heated, leading to the activation, purification of particle exterior surface and removing of oxide film, helping the rapid material diffusion and migration, promoting efficient heating and plastic deformation. When the pulse voltage is off, the temperature decreases rapidly through thermal diffusion and sintered material is subjected to quite short temperature exposure [21]. All of the phenomena and effects realize the consolidation of specific products with desired configuration and density at lower temperature and in shorter sintering time, as introduced in details in Fig. 10.

It is accepted that pulsed current concentrates on the junction of neighboring powder particles. Activated by pulsed electric current field, anode and cathode emerge in minute zones, where spark discharge and plasma are generated. The energetic particles from discharging knock on the contact part between adjacent particles, then the resulting local high temperature bring the superficial layer into a molten state, causing the emergency of 'neck' [22], as illustrated in Figs. 11. The transmission of the thermal energy from the particle center to the surface and its rapid dissipation result in a quick cooling in the 'neck's, prompting consolidation and material migration and transfer between neighboring powder particles through volume diffusion, surface diffusion and grain boundary diffusion [20], as Fig.12 shows. It is the simultaneous effect of electric current heating and uniaxial loading from the upper punch that reinforces volume diffusion and boundary condition so that the sintering synthesis and densification is accelerated and the sintered compacts of high quality can be gained in shorter duration of time and at lower sintering temperature.

However, how to verify the existence of discharged plasma has been a problem all the time. Besides, it has been proved on the basis of experimental studies that electric current cannot run through non-conductive powders, thus the view of generation of plasma in such powders does not work. Therefore, many scholars doubt the above-mentioned theory. For this reason, a 'self-adjusting' mechanism in microstructure evolution comes forward [23], as shown in Fig. 13. It is recognized that in the initial stage of sintering, many particles have come into contact under pressure. The electric current tends to be relatively larger in the particle who has contributed larger contact surface, that is $I_1 > I_2$. Thus, the sintered 'neck' forms firstly in the particle which I_1 passes through for the reason of the Joule heat effect. With the neck growing up, the contact surface increases further, then current I_1 ascends and neck temperature augments correspondingly, producing increased local resistivity and resistance. At the moment, current in I_1 particle is frustrated and commences to run through the particle with smaller contact surface, namely, $I_2 > I_1$. As a result, neck forms in the particle with smaller contact surface. And so on alternately until powder consolidation and densification process is completed. Such a mechanism can explain the cause why the sintered materials are always with fine, homogeneous and dense microstructures. But the explanation can apply only to conductive material with positive resistance-temperature coefficient. When it comes to non-conductive materials with no obvious resistance changing sensitivity to temperature, it does not work. It is generally believed that the rapid transportation of heat through mold and upper and lower punches, together with the large electric current, bring non-conductive powders directly into the high-temperature zone and make it possible to realize rapid synthesis.

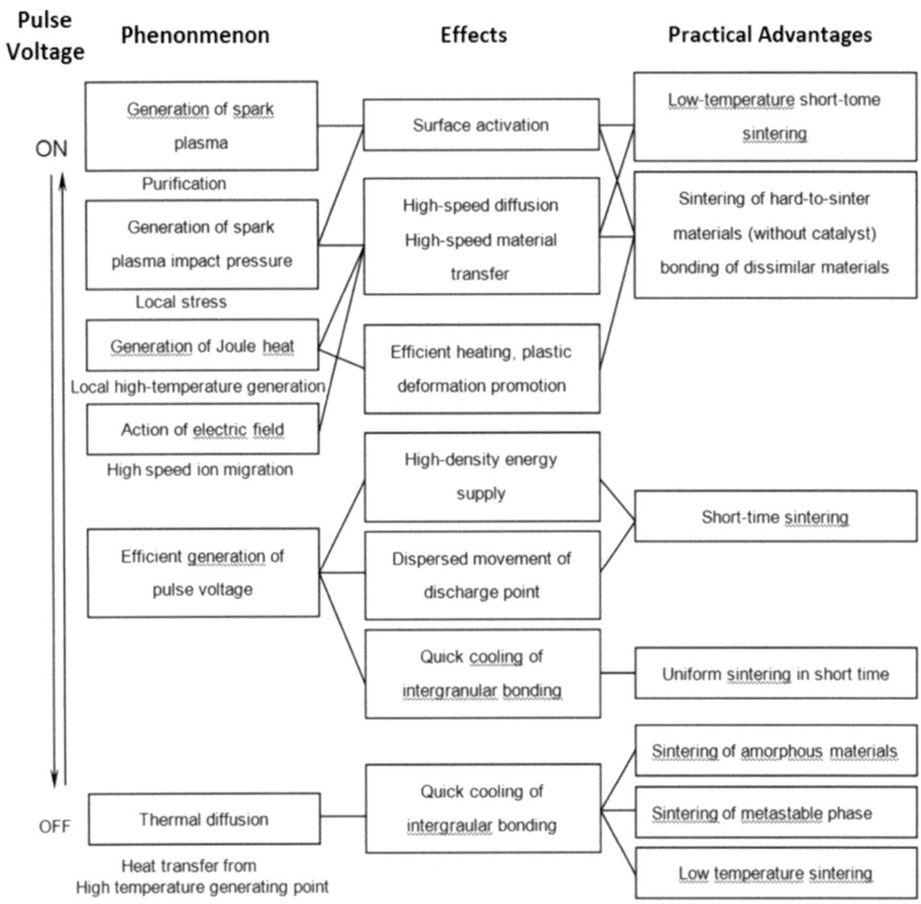

Figure 10. Effect of ON/OFF direct current pulse energizing

http://dx.doi.org/10.5772/53469

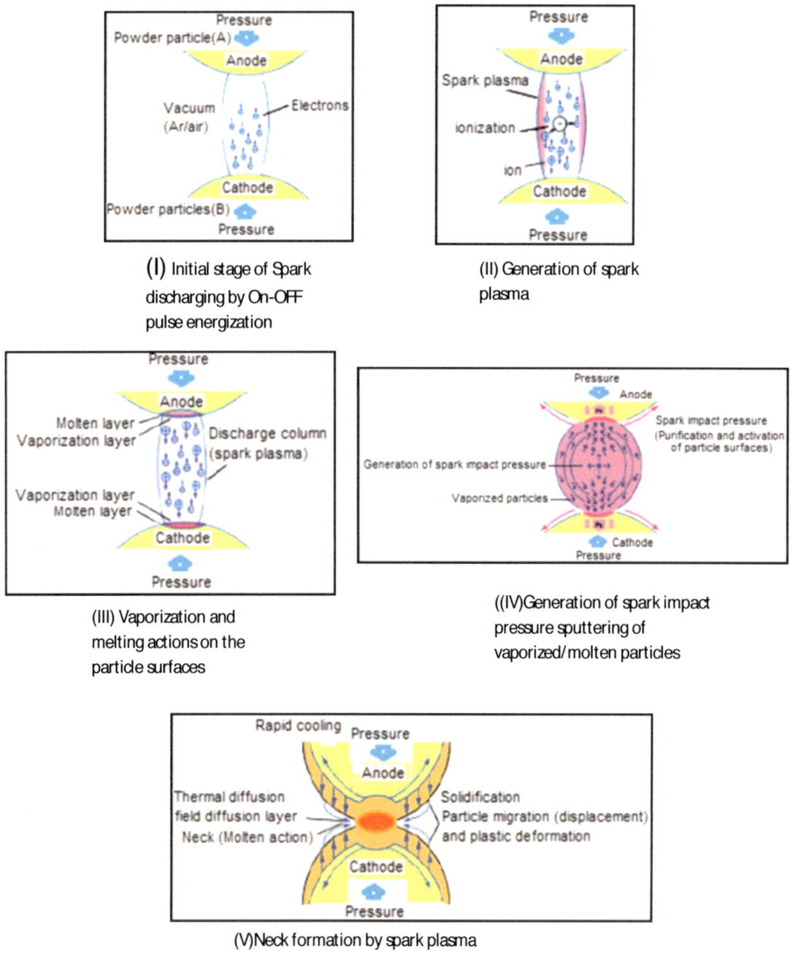

Figure 11. Mechanisms of neck formation between particles

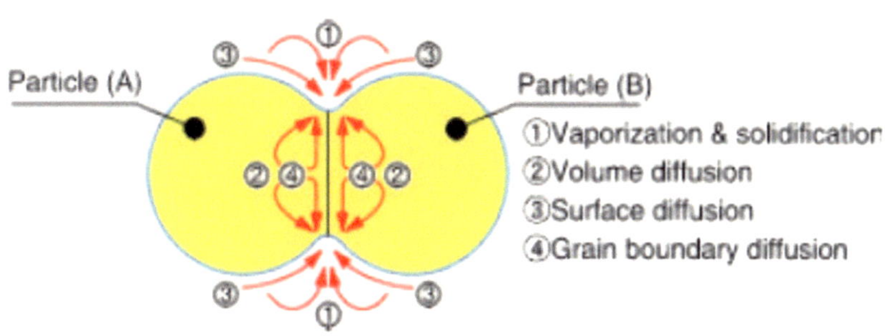

Figure 12. Material transfer paths during sintering process

Although the working mechanism of the technique is far from reaching a consensus, studies prove that electric pulse plays an importance effect on crystallization. Pulsed current possesses such a short relaxation time that it eliminates nucleation barrier and raises nucleation rate and refines sintered microstructures.

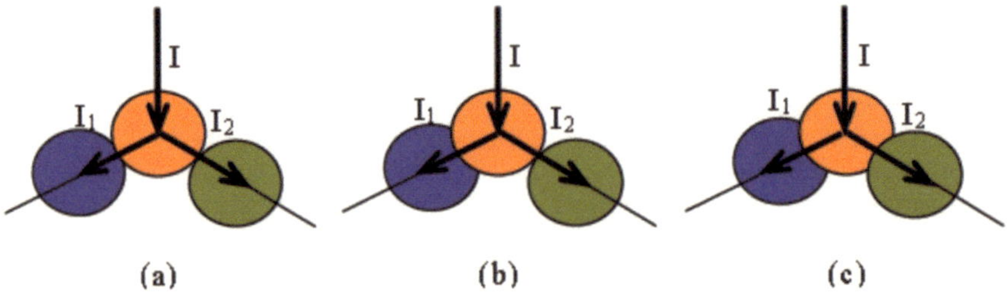

Figure 13. Change of contact areas between adjacent powders and distribution of electric current

1.4. characteristics and advantages over other sintering methods

This technology seems characterized by technological and economical advantages over conventional sintering methods, such as more efficient use of the heat input, faster heating rate, lower sintering temperature, shorter holding time, elimination of the need of sintering aids, no need of cold compaction, less sensitivity to initial powders characteristics, and marked comparative improvements in the properties of consolidated materials [24]. Especially when electrically insulating container is used and the electric current is applied for extremely short duration (down to few hundreds of microseconds), these advantages are more obvious.

For costly materials, this means considerable cost savings from reduced machining requirements and materials scrap. In addition, it should be noted that for engineering purposes shorter processing times usually result in productivity gains. As a consequence, sintered parts of higher quality may be expected to be obtained at lower processing costs.

On the contrary, in conventional powder metallurgy techniques, the powder container is typically heated by radiation from the enclosing furnace through external heating source. The resulting heating rate is then typically slow and the process can last hours. In the process, a lot of heat is wasted as the whole volume of space is heated and the compact receives heat indirectly from the hot environment.

2. Application to synthesis of γ-TiAl alloys

γ-TiAl alloys are of great interest for applications in aerospace industry due to their numerous advantages, such as low density, high specific Young's modulus and strength, good oxidation and burn resistance. However, such intermetallics suffer from having only poor machinability at room temperature and limited ductility at high temperatures, which limits their practical application [25]. In order to solve the problem, improving their room-temperature ductility and developing their superplasticity at high temperatures is of much importance.

Investigations have shown that fine grain size is beneficial for alloys to display good tensile behavior for the facility of grain sliding and diffusion [26]. However, it is difficult to directly obtain micron or submicron grains in γ-TiAl alloys through traditional cast or powder metallurgy route without subsequent heat treatment or complicated and costly mechanical processing. Thus, simplifying preparation route becomes an urgent need.

Up to now, electric current auxiliary sintering technique (ECAS) has been applied to synthesize γ-TiAl alloys and specimens with high densities and microstructures approaching the equilibrium state have been obtained [27-31]. The influence of operating parameters (i.e. temperature, processing time, etc.) on sintered γ-TiAl characteristics has been studied. However, heating rate effect was not widely explored. The typical heating rate is as low as1.7℃/s [27] or even less [32]. As a result, γ-TiAl alloys with coarse lamellar colonies were often obtained, even though the starting powders were managed to be refined. Thus, subsequent mechanical processing or other treatments become necessary for the purpose of improving the product properties. Currently, processed γ-TiAl alloys produced by electric current auxiliary sintering or other conventional processing routes, usually exhibit poor ambient ductility (in the range of 0.3~2% elongation).

In this context, dense γ-TiAl alloys with diverse microstructures through the controlling of sintering temperature, dwell time and heating rate are synthesized in PCAS apparatus from initially coarse powders (with particle size in the range of 5-350μm and the mean size of 103μm). When great care is taken, marked improvements in the tensile properties can be obtained both at room temperatures and at high temperatures. Above-mentioned experimental details are investigated and relevant results are given. Besides, an attempt was made to explain the effect of experimental parameters control, especially the role of heating rate, using thermodynamic theories, as well as nucleation and growth kinetics.

2.1. Experimental details

Pre-alloyed powders (Fig. 14) with the composition of Ti-42.5Al-2.3Nb-2.2Cr-0.28W- 0.15B (in at.%, mean grain size = 103μm, oxygen concentration ≈ 700ppm (wt.%)) were filled into a graphite mold sealed by two graphite punches at both ends. The loose powders in the graphite mold with a rectangular cross-section in the dimension of 10mm×30mm were subjected to pulsed direct current with different current densities to realize three groups of contrast experiments. The tested heating rate ranged from 2 to 9℃•s^{-1}; sintering optimal temperatures have been tested at 1200, 1250 and 1300℃; heating dwell time have been tried for 5min, 10min and 15min. In the experiments a gradual increasing pressure at the onset of the pulse was exerted until the optimal temperature and a specified pressure of 50MPa were attained nearly simultaneously. Then, the temperature and pressure were held constant for certain duration, followed by furnace cooling. A pulsed direct current with on/off cycles of 1.5ms (on-time) / 0.5ms (off-time) was applied. All the details for the three groups of contrast tests have been listed in Table 2.

Tensile specimens with a gauge length of 8mm and a section area of 3×1.2mm^2 were machined from the compacts. Tensile tests were carried out in air on American Instron 5500R testing machine at a strain rate of 2.083×$10^{-4}s^{-1}$ at ambient temperatures and high temperatures from 800 to 1000℃. SEM observations were operated on S4700 scanning electron microscopy (SEM), equipped with backscattered electron image. Conventional TEM (transmission electron microscope) observations were made on a Philips CM 12 operating at 120kv. Thin foils for TEM observation were cut parallel to the tensile axis from the gauge section. The foils were prepared by the standard jet polishing method using a solution of 30 vol.% nitric acid, 70vol.% methanol under ~15V and at -30℃.

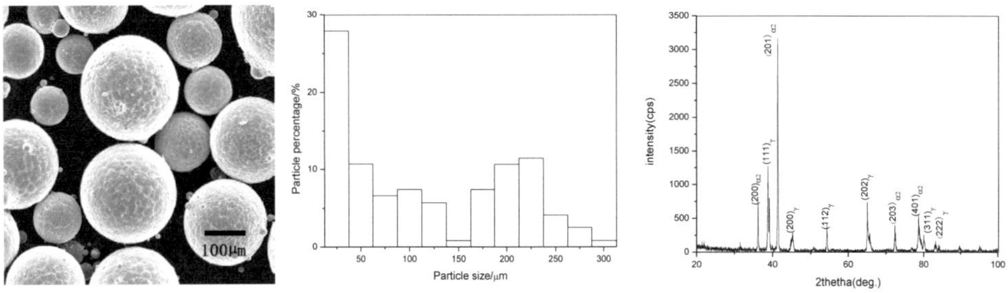

(a) (b) (c)

Figure 14. Basic information about pre-alloyed powders: (a) SEM micrographs, (b) particle size distribution (c) powders XRD analysis

Group	Heating rate ($°C \cdot s^{-1}$)	Sintering temperature (°C)	Dwell time (min)	Loading pressure (MPa)
1	9	1250	5	50
	4.5	1250	5	50
	3	1250	5	50
	2.5	1250	5	50
	2	1250	5	50
2	3	1200	10	50
	3	1250	10	50
	3	1300	10	50
3	3	1250	15	50

Table 2. Technological parameters adopted in sintering processes

2.2. Results and discussion

2.2.1. Dependence of as-sintered microstructures on sintering parameters

2.2.1.1. Heating rate

Typical back-scattered photomicrographs of four representative specimens sintered with heating rates of 2-9℃•s^{-1} are given in Fig. 15. It can be clearly observed that heating rate has a significant influence on the sintered microstructures. In the specimen sintered with heating-up rate of 9℃•s^{-1}, boundaries of original alloy powders can be clearly observed and a volume of pores between adjacent particles still exist. The compact is far from being consolidated. It can be remarked that inside individual powders, recovery and recrystallization have occurred, but grains have mal-developed. When powders were sintered with lower heating rate, 4.5℃•s^{-1}, grains developed fully and fine grains formed. When a local region is zoomed in, Fig. 16 can be obtained, presenting a dense micron-sized near-γ microstructure with irregular grayish-white α_2-Ti_3Al interspersed in the matrix of nearly equiaxial gray γ-TiAl. The $[01\bar{1}2]_{\alpha_2}$ and $[00\bar{2}]_{\gamma}$ zone diffraction patterns of the two phases are shown in Figs. 16 (c) and (d), respectively. A linear intercept method was used to reveal grain size d=4μm for γ and a volume fraction V_γ= 88.7%.

3℃•s^{-1} sintered materials were alloys with a duplex microstructure with nearly equi-axed gray γ grains dotted about α_2/γ lamellar colonies. TEM diffraction patterns of the two phases in lamellar region in Fig 15 (c) are shown by Fig.17, in which $[000\bar{2}]_{\alpha_2}//[\bar{2}\bar{2}\bar{2}]_{\gamma}$ is labeled, demonstrating $(111)_{\gamma}//(0002)\alpha_2$ and $[10\bar{1}]_{\gamma}//[11\bar{2}0]_{\alpha_2}$. The microstructure of specimens heated within 2-2.5℃•s^{-1} consists of fully regular patterns of α_2/γ laminates and no equiaxed γ grains. The slower the heating rate, the coarser the lamellar colony size is. Lamellar colony and grain sizes in samples consolidated with the heating rate of 2-4.5℃•s^{-1} are listed in Table 3. All the last four specimens have relative densities over 98% of the theoretical values, and microstructures are all uniform.

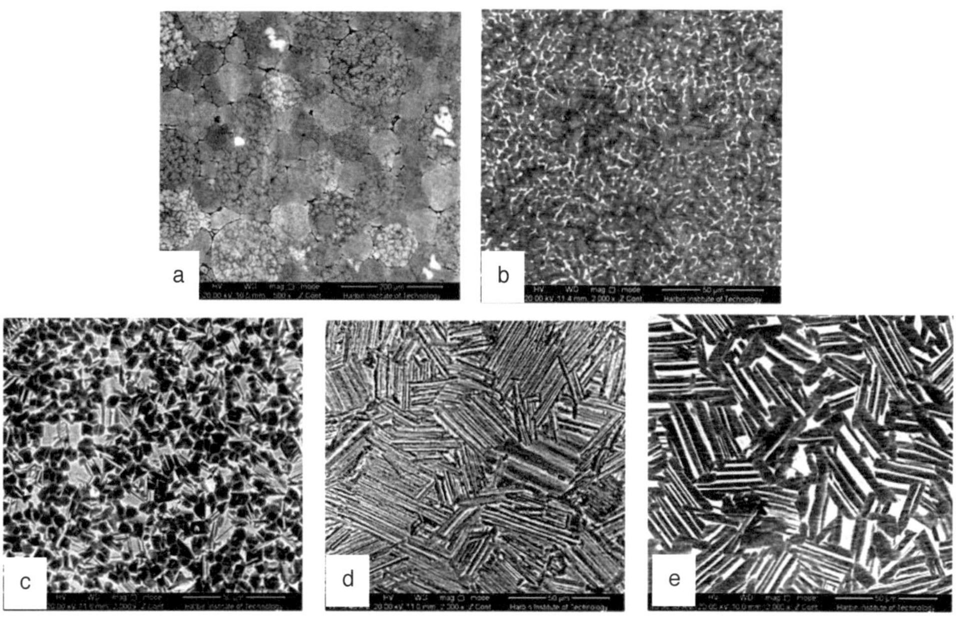

Figure 15. Back scattered SEM images of microstructures in alloys sintered at 1250°C for 5min with the heating-up rate of (a)9°C·s^{-1}, (b)4.5°C·s^{-1}, (c) 3°C·s^{-1}, (d)2.5°C·s^{-1}, (e)2°C·s^{-1}

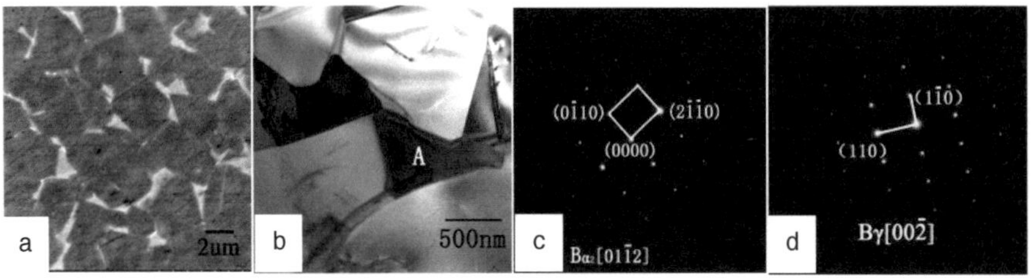

Figure 16. Microstructures in alloy sintered with heating rate of 4.5°C · s^{-1}: (a) Zoomed back scattering SEM image, (b) bright-field TEM image, (c) diffraction patterns of α_2 (grain A in (b)), (d) diffraction patterns of γ

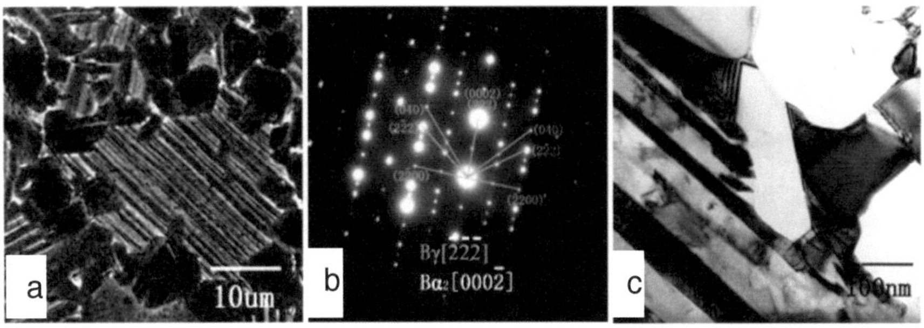

Figure 17. Microstructures in 3°C•s^{-1}-sintered alloy: (a) Zoomed back scattering SEM image; (b) bright-field TEM image; (c) diffraction patterns of α_2 and γ, in which $[000\bar{2}]_{\alpha_2}//[\bar{2}\bar{2}\bar{2}]_{\gamma}$

Heating rate (℃ · s^{-1})	Microstructure (μm)	Lamellar spacing (μm)	Lamellar Volume fraction (%)
4.5	NG(4)		0
3	DP(6~9)	0.21	22.3
2.5	FL(25)	0.89	100
2	FL (27)	2.53	100

Table 3. Controlled microstructures and corresponding values of the alloys sintered at 1250℃ for the duration time of 5min with different heating rates

2.2.1.2. Sintering temperature

When heating rate was fixed at 3°C · s^{-1}, duration time was lengthened to 10min, and loose powders were sintered at 1200°C, 1250°C and 1300°C, synthesized material display different microstructures, as shown in Fig. 18. In the compact sintered at 1200°C, original powder boundaries can still be seen although shapes of powders have changed from spheres to polyhedrons under pressure. Material migration was not finished fully enough to form grains and granular boundaries. When the sintering temperature was raised up to 1250°C, a uniform and full lamellar microstructure formed. The mean lamellar colony size is 30μm. When the material is sintered at 1300°C, a full lamellar microstructure with average colony size of 50μm was consolidated with obvious coarser lamellar spacing than the product obtained at 1250°C.

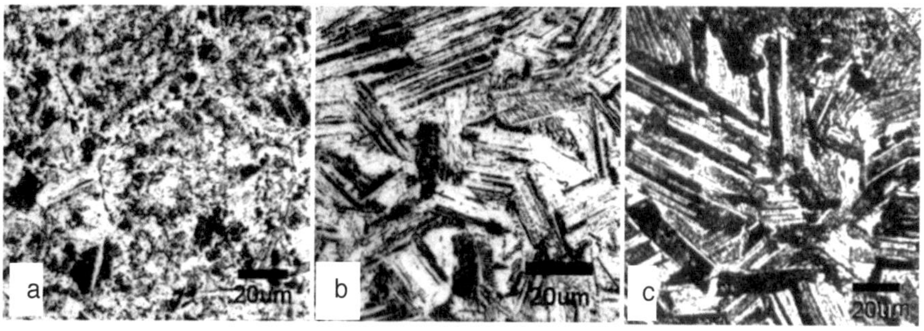

Figure 18. Microstructures of alloys heated with the rate of 3°C $\cdot s^{-1}$ and sintered at different temperatures for 10min :(a)1200°C, (b)1250°C, (c)1300°C

2.2.1.3. Dwell duration

The Fig. 19, combined with micrographs in Fig. 15(c) and Fig.18(b), the influence of duration time on the sintered microstructures can be deduced. In the sintered compact heated at the rate of 3°C $\cdot s^{-1}$ up to 1250°C and stayed at the temperature for 5min, the microstructure presented a duplex feature, composed of fine lamellar colonies and equiaxed γ grains. When the duration period is extended to 10min, a complete lamellar microstructure was generated. If the dwell time was further lengthened, a complete lamellar with coarser colony size was observed. The average lamellar colony size got to 120μm, the spacing augments to 2.2μm.

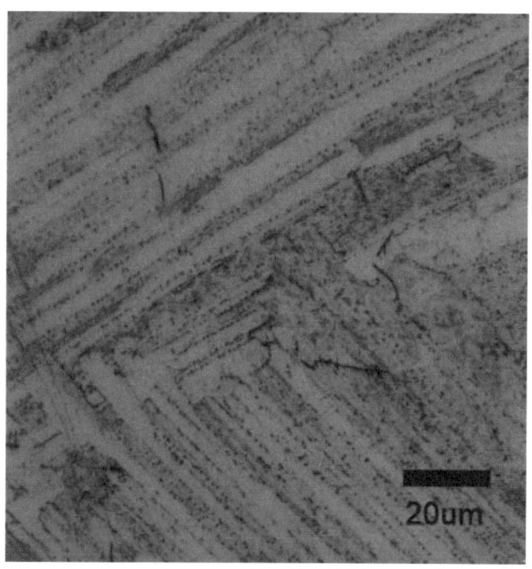

Figure 19. Microstructures of alloys heated with the rate of 3℃ $\cdot s^{-1}$ and sintered at 1250℃ for 15min

2.2.1.4. Discussion

Working mechanisms for the PCAS technique has been introduced in details in the first part of the chapter. But when the effect of sintering parameters on consequent microstructures is involved, the above-mentioned theory seems a little abstract to explain the corresponding evolution of microstructures. Hereinafter, the author tried to approach the effect of sintering parameters on resultant microstructures from the viewpoint of thermodynamics, as well as nucleation and growth kinetics.

Firstly, attention was paid to the effect of heating-up rate. If the joule heat generated by the pulsed electric current flowing through the tested material is denoted by Q_t, Q_t=Q_r+Q_h, where Q_r is the heat input during transient (i.e. temperature-rise) period, and Q_h is that supplied during steady-state sintering. Since four sintering processes about heating rate influence were all carried out at 1250℃ for 5min under the load of 50MPa, $Q_h$was the same. The only consideration is thus Q_r.

According to classical equation: $Q = I^2Rt$. I is the applied electrical current and $I = \varrho_I{}^*S$, where ϱ_I is the current density, which has been listed in table 4, and the cross-sectional area in tests $S = 30\times10mm^2$. The material's electrical resistance R can be derived in terms of the equation: $R = R_0(1+aT)$, where T is the thermodynamic temperature, R_0 the material resistance measured at 0℃, and 'a' is the temperature resistance coefficient. For the shortage of a suitable strong testing machine, it is sensible to choose R_0 as a constant. For simplicity, the averaging value of the coefficient 'a' was taken according to the equation: $a= (R_2-R_1)/R_1\Delta T$. $\Delta T = T_2 - T_1$, herein T_1 = 293k, T_2 = 1523k in experiments. In addition, R_1, R_2 respectively refer to the electrical resistance of initial loose powders and that of the alloy products. If all the alloy products are presumed to get to theoretical densification, it can be sure that R_2 must also be the same. Thus 'a' is identified as another constant term. The numeral value of the time 't' spent in temperature-rise period has been listed in table 4 as well. So the heat yielded by the current in temperature-rise period is calculated as follows:

$$Q_r = I^2Rt = \iint_{T_1 0}^{T_{2t}} I^2R_0(1 + aT)tdtdT = \int_{T_1}^{T_2} I^2R_0(1 + aT) = \frac{1}{2}t^2I^2R_0[(T_2 - T_1) + \frac{1}{2}a(T_2^2 - T_1^2)]$$

If the equation is reduced to Q_r= $(A+Ba)R_0$, table 5 will be obtained. Based on theory of electron and quantum, literature [33] confirms that resistance temperature coefficient for metals is about 0.4%. Then, 'a' in the above-equation must be a positive constant. Therefore, the result is straightforward that slower heating supplies a higher energy.

Usually the heat energy is consumed by three respects: Q_S, Q_T and Q_L, where Q_S is the energy for synthesis and densification of powders, Q_T is for morphologic transformation, and Q_L is for thermal losses. Here, Q_L can be neglected because the graphite mold used in the experiments was set inside a ceramic mold to prevent heat losses. Allowing for all the powders have been synthesized to compact alloys under the four sintered conditions, Q_S must be the same. Therefore, the more heat energy absorbed in 2℃•s^{-1}, 2.5℃•s^{-1}, 3℃•s^{-1}-sintering process than that in 4.5℃•s^{-1}-sintering period is used for morphological transformation. The slower the heating up rate, the more energy for such transformation will be.

http://dx.doi.org/10.5772/53469

Heading rate(°C·s^{-1})	ρ_I (A·mm^{-2})	t(h)
2	7.3	0.174
2.5	7.7	0.139
3	8.1	0.116
4.5	9.1	0.077

Table 4. Correlative numeral values in temperature-rise period

Heating rate(°C·s^{-1})	A(×10^6)	B(×10^9)
2	0.835	1.271
2.5	0.742	1.130
3	0.685	1.043
4.5	0.574	0.874

Table 5. Numeral values of A and B in Q_r

In the experiments in the current chapter, the more energy corresponds to more lamellar volume fraction in sintered alloys. Also it has been mentioned above that our starting metal powders consists of γ and α_2 (Fig.14(c)) with the approximate composition of Ti-42.5Al. In terms of Ti-Al binary phase diagram, Fig. 20, when experimented materials were heated up to 1250℃, the sintering temperature, the constituent phases γ and α_2 would both turn into disordered phase α. Thus it can be asserted that phase transformation in the sintering process followed the route: $\alpha \rightarrow \alpha+\gamma_{PPT} \rightarrow \alpha+\gamma_P\rightarrow$ Lamellar $(\alpha/\gamma) \rightarrow$ Lamellar$(\alpha_2+\gamma)$, where γ_{PPT} refers to disordered γ (equiaxed grains), $\gamma_P$ordered γ (γ plates).

When after heat preservation and the material was cooled down into $\gamma+\alpha$ two phase region, in α, the whole dislocation a/3<1120 > was discomposed into partial Shockley ones with simple stacking faults (SF): $\alpha \rightarrow \alpha^{SF}+\alpha \rightarrow \gamma_{ppt}+\alpha$ [34]. Such SFs would change local stacking sequences in $(0001)_\alpha$ in hcp α matrix, leading to a local change of crystal structure from hcp→LI_0. γ grains began to precipitate. Driven by surface tension, the precipitated disordered γ phase turned equiaxed. If at such time not enough energy can promote more equiaxed γ grains to precipitate further, such γ grains together with residual α would change into γ and ordered α_2, making up near γ microstructure. But the formation of equiaxed γ brought about differences in chemical free energy between γ and α matrix, and further redistribution of Ti and Al atoms. Thus it can be figured out that equiaxed γ is only a pioneer of lamellar γ/α_2 colonies. If the energy absorbed by the material can keep it in $\gamma+\alpha$ region for longer time, the internal structure tends to change into a more stable state.

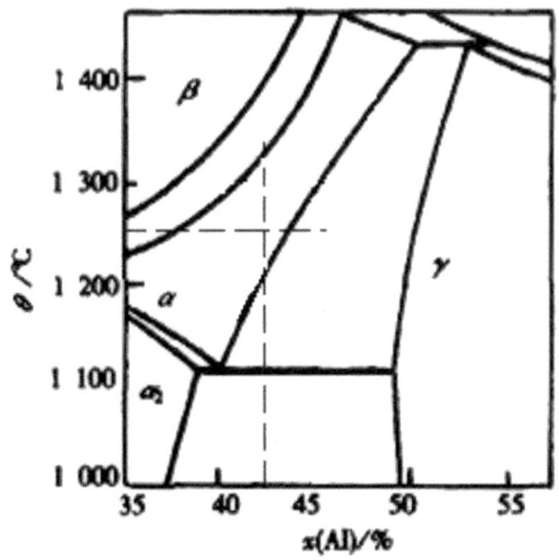

Figure 20. Central part of Ti-Al binary phase diagram

The surface of equiaxed γ grains is composed of many crystal faces with varied orientation, and such places can easily become nucleation location for γ lamellae. Extension of partial Shockley dislocations led to γ lateral growth and formation of "terrace" in phase boundaries. The formed kinking parts attracted atoms' diffusion and migration, resulting in γ growing laterally and γ plates formed. That is, γ plates form and grow by the "terrace-ledge-kink" mechanism [35]. In view of the fact that γ lamellae grew into neighboring α matrix interior with different rates and the growth rate in coherent α was greater than in incoherent ones, the differences in chemistry free energy would drive α on the side with incoherent boundaries to grow into γ interior. Then lamellar γ/α_2 colonies took shape[36,37]. γ plates can parallel align along $(0001)_{\alpha}$, the only habit plane in hcp α, while α lamellae can align along four habit planes in γ, including $(111)_{\gamma}$, $(\bar{1}11)_{\gamma}$, $(1\bar{1}1)_{\gamma}$ and $(11\bar{1})_{\gamma}$. As a result, the alignment of the two phases formed Blackburn phase relation [38, 39]. If excess energy existed at this time, all of equiaxed γ would be consumed and the whole microstructure in sintered alloys would be composed of lamellar γ/α_2 colonies. And there could be no doubt that more excess energy can thicken lamellae further. Then when the structures were cooled down to the room temperature, γ would remain and α would evolve into ordered α_2. Correspondingly, duplex or fine lamellar or coarse lamellar microstructure would be observed. Certainly, what microstructures can be obtained depends on how much energy the sintered material has absorbed.

The above-mentioned phase transformation process is in agreement with experimental results in this paper. Fig. 21 displayed such microstructure evolution process in TiAl alloys [40]. It should be pointed out that in the PCAS technique, electric field helps to improve nucleation rate and refine microstructures. As a result, the recrystallized α would be finer than that given in the figure, then more equiaxed γ would be separated out. When fine γ grains reached out to each other, triple junctions among boundaries would appear and near gamma would form. Therefore, the reason for the as-sintered microstructures changing from near gamma to duplex, fine lamellae, then coarse lamellae with the increment of heating-up rate can be clearly uncovered. The effect of lengthening duration time and raising sintering temperature is the case also.

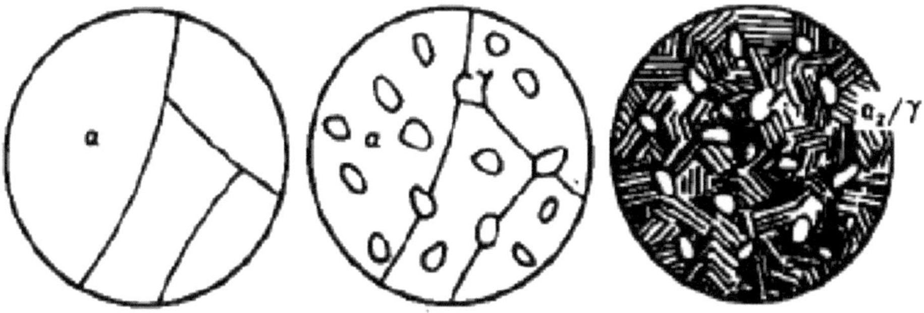

Figure 21. Schematic drawing of microstructure evolution process in TiAl alloys

2.3. Tensile properties of as-sintered alloys

Due to the fact that the alloys sintered at 1250 °C with heating rate of 2-4.5℃·s^{-1} and dwell time of 5 min present relative fine-grained microstructure. So, all these samples were subjected to room temperature tensile tests. 4.5°C·s^{-1}- and 3°C·s^{-1}- sintered samples were picked out to experience high temperature tensile tests. It was found that some alloys displayed improved tensile properties than results from other researchers.

2.3.1. Room temperature tensile properties

Tensile properties of the sintered alloys are listed in Table 6. As shown in the table, room-temperature mechanical properties of the materials are dependent upon the microstructure. Ultimate tensile strengths σ_b of all specimens fell between 300 and 700MPa. The ultimate tensile strength increases with a decrease in lamellar colony size. Whereas the two samples with lamellar structure exhibited ordinary tensile elongation, the ductility of samples with near γ and duplex structures were over the usual value of 0.3-2%.

Heating rate(°C·s^{-1})	Microstructure (μm)	σ_b(MPa)	δ(%)
4.5	NG(4)	578	3.59
3	DP(6~9)	632	4.51
2.5	FL(25)	416	2.21
2	FL(27)	367	1.58

Table 6. Results of ambient temperature tensile tests at 2.083×10^{-4} s^{-1}

It has been shown [41] that the refinement of grain size can enhance the mechanical properties of TiAl. Indeed tensile properties of the two coarse fully lamellar microstructures are lower than those of duplex microstructure (Table 6). The strengthening mechanism of TiAl alloys with refined microstructures may be attributed to the classical Hall-Petch equation $\sigma = \sigma_0 + KD^{-1/2}$, where σ_0 refers to a material intrinsic value, K a Hall-Petch constant, and D the average diameter of grains. Such equation can be used to explain higher ultimate tensile strength when α_2 / γ lamellae size decreases from 27 down to 9μm. In addition, fine microstructures are beneficial for the operation of grain boundary sliding and the compatibility of deformation, resulting in better strain [42-44]. Furthermore, allowing for lamellar microstructures possessing the anisotropic mechanical property, differently oriented lamellae in deformation will produce discontinuous strain [45], resulting in the generation of interfacial microcracks and brittle failure of the tested specimen. Thus, better room temperature ductility is suppressed.

However, the above-mentioned theory isn't applicable for all the experimental results here in that in the test a room temperature ductility peak was observed in the duplex TiAl alloy instead of finer equi-axed one. Specifically, the maximum tensile strength occurs also at the maximum tensile elongation. In order to explain such phenomenon, close attention is paid to sintering processes. During sintering, load and heat were applied simultaneously. Therefore, high heating rate corresponded to high pressurizing rate. A high heating rate is expected to inhibit grain growth. As a result, grain sizes are much refined when heating rate is high. However, high loading rate can quickly seal the venting channels among powder particles and prevent a complete outgassing. This, of course, was not in favor for specimen densification during sintering [46]. Thus, the relative density of specimens obtained at the heating rate in the range of 2-3℃•s^{-1} was over 99.5%, while the density of that one sintered at 4.5℃•s^{-1} was 98.6%. The lower density, thus more residual pores reduces the ductility and tensile properties. Also, near gamma microstructure is a product of unstable phase transformation, so vast defects would remain in internal structures, such as a large number of dislocations tangled up in grains, as shown in Fig. 22. So, even though grain boundary sliding and deformation compatibility ameliorated with grains being refined, numerous dislocations tangling up inhibited the alloy's capability of homogeneous deformation [45]. As a result, the tensile properties of fine-grained, near γ sample are inferior to those with a duplex structure.

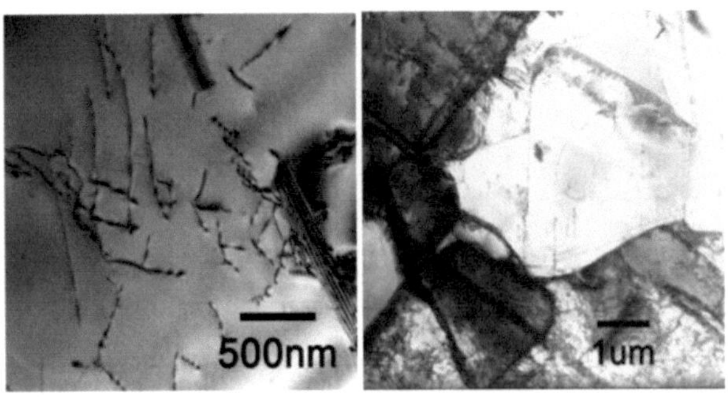

Figure 22. TEM micrographs of 4.5℃•s^{-1}-sintered alloy

2.3.2. *High temperature tensile properties*

Fig. 23(a) shows the true stress-strain curves of the 4.5℃•s^{-1}-sintered alloy for tensile deformation at temperatures ranging from 800 to 1000℃ at the strain rate of 2.083×10^{-4}s^{-1}. It can be seen that in the temperature range, plastic strain appeared, following the elastic deformation. Moreover, at above 900℃, steady flow took place. Besides, the equiaxial γ-TiAl based alloy under study is characterized by a remarkable elongation-to-fracture and a marked strain rate sensitivity m. As the temperature rises, the elongation exhibits monotonous character, increasing from 71% at 800℃, 165% 900℃, 241% 950℃ to the maximum elongation-to-failure δ=409% at 1000℃. The profile of the specimen fracture at 1000℃ is illustrated by the Fig. 23(b).

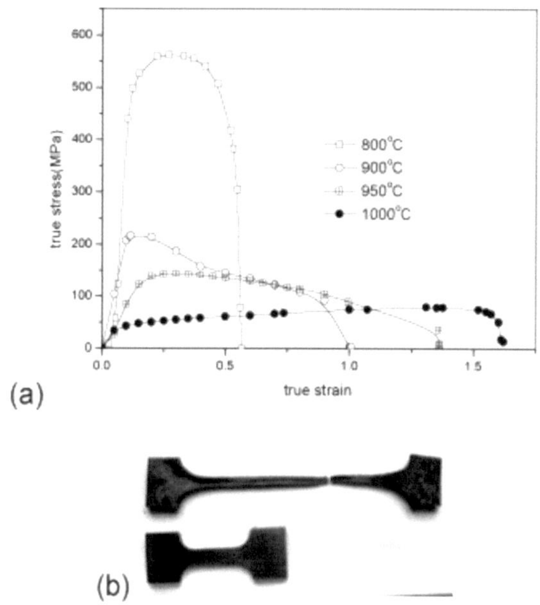

Figure 23. (a)True stress-strain curves of the 4.5℃•s^{-1}-sintered alloy for deformation at 800 -1000℃, (b) fracture sample deformed at 1000°C with 409% elongation and an as-prepared test piece (bottom) is also shown for comparison.

m value obeys the same tendency as the elongation, from 0.15, 0.33, and 0.45 up to 0.8 at 1000℃. The values of m and elongation-to-failure at temperatures above 900℃ are indicative of superplastic deformation behavior. Meanwhile, an increase of temperature leads to a rapid decrease in flow stress, decreasing from 565Mpa at 800℃ down to 79Mpa at 1000℃. This result is in accordance with the majority of γ-TiAl based alloys with micron-grained structure investigated to date [47-49].

The effect of testing temperature on the high temperature flow behavior of duplex microstructure is illustrated in Fig. 24(a). It can be seen that this alloy failed during the elastic stage at 750°C and the elongation reaches 10%.At 900°C, plastic strain began to appear, following the elastic deformation. At and above 950°C, steady flow took place. Moreover, an increase in the temperature led to an increase in the elongation, from 48% at 900°C to 99% at 950°C, with a maximum of 135% achieved at 1000°C (Fig. 24 (b)). m value reached 0.23 at 1000°C, lower than the critical value 0.3 for conventional superplastic deformation. Its flow stress followed the same tendency as 4.5°C•s^{-1}-sintered alloy, namely decreasing with the test temperature ascending. However, compared with the tensile behavior exhibited by 4.5°C•s^{-1}-sintered alloy, flow stress in 3°C•s^{-1}-sintered one is always higher at corresponding test temperature.

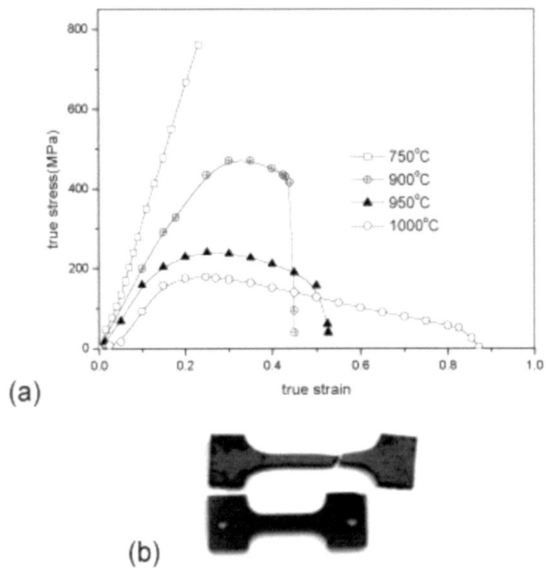

Figure 24. (a) True stress-strain curves of the 3°C•s^{-1}-sintered alloy for deformation at 750 -1000°C at the strain rate of $2.083\times10^{-4}s^{-1}$ (b) fracture sample deformed at 1000°C with 135% elongation and an as-prepared test piece (bottom) is also shown for comparison.

Superplastic behavior was found in nearly equi-axed TiAl alloy with the usual prerequisites of fine grain size [50]. It is obvious that a decrease in grain size to a micron level improves deformation homogeneity, facilitates grain boundary sliding and interaction of dislocation with grain boundaries, decreases flow stress, and suppresses deformation twinning, providing increased ductility at relatively low temperatures [51]. Fine microstructures are beneficial to the compatibility of deformation. Large number of grains joins in the deformation, resulting in better plasticity.This is also the reason for the non existence of superplasticity in 3°C•s^{-1}-sintered duplex alloy.

Also, a microcrystalline grain size can help the alloy to display Superplastic behavior and decrease the temperature for Superplastic behavior with extensive grain boundary diffusion and sliding. Based on calculation of the activation energy Q for Superplastic deformation being 212.6~252.2kJ•mol^{-1}, the Superplastic deformation mechanism for the alloy is determined as the grain boundary sliding accommodated by grain boundary diffusion [52].

http://dx.doi.org/10.5772/53469

3. Conclusion

In the first part of this chapter, the technique of PCAS was firstly introduced in detail, including its origin, development, classification, fundamental working mechanism and a lot of advantages over other traditional material preparation technique.

The second part paid special attention to the application of the technique to synthesis of γ-TiAl alloys. On self-built pulse current auxiliary sintering apparatus (PCAS), influences of different parameters, including heating-up rate, sintering temperature and duration time, on obtained microstructures in sintered products were studied. After systematic researches on heating rates of 2~9°C $\cdot s^{-1}$, sintering temperatures of 1200~1300℃, dwell times of 5~15min, it was determined that 1250℃ as the temperature and 5min as the duration time were more beneficial for generating dense, even and fine γ-TiAl based alloys.

When powders were synthesized in the oven at 1250℃ for 5min with heating rates of 4.5, 3, 2.5, 2℃ $\cdot s^{-1}$, near gamma, duplex, fine fully lamellar and coarse fully lamellar microstructures were obtained, correspondingly. After comparison studies among these alloys, it was found that grain size, lamellar volume fraction and lamellar spacing played an important effect on room temperature mechanical properties of the sintered products. In general, duplex alloys possessed better comprehensive behaviors. Their ductility displayed at room temperatures could get to 4.51%. Thus, brittleness problem was solved to some extent.

In tensile tests at various temperatures on sintered TiAl alloys with near γ and duplex microstructures, it was observed that the one with equi-axed fine grains possessed more superior elevated temperature tensile properties. In temperature range of 950℃~1000℃ and strain rate range of $2.083\times10^{-4}s^{-1}$,the obtained elongation was over 240%. The results demonstrated its ability to display superplastic behavior at relatively low temperatures. Based on calculation of the activation energy Q for Superplastic deformation being 212.6~252.2kJ•mol^{-1}, the superplastic deformation mechanism for the alloy was determined as the grain boundary sliding accommodated by grain boundary diffusion

Based on the success in the synthesis of dense γ-TiAl alloys with fine microstructures, improved room-temperature and ideal high-temperature tensile properties, more studies are under way for processing industrial parts with good quality and improved mechanical properties directly by PCAS technique. These relative studies will explore more prospects for the application of the technique.

Author details

Zhang Chunping and Zhang Kaifeng

National Key Laboratory for Precision Heat Processing of Materials, Harbin Institute of Technology, Harbin, People's Republic of China

References

[1] Orru R., Licheri R., A. Locci M., Cincotti A., Cao G. Materials Science and Engineering R 63 (2009) 127–287

[2] L.Gao. Spark plasma sintering technique. Jounal of inorganic material. 1997;12:129-133

[3] Gao W., Gao X., Asif M., Wu Z.W., Ling B.L., Li J.G.. The Study of MARFE during Long Pulse Discharge in the HT-7 Tokamak. Journal of Nuclear Materials. 2007;363-365(15):770-774

[4] Clyens S., Al-Hassani S.T.S., Johnson W. The Compaction of Powder Metallurgy Bars Using High Voltage Electrical Discharges. International Journal of Mechanical Sciences. 1976;18(1):37-40

[5] Omori M., Okubo A., Kang G.H., Hirai T. Preparation and Properties of Polyimide/Cu Functionally Graded Material. Functionally Graded Materials 1996. 1997;767-772

[6] Tracy M.J., Groza J.R. Consolidation of Nanocrystalline Nb-Al Powders by Plasma Activated Sintering. Nanostructured Materials. 1993;2(5):441-449

[7] Kalyanaraman R., Yoo S., Krupashankara M.S., Sudarshan T.S., Dowding R.J. Synthesis and Consolidation of Iron Nanopowders. Nanosructured Materials. 1998;10(8): 1379-1392

[8] Kim H.C., Yoon J.K., Doh J.M., Ko I.Y., Shon I.J. Rapid Sintering Process and Mechanical Properties of Binderless Ultra Fine Tungsten Carbide. Materials Science and Engineering A. 2006;435-436(5):717-724

[9] Clyens S., Williams D.J., Johnson W. The Compaction of Some Powdered Foodstuffs. International Journal of Mechanical Sciences. 1976;18(9-10):449-450

[10] Kimura H. Powder Processing and Mechanical Characteristics of Fully Dense Nano-Intermetallics. Nanostructured Materials. 1997;9(1-8):93-96

[11] Jaraszewicz J., Michalski A. Preparation of a TiB_2 Composite with a Nichkel Matrix by Pulse Plasma Sintering with Combustion Synthesis. Journal of the European Ceramic Society. 2006;26(13):2427-2430

[12] Song Z., Kishimoto S., Shinya N. Fabrication of Closed Cellural Nichel Alloy Containing Polymer by Sintering Method. Journal of Alloys and Compounds. 2003;335(1-2):166-170

[13] Xie G.Q., Ohashi O., Chiba K., Yamaguchi N., Song M., Furuya K., Noda T. Frequency Effect on Pulse Electric Current Sintering Process of Pure Aluminum Powder. Materials Science and Engineering A. 2003;359(1-2):384-390

[14] Anselmi-Tamburini U., Gennari S., Garay J.E., Munir Z.A. Fundamental Investigations on the Spark Plasma Sintering/Synthesis Process:II. Modellling of Current and

Temperature Distributions. Materials Science and Engineering A. 2005;394(1-2): 139-148

[15] Chen W., Anselmi-Tamburini U., Garay J.E., Groza J.R., Munir Z.A.. Fundamental Investigations on the Spark Plasma Sintering/Synthesis Process:I. Effect of Dc Pulsing on Reactivity. Materials Science and Engineering A. 2005;394(1-2):132-138

[16] Vanmeensel K., Laptev A., Hennicke J., Vleugels J., Van der Biest O. Modelling of the Temperature Distribution during Field Assisted Sintering. Acta Materialia. 2005;53(16):4379-4388

[17] Gennari S., Anselmi-Tamburini U., Maglia F., Spinolo G., Munir Z.A.. A New Approach to the Modeling of SHS Reactions: Combustion Synthesis of Transition Metal Aluminides. Acta Materialia. 2006;54(9):2343-2351

[18] Michalski A., Jaroszewicz J., Rosinski M., Siemiaszko D. NiAl-Al_2O_3 Composites Produced by Pulse Plasma Sintering with the Participation of the SHS Reaction. Intermetallics. 2006;14(6):603-606.

[19] Tokita M. Development of Large-size Ceramic/Metal Bulk FGM Fabricated by Spark Plasma Sintering. Materials Science Forum. 1999;308-311:83-88

[20] Kim H., Kawahara M., Tokita M. Specimen Temperature and Sinterability of Ni Powder by Spark Plasma Sintering. J.Japan Socoety Power. Powder Metallurgy. 2000;47: 887-891

[21] Groza J.R., Zavaliangos A. Sintering Activation by External Electrical Field. Materials Science and Engineering A. 2000;287(2):171-177

[22] Song X., Liu X., Zhang J. Evolution of microstructure in conductive powders and its mechanism during SPS sintering. Chinese Science. 2005;35:459-469

[23] Li B., Wang X., Han X., Liu X., Li S. Synthesis of nano-grained $BaTiO_3$ by SPS technique. Piezoelectricity and acousto-optic. 2005;27:43-46

[24] Khor K.A., Cheng K.H., Yu L.G., Boey F. Thermal Conductivity and Dielectric Constant of Spark Plasma Sintering Aluminium Nitride. Materials Science and Engineering A. 2003;347:300-305

[25] Vassiliou M.S., Rhodes C.G., Mitchell M.R., Graves J. Metastable Microstructure in Dynamically Consolidated γ Titanium Aluminide. Scripta Metallurgica. 1989;23(10): 1791-1794

[26] Liu C.T., Schneibel J.H., Maziasz P.J., Wright J.L., Easton D.S.. Tensile Properties and Fracture Toughness of TiAl Alloys with Controlled Microstructures. Intermetallics. 1996;4(6):429-440

[27] Couret A., Molenat G., Galy J., Thomas M. Microstructures and Mechanical Properties of TiAl Alloy Consolidated by Spark Plasma Sintering. Intermetallics. 2008;(16): 1134-1141

[28] Liu Y.D., Liu W. Mechanical Alloying and Spark Plasma Sintering of the Intermetallic Compound $Ti_{50}Al_{50}$. Journal of Alloys and Compounds.2007;440:154-157

[29] Matsugi K., Hatayama T. Yanagisawa O. Impact Properties of Spark Sintered Titanium Aluminides at Elevated Temperatures. Intermetallics. 1999;7:1049-1057

[30] Morris M.A., Leboeuf M. Analysis of Thermal and Athermal Deformation Mechanisms during Creep of γ-TiAl Alloys. Material Science and Engineering A. 1997;239-240: 429-437

[31] Lee W.B.. Yang H.S, Kim Y.W., Mukherjee A.K. Superplastic Behavior in a Two-phase TiAl Alloy. Scripta Metallurgica et Materialia. 1993;11:1403-1408

[32] Liu Y., Lu W.J., Qin J.N., Zhang D. A New Route for the Synthesis of NdB6 powder from Nd_2O_3-B_4C System. Journal of Alloys and Compounds 2007;431:337-341

[33] Liu H.Z., Li S.E. Discussion of Metals' Resistance Temperature Coefficient. Journal of Jining Teachers College. 2001;3:12-13

[34]]Valencia J.J., McCullough C., Levi C.G., Mehrabian R. Solidification Microstructure of Supercooled Ti-Al Alloys Containing Intermetallic Phases. Acta Metallurgica. 1989;37(9):2517-2530

[35] Denquin A., Naka S. Phase Transformation Mechanisms Involved in Two-Phase TiAl-based Alloys—II. Discontinuous Coarsening and Massive-type Transformation. Acta Materialia. 1996;44(1):353-365

[36] Kumagai T., Abe E., Takeyama M., Nakamura M. Microstructural Evolution of Massively Transformed γ-TiAl during Isothermal Aging. Scripta Materialia. 1997;36(5): 523-529

[37] Li Z., Cao C. $\gamma \rightarrow \alpha$ precipitation and transformation in Ti-Al alloy. Journal of Chinese nonferrous metal. 2003;13(4):827-834

[38] Jones S.A., Kanfman M.J.. Phase Equilibria and Transformation in Intermediate Titanium-aluminum Alloys. Acta Metallurgica et Materialia. 1993;41(2):387-398

[39] Hao S., Wu W., Han C. Hisological analysisof phase transition in TiAl two-phase alloys. Acia Metallurgica Sinica. 1993;29(2):49-55

[40] Kong F., Xiao S., Chen Y., Li B. Phase transition in Ti-45Al-5Nb(-0.3Y) in the process of continous cooling. Rare metals and engineering. 2009;38(1):25-28

[41] Chen X.Q., Huang B.Y., He Y.H., Qu X.H., Lei C.H. Microstructure Characteristics of a High Ductility γ-TiAl Alloy. Acta Metallurgica Sinica. 1997;7:683-689

[42] Venkateswara K.T., Kim Y.W., Muhlstein C.L., Ritchie R.O. Fatigue-crack Growth and Fracture Resistance of a Two-phase (γ+α_2) TiAl Alloy in Duplex and Lamellar Microstructures. Materials Science and Engineering A. 1995;192-193:474-482

[43] Tang J.C., Huang B.Y., He Y.H., Liu W.S., Zhou K.C., Wu A.H. Hall-petch Relationship in Two-phase TiAl Alloys with Fully Lamellar Microstructures. Materials Research Bulletin. 2002;37:1315-1321

[44] Bohn R., Klassen T., Bormann R. Room Temperature Mechanical Behavior of Silicon-doped TiAl Alloys with Grain Size in the Nano- and Submicron-Range. Acta Materialia. 2001;49:299-311

[45] Inui H., Oh M.H., Nakamura A., Yamaguchi M.. Room-temperature Tensile Deformation of Polysynthetically Twinned (PST) Crystals of TiAl. Acta Metallurgica et Materialia. 1992;40:3095-3104

[46] Liu J., She Z.G. in: Powder Metallurgy and Ceramic Forming Technology. Chemical Industry Publishing House. 2005;116

[47] Wang J.N., Y. W. An Investigation of the Origin of the Superplasticity of Cast TiAl Alloys. International Journal of Plasticity. 2006;22(8):1530-1548

[48] Lin D.L., Sun F. Superplasticity in a Large-grained TiAl Alloy. Intermetallics. 2004;12(7-9):875-883

[49] Hsiung L.M., Nieh T.G. Microstructures and Properties of Powder Metallurgy TiAl Alloys. Materials Science and Engineering A. 2004;364(1-2):1-10

[50] Mishra R.S., Stolyarov V.V., Echer C., Valiev R.Z., Mukherjee A.K. Mechanical Behavior and Superplasticity of a Severe Plastic Deformation Processed Nanocrystalline Ti-6Al-4V Alloy. Materials Science and Engineering A. 2001;298(1-2):44-50

[51] Bohn R., lassen T.K., Bormann R. Mechanical Behavior of Submicron-Grained γ-TiAl-based Alloys at Elevated Temperatures. Intermetallics. 2001; (9):559-569

[52] Nieh T.G., Wang J.N., Lsiung L M., Wadsworth J., Sikka V. Low Temperature Superplasticity in a TiAl Alloy with a Metastable Microstructure. Scripta Materialia. 1997;37:773-779

Chapter 10

Development of a Stress Insensitive MgCuZn-NiCuZn Composite Ferrite Useful for Microinductors Applications

N. Varalaxmi and K. Sivakumar

Additional information is available at the end of the chapter

http://dx.doi.org/10.5772/52857

1. Introduction

Ferrite materials are recognized as more important and essential for the further development of electronics than before, and it is believed that the production of ferrites will increase year by year as their applications become more diverse.

A tendency to miniaturize electronic components began in the 1990s. Concurrently, progress also occurred in surface-mounting technology, and attempts have been made to accomplish high density, incorporation of ferrite inductors into a printed circuit board. This has, as a result, allowed development of various types of multilayer ferrite chip inductors [1]. Generally, this has a length of 10mm and a width of 0.5 mm, to be used in television receivers, video equipment, headphone stereos, hard disk device systems, personal computers, cordless telephones, automobiles parts etc. Multilayer ferrite chip inductors are manufactured using the thick film printing method, and the latest chip inductors are successfully constructed as a single monolithic structure that combines the inductor with other passive elements such as capacitors and resistors.

The present chip inductor features make the miniaturization process very easy. The flux is entirely free from leakage because the coil is shielded with ferrite material. Hence it is expected that the demand for the chip inductors will increase in future.

Chip inductors are one of the passive surface mounting devices (SMD). They are important components for the latest electronic products such as cellular phones, video cameras, note book computers, hard and floppy drives etc., and those that require small dimensions, light weight and better functions [2-3]. The traditional wire wound chip inductors can only be miniaturized to a certain limit and lack of magnetic shielding leads to the development of

INTECH
open science | open minds

© 2013 The Author(s). Licensee InTech. This chapter is distributed under the terms of the Creative Commons Attribution License http://creativecommons.org/licenses/by/3.0), which permits unrestricted use, distribution, and reproduction in any medium, provided the original work is properly cited.

new materials for the multilayer chip inductors. In this direction only NiCuZn ferrites were developed as the core material used in the chip inductors [4-5]. But, it was found that these ferrites are comparatively sensitive to stress and magnetic properties are easily changed or deteriorated by the stress caused at the electrode.

Silver is generally used as the material for the internal electrode of the multilayer chip inductors due to its low resistivity, resulting in the components with high quality factor (Q) [4]. In addition to this, Ag paste is commercially available at lower cost than Ag-Pd paste. Since the melting point of silver is 961 ºC, the sintering temperature of ferrite which is used for the manufacture of chip inductor should be below 950 ºC. This is because of the need to prevent Ag diffusion into the ferrite that would increase the resistivity of the internal conductor. Further, the segregation of Cu^{2+} from the ferrite induced by the diffused Ag can be avoided and thus no deterioration in magnetic properties of the material. In order to overcome these problems, MgCuZn ferrites were found to be suitable [6-8]. Normally, MgCuZn ferrites were sintered at temperatures higher than 1100 ºC [6-7, 9]. In order to use these ferrites in multilayer chip components the sintering temperature must not be more than the melting point of silver.

With the rapid development and advancement of mobile communication and information technology, the electronic components with small size, high efficiency and low cost are urgently needed [10]. Multilayer chip inductors (MLCI) as key component of electronic devices confront new challenges. Better magnetic properties, especially high initial permeability are required for reducing the number layers of multi layer chip inductors. At present NiCuZn ferrites have been used extensively for the production of the MLCI [11]. But NiCuZn ferrites are highly stress sensitive. The MgCuZn ferrite is also a pertinent magnetic material for wide range of applications owing to its high resistivity, fairly high Curie temperature, low magnetostriction, environmental stability and low cost [6-16].

In search of suitable ferrite materials for microinductor applications, an attempt has been made, in the present work, to prepare ferrite composites containing two pure ferrite components viz., NiCuZn and MgCuZn ferrites with high initial permeability viz., 1-x $(Mg_{0.25}Cu_{0.25}Zn_{0.5}Fe_2O_4)$ + x $(Ni_{0.35}Cu_{0.05}Zn_{0.6}Fe_2O_4)$, (where x varies as 0.0, 0.2, 0.4, 0.5, 0.6, 0.8 and 1.0) with a view to examine whether there would be any enhancement of magnetic properties in this system. In this system both the end members have fairly high permeability. This attempted is made to reduce the stress sensitivity of the ferrites since the magnetostriction constant of MgCuZn ferrite is lower than that of NiCuZn ferrite [11]. It was also expected that MLCI using MgCuZn ferrites would show better magnetic properties [7] comparable to that of NiCuZn ferrites and can realize low cost MLCIs with MgCuZn ferrites. Therefore in the present investigation along with the MgCuZn and NiCuZn ferrites as pure components and their composites were chosen. Since, the author has been studying the modification of soft magnetic ferrites, for different applications, she thought that, the ferrites composites would produce dense microstructure and excellent magnetic properties. To the best of the author's knowledge no systematic investigation on the composites containing (NiCuZn-MgCuZn ferrites) were made, hence the author has under taken a detailed study of initial permeability in these composites.

In this paper, different mole fractions of the ferrites composites with generic formula 1-x ($Mg_{0.25}Cu_{0.25}Zn_{0.5}Fe_2O_4$) +x ($Ni_{0.35}Cu_{0.05}Zn_{0.6}Fe_2O_4$) where 'x' varies from 0.0 to 1.0 were prepared and an attempt was made both to promote the high initial permeability ferrite composites and having stress insensitivity.

2. Experimental

2.1. Preparation of ferrite composites

Ferrite composites with ferrimagnetic phases have been prepared by employing conventional ceramic method.

In the present investigation the ferrite composites contain two ferrite components viz.,

i. $Mg_{0.25}Cu_{0.25}Zn_{0.5}Fe_2O_4$ (MgCuZn ferrite)

ii. $Ni_{0.35}Cu_{0.05}Zn_{0.6}Fe_2O_4$ (NiCuZn ferrite)

The pure components, viz., $Mg_{0.25}Cu_{0.25}Zn_{0.5}Fe_2O_4$ and $Ni_{0.35}Cu_{0.05}Zn_{0.6}Fe_2O_4$ compositions were prepared using the conventional ceramic double sintering method.

First, the composition $Mg_{0.25}Cu_{0.25}Zn_{0.5}Fe_2O_4$ was prepared by employing conventional ceramic double sintering method using analytical grade MgO, CuO, ZnO and Fe_2O_3 in their respective proportions. These oxides were weighed and intimately mixed and then ball-milled with acetone using polyethylene jars and iron balls (ϕ = 5 mm) media for 24 hrs. The ground powders were presintered at 800 ºC in the form of cakes. After pre-sintering, these cakes were crushed, ground again to obtain fine particle size and finally these powders were sieved to get uniform particle size. The $Ni_{0.35}Cu_{0.05}$ $Zn_{0.6}Fe_2O_4$ composition was also prepared in the same method by employing conventional ceramic double sintering method using analytical grade NiO, CuO, ZnO and Fe_2O_3 in their respective proportions.

The ferrite composites were prepared by mixing the two individual green powders in molar proportions 1-x ($Mg_{0.25}Cu_{0.25}Zn_{0.5}Fe_2O_4$) + x ($Ni_{0.35}Cu_{0.05}$ $Zn_{0.6}Fe_2O_4$) in which 'x' varies from 0.0, 0.2, 0.4, 0.5, 0.6, 0.8 and 1.0. These compositions were taken in the respective proportions as mentioned above and they were ultimately mixed and ground further to obtain fine powder.

These presintered powders were mixed with 2% PVA as a binder and were compacted in the form of toroids of 12mm outer diameter; 8mm inner diameter; and 4mm height to measure the initial permeabilities and their Curie temperature, and in the form of cylinders of diameter 10 mm and length nearly 20mm to study the effect of external stress, using hydraulic press by applying a pressure of 150 MPa. These compacted bodies were finally sintered at 1250 ºC for 2 hrs in a programmable furnace and were cooled to room temperature at the rate of 80º/hr. Sufficient care was taken to avoid the zinc loss during the sintering process.

All the samples were structurally characterized by X-ray powder diffraction analysis using a Philips diffractometer PW-1710 with monochromatised CuK_α radiation at room temperature. Infrared spectra of pure components and the equimolar composition were recorded using Perkin-Elmer FT–IS spectrometer Model 1605 in the wave number range 400-1100 cm^{-1} at room temperature. The scanning electron micrographs of samples were obtained using (Philips XL30ESEM) instrument. The samples were laped and polished using 0.3 μm alumina powder and later they were thermally etched. The thermally etched and gold coated samples were used for obtaining the SEM photographs.

2.1.1. Analysis of X-ray diffractograms of ferrite composite system

1-x ($Mg_{0.25}Cu_{0.25}Zn_{0.5}Fe_2O_4$) + x ($Ni_{0.35}Cu_{0.05}Zn_{0.6}Fe_2O_4$)

Typical X-ray patterns obtained in the present investigation for ferrimagnetic composites with generic formula 1-x ($Mg_{0.25}Cu_{0.25}Zn_{0.5}Fe_2O_4$) + x ($Ni_{0.35}Cu_{0.05}Zn_{0.6}Fe_2O_4$) where x = 1.0, 0.5 and 0.0 are presented in Figs 1a, 1b and 1c respectively. An examination of the x-ray diffractograms reveals that both pure ferrites and composite show the formation of single phase cubic spinel structure showing well-defined peaks.

2.1.2. Scanning electron micrographs

Typical microstructures of the pure components and a few composites are obtained using SEM. The scanning electron micrographs of pure components and equimolar mixture along with x = 0.4 were taken with the help of Philips XL30ESEM. Thermally etched samples were taken for making the SEM specimens. The samples were mounted on brass studs with double sided adhesive tape and coated with Au-Pd alloy of 120-150 Aº thickness in argon ambient atmosphere of 8-10 Pascal. Then the samples were scanned for various surfaces and the best photographs were recorded. These photographs are shown in Figs.2a, 2b and 2c. An examination of the figures presented shows clearly the uniform distribution of the ferrite phases.

2.1.3. Infrared analysis

The infrared spectra of MgCuZn and NiCuZn ferrites and their equimolar mixture are shown in Figs.3a, 3b and 3c All the samples show only two absorption bands, viz., high frequency band ν_1 and second absorption band ν_2 which are found to be in the range 571 to 563 $cm^{-1}$and 400 to 398 cm^{-1} respectively. The difference is expected because of the difference in Fe^{3+}- O^{2-} distance of the octahedral and tetrahedral sites. The ν_1 band is assigned to the tetrahedral co-ordination, ν_2 and ν_3 to octahedral co-ordination compounds where as ν_4 is attributed to some sort of lattice vibrations involving displacement of tetrahedral cations [17]. It is reported that the systems in more disorder state gives rise to broader and less intense bands in IR spectra [18].

http://dx.doi.org/10.5772/52857

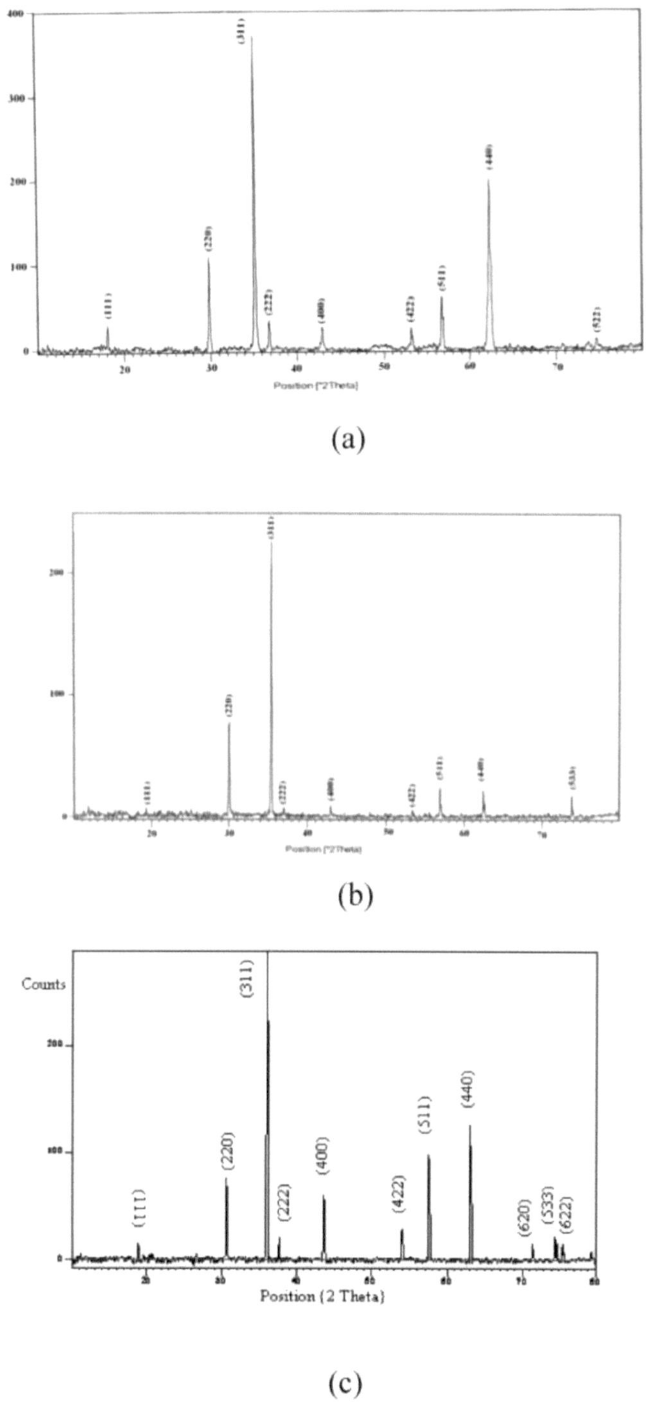

Figure 1. X-ray diffractograms of pure components and ferrite composite (a) X = 0.0, (b) X = 0.5 and (c) X =1.0.

(a) X = 0.0

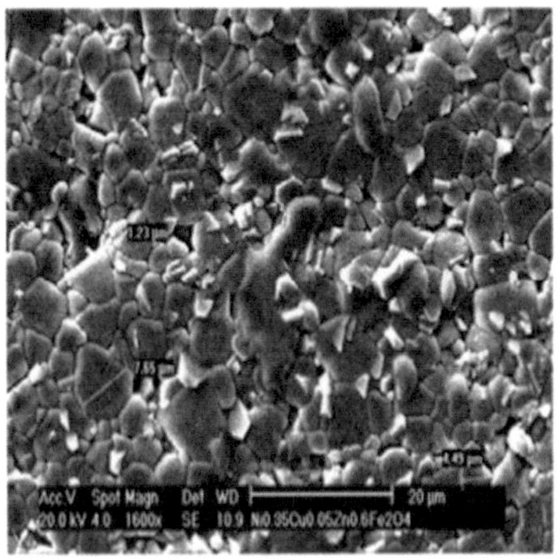

(b) X = 1.0

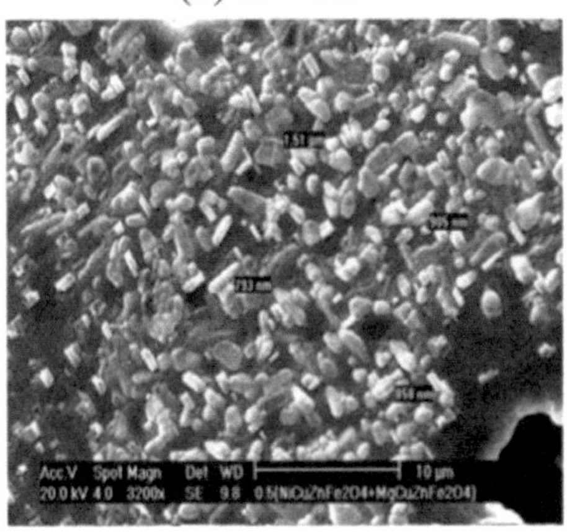

(c) X = 0.5.

Figure 2. SEM photographs of ferrites composites (a) X = 0.0, (b) X = 1.0 and (c) X = 1.0.

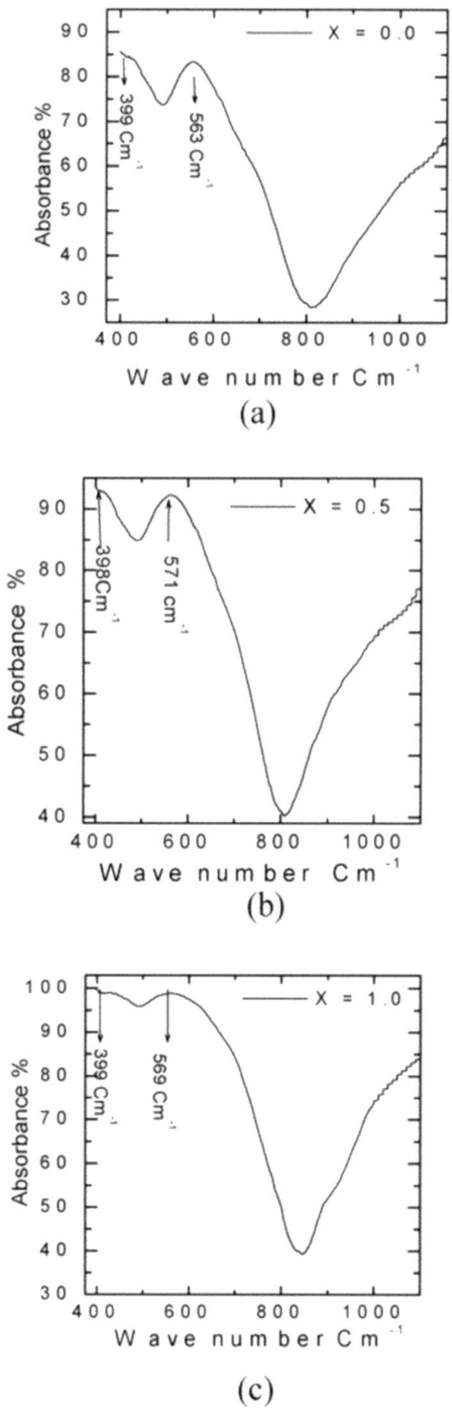

Figure 3. Infrared spectra of ferrite composites (a) X = 0.0, (b) X = 0.5 and (c) X = 1.0.

3. Experimental technique

a. Initial permeability

For the determination of the initial permeability (μ_i) the samples were prepared in the form of toroids with the help of suitable die having an inner diameter of 8mm outer diameter of 12mm and 4mm height and 30 to 35 turns of insulated copper wire (24SWG) was wound on the toroid and was used as a test specimen. The initial permeability, $\mu_{i,}$ of these ferrite toroids were evaluated using the standard formulae from the inductance measurements carried out at 1 kHz using impedance analyzer (Hioki Model 3532-50 LCR HiTester). The values of inductance were recorded noted in the temperature range 25 ºC to 250 ºC in the frequency of 500 Hz to 1000 kHz. The permeability was calculated by using the relation

$$L = 0.0046N^2 * \mu_i * h * \log_{10}\left(\frac{D_1}{D_2}\right) \qquad (1)$$

Where L is the inductance in micro henrys, h is the height of the core in inches, D_1 and D_2 are the outer and inner diameters in inches μ_i is the permeability of the core and N is the number of turns.

b. Stress sensitivity

In order to study the effect of external stress, uniaxial compressive stress parallel to the magnetizing direction was applied to the cylindrical shape ferrite cores using uniaxial press system. The stress magnitudes were varied from 0 to 10 MPa. The change in inductance was measured using the above mentioned LCR HiTester by employing 120 turns coil wound on each cylinder.

c. Excess permeability and excess volume

From the experimental measurements of molar volume (molecular weight/density) and initial permeability the excess parameters V^E and μ_i^E are evaluated using the formulae

$$V^E = V_{exp} - \left[V_1(1\text{-}x) + V_2 x\right] \qquad (2)$$

and

$$m_i^E = m_{iexp} - [m_{i1}(1\text{-}x) + m_{i2}x] \qquad (3)$$

where V^E and μ_i^E are excess molar volume and excess initial permeability. The quantities in the brackets are the ideal values. The excess parameters obtained in the present work are presented as function of mole fraction x in Fig.5 at room temperature.

http://dx.doi.org/10.5772/52857

4. A brief review of initial permeability work carried out on MgCuZn and NiCuZn ferrites

Development of nickel copper zinc ferrites for use in high performance multilayer chip ferrites were investigated by Suzuki et.al., [19]. Kanada et.al., [20] studied the replacement of Fe_2O_3 by Mn_2O_3 and replacement of CuO for MgO in MgZn ferrite in order to achieve low core loss, focusing on the microstructure. They have also investigated the initial permeability of MnZn, MnMgZn and MnMgCuZn ferrites and concluded that the initial permeability of both MnMgZn and MnMgCuZn ferrites are higher than those of MgZn ferrite.

Yan et.al., [21] studied the effect of CuO and V_2O_5 additions and the particle size of precursor materials on the microstructure and relative initial permeability of low temperature fired NiCuZn ferrites and found that the additions of CuO and V_2O_5 contribute to the grain growth and densification of matrix in the sintering process. Addition of 10 mol% CuO and 0.20 mol% V_2O_5, favoured the low firing temperature of NiCuZn ferrite, and showed very high relative initial permeability of 1417 at 1MHz. Zhang et.al., [11] carried out the preparation of $(Ni_{0.15}Cu_{0.25}Zn_{0.60})Fe_{1.96}O_4$ ferrite with very high initial permeability from self-propagated powders. Caultun et.al., [22] studied the magnetic properties of high frequency Ni-Zn ferrites doped with CuO. Nakamura et.al., [23] studied the frequency dispersion of permeability in ferrite composite materials and observed that permeability spectra of Ni-Zn ferrite composite materials. They observed that in the sintered ferrite (volume fraction 1.0), the spin resonance was around 9 MHz and the static permeability was about 1400. As the ferrite content decreases in composite materials, the static susceptibility of the spin component decreases and the spin resonance frequency shifts to higher frequencies. The real part of the permeability in the ferrite composite materials becomes larger than that of the sintered ferrite in the rf frequency region.

Oh [24] studied the effect of MnO addition on sintering behaviour and magnetic properties of NiCuZn ferrites for multilayer chip inductors. Su et.al., [25] studied the effect of composition and sintering temperature on grain size, porosity and magnetic properties of the NiZn and NiCuZn ferrites. It was found that the lowest power loss could be obtained with the equimolar composition for both NiZn and NiCuZn ferrites, which could be attributed to the lowest porosity.

High strength NiCuZn ferrite used for surface mounting devices was investigated by Murayama et.al., [26]. Kanada et.al., [27] studied the effects of chemical composition and microstructure on the stress sensitivity of initial permeability of NiMgCuZn ferrites. When the external stress is applied to the ferrites, the reduction of initial permeability due to the deficiency of Fe_2O_3 was observed. Furthermore, they reported that a finely dispersed grain size in microcrystalline structure is highly effective in stabilization of initial permeability. As the compressive stress is applied parallel to the magnetic field, the maximum initial permeability of NiCuZn ferrite could be observed at 4 MPa. On the other hand, in the case of MgCuZn ferrite the maximum was observed at around 0 MPa. It seems possible to stabilize the initial permeability against the external stress by adjusting composition ratio of NiO/ (NiO + MgO).

Bhosale et.al., [28] presented a novel route for the preparation of high density, high permeability Cu-Mg-Zn ferrites using oxalate precursor method, with various Mg^{2+} contents with a generic formula $Cu_{(0.5-x)}Mg_xZn_{0.5}Fe_2O_4$ where x = 0.00, 0.20, 0.25 and 0.40. They investigated the ferritization temperature of this system and observed the initial permeability of all the samples increased with increase in sintering temperature and Mg^{2+} content for samples with $x \leq 0.20$.

Yan and Hu [29] studied the $(Ni_{0.20}Zn_{0.60}Cu_{0.20})Fe_{1.98}O_4$ ferrite which was sintered using microwave sintering and conventional sintering techniques, respectively and found that microwave sintering technique could effectively promote the forward diffusion of ions and thus accelerate the sintering process, resulting in the grain growth and the densification of the matrix. They found that at 100 kHz, the magnetizing contribution of domain wall motion is predominant. Compact and coarse matrices are favorable for domain wall motion, giving rise to improvement of relative initial permeability and loss of ferrites. With microwave sintering technique, $(Ni_{0.20}Zn_{0.60}Cu_{0.20})Fe_{1.98}O_4$ ferrite with the relative initial permeability μ_i of about 2000 and the relative loss factor $\tan\delta/\mu_i$ of about 8.7×10^{-6} at 100 kHz were achieved at 980 °C sintering temperature. In addition, the sintering time of ferrites was reduced from 5 to 0.5 h by using microwave sintering technique.

Ghodake et.al., [30] studied the magnetic properties of NiCuZn ferrites with generic formula $Ni_xCu_{[(1-t)-x]}Zn_tFe_2(C_2O_4).n.H_2O$ where t = 0.45, 0.50, 0.60 and x = 0.0, 0.05,0.10, 0.15, 0.20, 0.30, 0.40,0.50 and 0.55 were synthesized by oxalate precursor method and observed that the initial permeability (μ_i) showed an increase when Ni^{2+} was added up to x = 0.15 while for x > 0.15, it decreased. The increase in initial permeability (μ_i) was attributed by them to monotonic increase in M_s, and K_1 on addition of Ni^{2+}. However, the microstructure and density (porosity) also influence μ_i variations. The decrease in μ_i is attributable to increase of $K_{1.}$

Xiao et.al., [31] studied the fabrication of solenoid type inductor with electro pated NiFe magnetic core, the inductance and quality factor were 0.42-0.345 μH and 1.8-5.3 in the frequency range 1-10 MHz respectively. Bhosale et.al., [32] studied a novel route for the preparation of high density, high permeability Cu-Mg-Zn using oxalate precursor method for various Mg^{2+} contents with generic formula $Cu_{(0.5-x)}Mg_x\ Zn_{0.5}\ Fe_2O_4$ where x = 0.0, 0.2, 0.25 and 0.4 have been carried out in the temperature range 599 to 743 K and observed that initial permeability of all the samples increases with increase in temperature and Mg^{2+} content upto $x \leq 0.2$.

Densification and magnetic properties of low-temperature fired NiCuZn ferrites was studied by Wang et.al., [33]. Effects of presintering temperature on the property of high permeability of NiCuZn ferrite was studied by Su et.al., [34]. Mukesh et.al., [35] prepared fine powders of $Ni_{0.6-x}Cu_xZn_{0.4}Fe_2O_4$ where $0 \leq x \leq 0.4$ by citrate precursor method under investigation and they concluded that the ferrite with Cu concentration of x = 0.4 showed highest initial permeability. Enhancement initial permeability of low temperature sintered NiCuZn ferrites was investigated by Oi et.al., [36]. The study of low temperature sintered MgCuZn ferrites were carried out by Nakahata et.al., [37].

Ramamanohar Reddy et.al., [38] studied the stress sensitivity of inductance in NiCuZn ferrites and the initial permeability measurements were reported in the temperature range 30-400 ºC. Nam et.al., [39] studied the effect of Cu substitution on the properties of NiZn ferrites sintered at low temperature in various compositions, they concluded that the densification of NiCuZn ferrite is dependent on Cu content in these ferrites, Dispersion characteristics of complex permeability in $(Ni_{0.5-x}Cu_xZn_{0.5})(Fe_2 O_3)_{0.98}$ was reported by them and They also showed that the relaxation frequency increases with increasing temperature. Caultn [40] studied the dependence of permeability spectra on the frequency for samples of Ni-Zn-Cu ferrite sintered at different temperatures. The effect of Cu substitution on the properties of magnesium zinc ferrites sintered at low temperature were investigated by Rezlescu et.al., [14]. Densification and properties of flux added NiCuZn ferrites were carried out by Wang et.al., [41].

Wang et.al., [42] also studied the densification and magnetic properties of low temperature fired NiCuZn ferrites. Murthy [43] studied the electrical and magnetic properties of low temperature sintered MgCuZn ferrites. Wang et.al., [44] also studied the effects of processing on the densification and properties of low-temperature fired NiCuZn ferrites. Bhosale et.al., [45] studied the synthesis of high permeability Cu-Mg-Zn ferrites using oxalate precursor method and concluded that for the sample with x = 0.25 in $Mg_xCu_{0.5-x}Zn_{0.5}Fe_2O_4$ has the highest permeability of (μ_i = 2158) which was sufficiently higher than that reported for ETMG-5 ferrite (μ_i = 550). Magnetic and structural properties of ultra fine Ni-Zn-Cu ferrite prepared by a sol-gel method were studied by Lee et.al., [46].

Effects of composition and sintering temperature on grain size, porosity and magnetic properties of the NiZn and NiCuZn ferrites were investigated by Hua Su et.al., [47] and they found that the lowest power loss could be obtained with the equimolar composition for both NiZn and NiCuZn ferrites, which could be attributed to the lowest porosity and a slight deficiency or excess of Fe_2O_3 content which no pronounced influence on saturation magnetic flux density (B_s). However, a slight excess of Fe_2O_3 was effective to improve the initial permeability, which could be attributed to decrease of the magnetocrystalline anisotropy. With the increase of sintering temperature, the initial permeability and power loss of the NiZn and NiCuZn ferrites had different trend, which was explained by the grain size and porosity. They reported that the power losses of the NiCuZn ferrite samples were lower than that of the NiZn ferrite samples at any sintering temperature and the NiCuZn ferrites had a better performance than the NiZn ferrites. Bhosale et.al., [48] carried out the initial permeability studies on high density Cu-Mg-Zn ferrite and based upon the results concluded that variation of initial permeability (μ_i) with composition was mainly affected by variation of magnetization (M_s) and average particle size. They showed that the initial permeability increased with increase of magnesium content. Effect of copper on the electromagnetic properties of Mg-Zn-Cu ferrites prepared by sol-gel auto combustion method were investigated by Yue et.al., [49] they observed that low temperature sintered Mg-Zn-Cu ferrites possess good electro magnetic properties as well as fine grained microstructure.

Yue et.al., [50] studied the Mn-doped NiCuZn ferrites with the generic formula $(Ni_{0.2}Cu_{0.2}Zn_{0.6})O(Fe_{2-x/}Mn_xO_3)_{0.98}$ where x = 0, 0.02, 0.04 and 0.06 prepared by a novel sol-gel auto combustion process, and found that with increase of Mn content, the initial permeability is significantly increased. Ramana Muthy [51] also studied the low temperature sintered of NiCuZn ferrites with generic formula $Ni_{0.65-x}Cu_xZn_{0.35}Fe_{2-x}O_4$ where x = 1 to 14% its electrical, magnetic and elastic properties were investigated, The temperature variation of initial permeability was measured in the temperature range 300-800 K at a frequency of 1 kHz in a field of 4 A/m, and found that for higher Cu content initial permeability increases continuously with temperature. Influence of PbO on some properties MgCuZn ferrites were investigated by Sachelarie et.al., [16]. Rezlescu et.al., [15] also studied the influence of PbO and Ta_2O_5 on some properties of MgCuZn ferrites. Rezlescu et.al., [14] studied the copper ions influence on the physical properties of MgZn ferrites. Qi et.al., [52] studied a series of Mn substituted MgCuZn ferrites with generic formula $(Mg_{0.2}Cu_{0.2}Zn_{0.6}O)$ $(Fe_{2-x}Mn_xO_3)_{0.97}$ with x = 0.00, 0.01, 0.03, 0.05 and 0.07 prepared with nanosized precursor powders synthesized by a sol-gel auto-combustion method. All the ceramic samples could be sintered at low temperature (930 °C) (below the melt point of Ag (961°C)) and their results showed that low temperature sintered MgCuZn ferrites doped with Mn posses higher initial permeability and better grain structure than that of low temperature sintered NiCuZn ferrites prepared by the same method. Therefore, Mn doped MgCuZn ferrites should be ideal materials for high inductance multilayer chip inductor. It is thought that the variation of initial permeability of MgCuZn ferrites with the Mn substitution was attributed to the decrease of magnetostriction constant.

Preparation of low-power loss MgCuZn ferrites using the microwave sintering method were studied by Bhaskar et.al., [53]. Hyie et.al., [54] carried out the studies on the electromagnetic properties of various compositions of magnesium-copper-zinc ferrites and suggested that good electromagnetic properties of MgCuZn ferrites were attributed to low MgO, moderate CuO and high ZnO contents. Roy and Bera [55] studied the effect of Mg substitution on electromagnetic properties of $(Ni_{0.25-x}Mg_xCu_{0.2}Zn_{0.55})Fe_2O_4$ with x = 0.0, 0.07, 0.13, 0.18, and 0.25 synthesized through nitrate-citrate auto-combustion method. The initial permeability was measured in the frequency range 10 Hz-10 MHz and the permeability were found to increase and the magnetic loss decreased with Mg substitution for Ni, up to x = 0.18 and also observed a very high permeability in the composition x = 0.18, which was due to better densification, lower magnetostriction constant and inner stresses, etc., and concluded that this composition would be better than NiCuZn-based material for more miniaturization of multi layer chip inductors.

Low and Sale et.al., [56] studied the development and analysis of property composition diagrams on gel-derived stoichmetric NiCuZn ferrite. As the sintering temperature increases, the reduction in resistivity is attributed to the formation of $BiCu_2O_4$ at grain boundaries which has lower resistivity compared to that of ferrite materials. The effects of Cu substitution for Mg on the properties of MgZn ferrite sintered at low temperature was investigated by Rezlescu et.al., [57]. Park et.al., [58] studied the effect of WO_3 addition on the electrical and magnetic properties of $(Ni_xCu_{0.2}Zn_{0.8-x}O)_{1.02}(Fe_2O_3)_{0.98}$ sintered at low temperature. They found that the addition of small amount (below 1wt%) of WO_3 improves their electrical and magnetic properties. Zhang et.al., [59] carried out the investigation on structure and properties of low temperature sintered composite ferrites: viz., $(Ni_{0.60}Zn_{0.20}Cu_{0.20})Fe_2O_4$ spinel ferrite powder and Co_2Z were synthesized by self propagating method. They observed that the composite ferrites have excellent electromagnetic properties in hyper frequencies more than 1.5 GHz.

Qi et.al., [60] investigated the magnetic properties of composite materials with a ferrimagnetic $Ni_{0.2}Cu_{0.2}Zn_{0.6}Fe_{1.96}O_4$ phase and a ferro electric $BaTiO_3$ phase with generic formula 1-x $(Ni_{0.2}Cu_{0.2}Zn_{0.6}Fe_{1.96}O_4)$ – x $(BaTiO_3)$ composites in which x varies from 0 to 1. They found that the saturation magnetization and initial permeability decreases with increasing ferroelectric content $BaTiO_3$ phase. Aoki et.al., [61] studied the bending strength of sintered ferrites for MgCuZn-NiCuZn and MnZn ferrites and observed that residual carbon content affects bending strength of these sintered ferrites and the sintered ferrites containing much residual carbon lower the bending strength. Nakano and Nomura [62] carried out the studies on low temperature sintered NiCuZn ferrites used for multilayer ferrite chips.

Effect of manganese impurity on the magnetic properties of $Ni_{0.3}Mn_xZn_{0.7-x}Fe_2O_4$ in which x varies from 0.1 to 0.5, were studied by Amarendra et.al., [63] and they concluded that the initial permeability continuously decreases with increase in manganese content. Srinivasan et.al., [64] carried out the studies on high density nickel zinc ferrites $Ni_{1-x}Zn_xFe_2O_4$ and its magnetic properties using novel hydrazine precursor method. Low-temperature sintering method for NiCuZn ferrite and the effect of Mn addition on electromagnetic properties were carried out by Ju and Pian [65]. Zhang et.al., [66] studied the preparation and investigation of $(Ni_{0.15}Cu_{0.25}Zn_{0.6})Fe_{1.96}O_4$ ferrite with very high initial permeability from self-propagated powders and found that the sintered ferrite specimens have very high initial permeability. They also found that densification of the ferrites and homogeneous grain growth were the main factors affecting high initial permeability. Kim et.al., [67] studied structure and mossbauer studies of Cu doped NiZn ferrites.

Tsay et.al., [68] studied the microwave sintering technique of NiCuZn ferrites materials with generic formula $(Ni_{1-x-y}Zn_xCu_yFe_2O_4)$ and multi layer chip inductors. They concluded that the benefit of using microwave sintering is in densifying multilayer chip inductors. They could achieve high-impedance multilayer chip inductors by microwave sintering of the materials at 800-850 ºC for 20 minutes. Wang et.al., [69] studied the effects in processing in the densification and properties of low temperature sintered NiCuZn ferrites.

Nakamura [5] reported low temperature sintering of NiZnCu ferrite and its permeability spectra and concluded that the post sintering density and complex permeability of NiZnCu ferrite ceramics can be controlled by altering the particle size of the starting oxide materials and the calcination temperature. The complex permeability of sintered ferrite is described as the summation of the spin rotational contribution and domain wall motion. The permeability in 100 MHz region is determined mainly by the spin rotation magnetization mechanism and it depends only on the ferrite volume loading.

Wu et.al., [70] studied the preparation of NiZn ferrites/SiO_2 nano composite powders by sol-gel auto combustion method. Ladgaonkar [71] investigated the initial permeability studies of Nd^{3+} substituted Zn-Mg ferrite system and explained that on substitution of Nd^{3+} ion causes reduction in the magnetic moment and curie temperatures. Substituted Nd^{3+} ion shows its occupancy on tetrahedral (B) site, and explained the positive contribution to the anisotropy constant K_1.

Bhosale et.al., [72] studied the initial permeability on high density CuMgZn ferrites and explained the variation in initial permeability was mainly affected by variations of magnetization (M_s) and average particle size. The μ_i decreases with increasing Mg^{2+} content. This is attributed to a lower value of the anisotropy constant (K_1) for $MgFe_2O_4$ than that of $CuFe_2O_4$. Ghodake et.al., [73] studied the magnetic properties of Mn substituted Mg-Zn ferrites and they explain the decrease in Curie temperature (T_c) with addition of manganese in terms of the exchange interaction. The decrease in permeability with the addition of Ti^{4+} in Mg-Zn ferrite system was reported by Suryavanshi et.al., [74]. A variation of K_1 has been explained on the basis of cation distribution. The relaxation phenomenon and frequency dispersion of complex permeability have been investigated in system of MgCuZn ferrites by Umeda et.al., [75]. Krishnaveni et.al., [76] studied the microwave hydrothermal synthesis and characterization of nanosized NiCuZn ferrites.

Ladgaonkar et.al., [77] studied the influence of Nd^{3+}substitution on the permeability spectrum in Zn-Mg ferrites with generic formula $Zn_xMg_{1-x}Fe_{2-y}Nd_yO_4$ where x = 0.00, 0.20, 0.40, 0.60, 0.80 and 1.00, y = 0.00, 0.05 and 1.00. The frequency dependence of initial permeability was measured by them in the frequency range from 100Hz to 1MHz, Their study reveals dispersion in initial permeability and the dispersion frequency was found to be lower for the high permeability compositions than the lower permeability compositions. Atassi and Tally [78] studied the low temperature sintering of MgCuZn ferrites prepared by citrate precursor method and found that sintering process increases the crystallinity of the solid and its domain size.

Effects of composition and sintering temperature on grain size, porosity and magnetic properties of the NiZn and NiCuZn ferrites were investigated by su et.al., [79] and they found that the lowest power loss could be obtained with the equimolar composition for both NiZn and NiCuZn ferrites, which was attributed to the lowest porosity in those samples. A slight deficiency or excess of Fe_2O_3 content had no pronounced influence on saturation magnetic flux density (B_s) in their testing range. However, a slight excess of Fe_2O_3 was effective to improve the initial permeability, which could be attributed to decrease of the magnetocrystalline anisotropy. With the increase of sintering temperature, the initial permeability and power loss of the NiZn and NiCuZn ferrites showed increasing trend, which was explained by the variation of the grain sizes and porosities. Their observation shows that the power losses of the NiCuZn ferrite samples were lower than that of the NiZn ferrite samples at any sintering temperature. The NiCuZn ferrites had a better performance than the NiZn ferrites in power applications. Bhosale et.al., [80] studied the synthesis of Cu-Mg-Zn ferrites and correlation of magnetic properties with microstructure with a generic formula $Cu_{0.5-x}Mg_xZn_{0.5}Fe_2O_4$ where x = 0.00, 0.20, 0.25 and 0.40 and observed that initial permeability of all the samples increases with increase in temperature and Mg^{2+} content upto ($x \leq 0.20$).

Zahi et.al., [81] investigated the synthesis, magnetic properties and micro structure of Ni-Zn ferrite with generic formula $Ni_{0.3}Zn_{0.7}Fe_2O_4$ by sol-gel technique using metal acetates at low temperature and they observed that initial permeability increases in the large frequency band (31MHz). Ghodake et.al., [82] studied the magnetic properties of NiCuZn ferrites synthesized by oxalate precursor method with generic formula $Ni_xCu_{[(1-t)-x]}Zn_tFe_2(C_2O_4).nH_2O$ (where t = 0.45, 0.50, 0.55, 0.60 and x = 0.00, 0.05, 0.10, 0.15, 0.20, 0.30, 0.40, 0.50, and 0.55) in the temperature range 30 to 240 ºC and in the frequency range from 20 Hz to 700 kHz. Initial permeability (μ_i) shows increase with addition of Ni^{2+} up to x = 0.15 while for (x > 0.15), it decreases. The increase in initial permeability (μ_i) is attributed to monotonic increase in M_s, and K_1 on addition of Ni^{2+}. However, the microstructure and density (porosity) also influence the μ_i variations. The decrease in (μ_i) is attributable to increase of K_1.

Hu et.al., [83] carried out the studies on the effect of magnetic permeability of $(Ni_{0.21}Zn_{0.58}Cu_{0.23})Fe_{1.95}O_4$ prepared by sol-gel self-propagating method. With this method of low temperature fired NiCuZn ferrite obtained as an initial permeability of 1120 and quality factor Q of 84 at 100 kHz. Byu et.al., [84] studied the factors effecting the initial permeability of Co-substituted Ni-Cu-Zn ferrites in the iron deficient composition of $(Ni_{0.2}Cu_{0.2}Zn_{0.6})_{1.02-x}Co_xFe_{1.98}O_4$ where x varies between 0 and 0.05 prepared to investigate the initial permeability dependence on cobalt contents. The compositional variation of initial permeability in cobalt substituted Ni-Zn-Cu ferrites could not be explained fully by intrinsic factors such as saturation magnetization and magnetic anisotropy based on a one-ion model. There results show that in addition to the magnetocrystalline and magnetoelastic anisotropy, the increase in induced anisotropy with increasing Co content results in the decrease of initial permeability.

Yue et.al., [85] studied the low sintered Mn-doped NiCuZn ferrites with compositions of $(Ni_{0.25-x}Mn_xCu_{0.25}Zn_{0.5})O(Fe_2O_3)_{0.98}$ where x = 0.02, 0.04, 0.06, 0.08, 0.10 were prepared by usual ceramic method. Initial permeabilities (μ'_i and μ''_i), were measured as a function of frequency and temperature. They reported that with increase in MnO_2 content, the resonant frequency and the Curie temperature decreased, whereas the μ'_i increases at low concentrations and decreases for higher concentration showing a maximum at MnO_2 for x = 0.06.

Hu and Yan [86] studied the preparation of high permeability NiCuZn ferrite with generic formula $(Ni_{0.17}Zn_{0.63}Cu_{0.20})Fe_{1.915}O_4$ by sintering at 930°C it showed an initial permeability of 1700 and relative loss coefficient $\tan\delta/\mu i$ of 9.0×10^{-6} at 100 kHz, due to the appropriate addition of CuO/V_2O_5 and showed a decrease in granularity of particles. Tantalum oxide added MgCuZn ferrite powders [87] were synthesized by a co-precipitation method using NaOH in a microwave-hydrothermal method. They showed the variation of initial permeability as a function of additive concentration at room temperature. Bhaskar et.al., [88] carried out studies on low temperature sintered Mn added MgCuZn ferrites were prepared using microwave sintering method and room temperature property of quality factor and initial permeability have been measured in the frequency range of 100kHz-100 MHz. The initial permeability was also measured in the temperature range of 300-650 K. Murase et.al., [89] studied the state of impurities, grain growth behavior and magnetic characteristics in MgZn ferrites.

Krishnaveni and Ramana Murthy [90] studied the microwave hydrothermal synthesis of nanosized Ta_2O_5 added Mg-Cu-Zn ferrites. Its initial permeability as a function of additive at room temperature was investigated and observed that the prepared microinductors posses better electrical properties than the microinductors prepared by using the NiCuZn ferrites. Mukesh et.al., [91] studied the effect of substitution of cations by cobalt in small concentration on the magnetic properties at low and high frequencies, in the polycrystalline ferrite samples having the composition $(Ni_{0.2}Cu_{0.2}Zn_{0.6})_{1-x}Co_xFe_2O_4$, where x = 0, 0.01, 0.03 and 0.05, they were prepared by citrate precursor method. Complex permittivity and permeability were measured at microwave frequencies (X-band) using the cavity perturbation method, which is a non-contact method.

Hsu et.al., [92] verified the addition of nanocrystalline ferrite particles (NFPs) was effective in improving the densification behavior and magnetic properties of the NiCuZn ferrites for multilayer chip inductor (MLCI) applications. The NFPs, which have high surface free energy, spread around the micron-sized ferrite particles (MFPs) and increased contacting area and inter-diffusion of the particles. They observed that the sample with 30 wt% NFPs additive exhibited the highest initial permeability when sintered at 900 ºC, than that made by the ferrite with 1.5 wt% Bi_2O_3 due to less diffusion of Ag electrode to ferrite and higher initial permeability. It was mainly attributed to relatively larger grain size and higher sintering density.

Su et.al., [93] carried out the studies on high-permeability and high-Curie temperature NiCuZn ferrite. In this process, attention was focused on the amount of CuO contained in NiZn ferrite and a small additive of MoO_3 added in the NiCuZn ferrite and observed that the initial permeability of the core increased and the Curie temperature only decreased a little with 4 mol% CuO content. This was mainly attributed to the presence of Cu ions activating the sintering processes in ferrites and leading to increase in density. The decrease of magnetocrystalline anisotropy constant also does some contribution, by optimizing the MoO_3 as additive, the initial permeability could get a sharp increase because of enhancement of grain size. So in the NiCuZn ferrite with a permeability of 2480 and Curie temperature of around 118 ^{0}C was obtained and by optimizing the CuO content, the initial permeability of the cores was increased and the Curie temperature decrease was only a little. This was attributed to the appropriate MoO_3 having the effect of promoting crystalline grain growth and not forming the abnormal grain growth and closure of pores.

Wang et.al., [94] prepared nano-sized NiCuZn ferrite powders with a composition of $(Ni_{0.15}Cu_{0.2}Zn_{0.65}O)(Fe_{2-x}Mn_xO_{3+0.5x})_{0.99}$ where x = 0 to 0.04, and studied the effect of Mn dopant on the microstructures and the initial permeability. They observed that introduction of Mn into NiCuZn ferrite has a great influence on its electromagnetic properties. i.e., initial permeability of NiCuZn ferrite increases greatly with a small amount of Mn addition, and then further increase of Mn content leads to a decrease in the initial permeability.

The effect of Mn substitution on the magnetic properties and microstructure of the $(Ni_{0.2}Cu_{0.2}Zn_{0.6})_{1.03}(Fe_2O_3)_{0.97} + 0.97xMnO_2$ where x = 0.1 to 0.5, were investigated by Li et.al., [95]. They observed that Mn content increases the initial permeability. Gao et.al., [96] studied solenoid type inductor with ultra low profile was fabricated by micro electromechnical system (MEMs) in which NiFe was electroplated as the magnetic core, and measured the value of inductance and quality factor. They observed 0.42 μH and 1.8 at 1 MHz, while the quality factor is 5.3 with inductance value of 0.345 μH at 10 MHz.

Micro structural dependence of permeability and permittivity spectra in Ni-Zn ferrites with generic formula $Ni_{0.4}Zn_{0.6}Fe_2O_4$ doped with V_2O_5 and Bi_2O_3 were studied by Pyun and Beak [97] and they explained that the initial permeability increases linearly with the grain size up to 5.9 μm and then decreases for undoped ferrite specimens, where as it increases linearly with the grain size for V_2O_5 doped ferrite specimens. The decrease in permeability with increase in grain size for undoped ferrite specimens was discussed in terms of the formation of excess Fe^{2+} due to higher equilibrium oxygen pressure.

5. Review of the previous experimental work on the stress sensitivity of ferrites

Stress sensitivity of inductance in NiCuZn ferrites was carried out by Ramamanohar Reddy et.al., [38] and observed that stress sensitivity is more in the case of iron rich and stoichiometric iron samples while in the case of iron deficient samples the stress sensitivity was found to be less. The stress sensitivity in Ni-Mn, Ni-Mn-Co and Ni-Mn-Co-Cu ferrites was studied by Kaczkowski [98] and he concluded that the greatest stress sensitivity was observed for the ferrites with the smallest magnetocrystalline anisotropy.

The effects of chemical composition and micro structure on stress sensitivity of magnetic properties to applied stress of Ni-Mg-Cu-Zn ferrite materials were investigated by Kanada et.al., [27] and they found that the application of external stress reduces the permeability Mechanical and electrical properties of $Ni_{0.65}Zn_{0.35}Cu_xFe_{2-x}O_4$ ferrites were carried out by Abd.El-Ati and Tawfik [99]. Enokido et.al., [100] studied the effects of additives on the stress variation and temperature dependence of inductance in NiCuZn ferrite. Effect of glass addition and quenching on the relation between inductance and external compressive stress in Ni-Cu-Zn ferrite-glass composites were reported by Yamaguchi and Shingawa [101]. Developments of stress-insensitive ferrite were carried out by Ikeda and Kumagi [102] and concluded that the sensitivity of the magnetic property to the external stress in NiCuZn ferrite was examined from the relationship between stress and permeability. They found that in the mass production process of inductors, when external stress is applied to the NiCuZn ferrite core by moulding them in resin, which is a protection against humidity or shock. It reduces inductance magnitudes as the permeability of NiCuZn ferrites is very sensitive to the external stress, the inductance of inductors changes very easily by the applied stress. This change has been recognized as the barrier to produce the inductors with a narrow inductance tolerance and found the best way to control the stress-sensitivity of permeability in NiCuZn ferrite was the addition of oxides such as SiO_2, the change in the inductance could be completely explained from the residual stress caused by the difference of coefficients of linear thermal expansion between spinel phase (magnetic) and second phase (non-magnetic). By this process, they succeeded in developing "stress-insensitive ferrite". Nakano et.al., [18] studied the development of low temperature fired NiCuZn ferrites, they studied the high performance for multilayer chip inductors and concluded that controlling stress by the internal Ag-conductor and CuO_{1-x}/ Ag on ferrites grain boundary is most important key point for high performance multilayer chip ferrites as well as the chemical composition of ferrite. The stress insensitive ferrite for micro inductors was carried out by Kumagi and Ikeda. [103]. Nakano et.al., [104] investigated the electro magnetic properties of low temperature sintered Mg-Cu-Zn ferrites and observed that under a compressive stress, the change in permeability of low temperature sintering MgCuZn ferrite was lower than that of low temperature sintering NiCuZn ferrite, also the multilayer chip inductor using low temperature sintering MgCuZn ferrite would show a higher inductance than the chip using low temperature sintered NiCuZn and low temperature sintering MgCuZn ferrite has the high potentialities of useful materials in the multilayer chip ferrite components.

The effect of SiO_2 on the bending strength of NiCuZn ferrite was investigated by Aoki et.al., [105] and they found that the bending strength of NiCuZn ferrite was increased by the addition of SiO_2. Also, the NiCuZn ferrite with SiO_2 addition showed a structure with smaller and more uniform grain size than non-addition. The improvement in the bending strength of NiCuZn ferrite was attributed to the control of grain growth by Zn_2SiO_4, and the residual stress caused by the difference of coefficients of linear thermal expansion between ferrite phase and Zn_2SiO_4 phase.

Aoki et.al., [106] studied the bending strength of sintered ferrites for MgCuZn-NiCuZn and MnZn ferrites. They found that the residual carbon content affects bending strength of these ferrites more over, sintered ferrites, containing much residual carbon lowers the bending strength. Fujimoto [107] studied the inner stress induced by Cu metal precipitation at grain boundaries in low temperature fired Ni-Zn-Cu ferrite.

From the above review, the author noticed that the experimental work carried out on ferrite composites is scanty. In view of this, the author has taken up a detailed study of initial permeability in pure components $Mg_{0.25}Cu_{0.1}Zn_{0.5}Fe_2O_4$ and $Ni_{0.35}Cu_{0.05}Zn_{0.6}Fe_2O_4$ and their composites with x = 0.2, 0.4, 0.5, 0.6 and 0.8 at a frequency of 1 kHz and temperature range 25 to 250 ºC. The stress sensitivity studies on the pure components and equimolar mixture of these components were also studied and the results are presented in the following section.

6. Results and discussion

a. X-ray diffractogram analysis

Pure ferrites and their composites sintered at 1250 ºC confirm the formation of single phase cubic structure.

http://dx.doi.org/10.5772/52857

An observation from the Fig.1 shows that the (222), (111) (511) and (440) peaks which belong to octahedral and tetrahedral sites are maximum suppressed in the equimolar composition of ferrite composite i.e for x = 0.5, when compared to pure ferrite components. And also it can be noticed that the intensity of (311) peak is maximum in equimolar composition when compared to pure ferrite components.

Pure ferrites and their composites sintered at 1250 ºC confirm the formation of single phase cubic structure.

An observation from the Fig.1 shows that the (222), (111) (511) and (440) peaks which belong to octahedral and tetrahedral sites are maximum suppressed in the equimolar composition of ferrite composite i.e for x = 0.5, when compared to pure ferrite components. And also it can be noticed that the intensity of (311) peak is maximum in equimolar composition when compared to pure ferrite components.

b. Scanning electron micrographic (SEM) studies

The scanning electron micrograph for $Mg_{0.25}Cu_{0.25}Zn_{0.5}Fe_2O_4$ shows a systematic grain structure (shown in Fig.2) having clear grain boundaries. The initial permeability (μ_i) is comparatively low. For x = 0.4 (not shown in Fig) composition shows a discontinuous grain structure with out any voids having large grain size.

On the other hand, the micrograph of ferrite composite with x = 0.5 (shown in Fig.2b). It is interesting to note that the morphology of grains is different from That of pure components. However, the initial permeability obtained in this composite is fairly high.

The micrograph of $Ni_{0.35}Cu_{0.05}Zn_{0.6}Fe_2O_4$ composition (shown in Fig.2c). it can be very clearly seen that the morphology o fgrains in this sample is different from that $Mg_{0.25}Cu_{0.25}Zn_{0.5}Fe_2O_4$

6.1. Initial permeability

a. Composition dependence of initial permeability

The initial permeability (μ_i) as function of composition at room temperature (30 ºC) for various mole fractions of ferrite composites along with their curie temperatures (T_c) is graphically represented in Fig.4. It can be noticed from the figure that with increasing x the initial permeability magnitudes reaches a maximum value at equimolar composition. i.e., 0.5 ($Mg_{0.25}Cu_{025}Zn_{0.5}Fe_2O_4$) +0.5 ($Ni_{0.35}Cu_{0.05}Zn_{0.6}Fe_2O_4$) and there after it decreases. The initial permeability values of pure ferrites are 1040 and 1090 in $Mg_{0.25}Cu_{0.25}Zn_{0.5}Fe_2O_4$ composition and $Ni_{0.35}Cu_{0.05}Zn_{0.6}Fe_2O_4$ composition respectively. The permeability of equimolar mixture shows a magnitude of 9619. However, the Curie temperature of these composites linearly decreases with in increase in x i.e., $Ni_{0.35}Cu_{0.05}Zn_{0.5}Fe_2O_4$ composition. Fig.4 shows the variation of excess initial permeability μ^E and excess volume V^E for ferrite composites

The permeability and molar volume show high values of at equimolar composition. This property can be explained to achieve high permeability ferrites. This is the excess property of the mixtures.

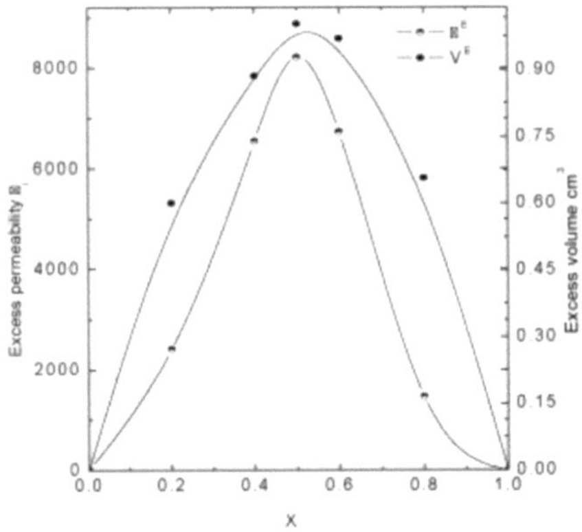

Figure 4. Variation of excess initial permeability (μ_i^E) and excess volume (V^E) of the system with increasing X.

b. Temperature dependence of initial permeability

The plots of initial permeability (μ_i) with temperature in the temperature range 30 ºC to 250 ºC for various compositions of the ferrite composite system is shown in the Fig.5. As the temperature increases the initial permeability remains flat upto a particular temperature and suddenly falls to a minimum value in all the composites except in equimolar composition. In equi molar composition the initial permeability gradually decreases upto Curie temperature and there after it suddenly falls to zero value.

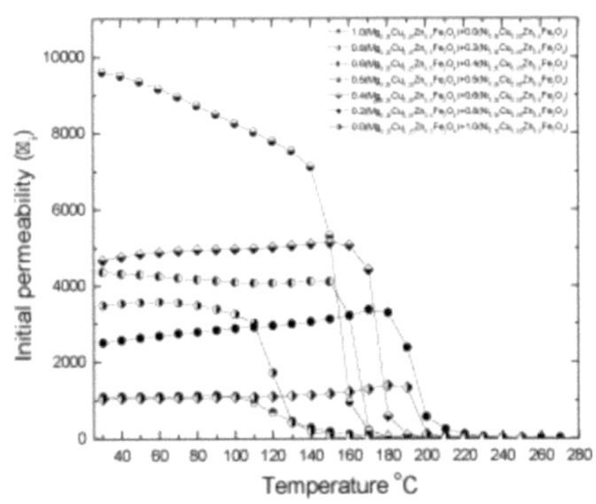

Figure 5. Variation of initial permeability (μ_i) with temperature at room temperature for the ferrite composites.

The increase of μ_i with temperature is explained as follows: The anisotropy field usually decrease with temperature much faster than M_s. The high value of μ_i corresponds to zero point in anisotropy, caused by presence of Fe^{2+} ions in the sample.

6.2. Stress sensitivity

In pure component ($Mg_{0.25}Cu_{0.25}Zn_{0.5}Fe_2O_4$ and $Ni_{0.35}Cu_{0.05}Zn_{0.6}Fe_2O_4$) and their equimolar composite i.e., 0.5 ($Mg_{0.25}Cu_{0.25}Zn_{0.5}Fe_2O_4$) + 0.5 ($Ni_{0.35}Cu_{0.05}Zn_{0.6}Fe_2O_4$) were chosen to examine the stress sensitivity of inductance in them. The pure composition $Mg_{0.25}Cu_{0.25}Zn_{0.6}Fe_2O_4$ has an initial permeability (μ_i = 1041 and the other $Ni_{0.35}Cu_{0.05}Zn_{0.6}Fe_2O_4$ has initial permeability value μ_i =1090. where as their equimolar i.e.,0.5 ($Mg_{0.25}Cu_{0.25}Zn_{0.5}Fe_2O_4$) + 0.5 ($Ni_{0.35}Cu_{0.05}Zn_{0.6}Fe_2O_4$) mixture showed high initial permeability value. It is thought worth while to study the stress sensitivity behaviour in this composite. In addition to this stress sensitivity of pure components were also under taken, the ratio of change in inductance with external compressive stress for all the three samples are shown in the Fig.6.

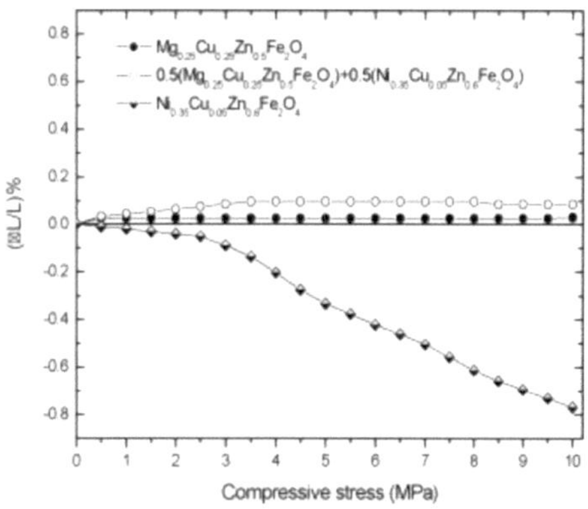

Figure 6. Variation of ratio of change in inductance (ΔL/L)% as a function of applied compressive stress for the x = 1.0, 0.5 and 0.0 ferrite composites.

Fig.6 shows the values of the ratio of inductance change (ΔL/L)% as a function of applied compressive stress for $Mg_{0.25}Cu_{0.25}Zn_{0.5}Fe_2O_4$, $Ni_{0.35}Cu_{0.05}Zn_{0.6}Fe_2O_4$ and 0.5 ($Mg_{0.25}Cu_{0.25}Zn_{0.5}Fe_2O_4$) + 0.5 ($Ni_{0.35}Cu_{0.05}Zn_{0.6}Fe_2O_4$) composition. From an examination of the Fig.6 it is clear that these figures indicate that the stress sensitivity is different for pure components and their equimolar mixture.

It is clear that (ΔL/L)% decreases with increasing applied stress for $Ni_{0.35}Cu_{0.05}Zn_{0.6}Fe_2O$. But it is found to be more or less insensitive in $Mg_{0.25}Cu_{0.25}Zn_{0.5}Fe_2O_4$. In $Mg_{0.25}Cu_{0.25}Zn_{0.5}Fe_2O_4$ and in equimolar mixture the (ΔL/L)% values are positive in the entire region of applied stress, while it is negative for $Ni_{0.35}Cu_{0.05}Zn_{0.6}Fe_2O_4$ composition.

This is in agreement with the observations made by Kumagai and Ikeda [103]. For small compressive stresses, the stress raises initial permeability with negative magnetostriction and for large tensile stresses the permeability decreases [108]. The variations of permeability with applied stress can be attributed to the magnetostrictive contributions of varied amounts of nickel and iron present in these samples. It can be seen from the above figures that in $Mg_{0.25}Cu_{0.25}Zn_{0.5}Fe_2O_4$ the stress sensitivity is found to be less, because magnesium ferrite has low magnetostriction. The eqiumolar mixture shows a broad and blunt peak of (ΔL/L)% with external compressive stress. A similar trend was observed by kanada et.al., [27].

The variation with external compressive stress of (ΔL/L)% in equimolar mixture is almost negliable and hence the ferrite composite material can be used as stress insensitive core material for microinductor applications.

7. Conclusions

S.No	Author Name	Ref. No.	Studies Carried Out	Observations/Conclusions
			Initial Permeability Studies	
1.	Kanada et.al.,	[20]	The replacement of Fe_2O_3 by Mn_2O_3 and replacement of CuO for MgO in MgZn ferrite in order to achieve low core loss, focusing on the microstructure. They have also investigated the initial permeability of MnZn, MnMgZn and MnMgCuZn ferrites	The initial permeability of both MnMgZn and MnMgCuZn ferrites are higher than those of MgZn ferrite.
2.	Yan et.al.,	[21]	The effect of CuO and V_2O_5 additions and the particle size of precursor materials on the microstructure and relative initial permeability of low temperature fired NiCuZn ferrites	The additions of CuO and V_2O_5 contribute to the grain growth and densification of matrix in the sintering process. Addition of 10 mol% CuO and 0.20 mol% V_2O_5, favoured the low firing temperature of NiCuZn ferrite, and showed very high relative initial permeability of 1417 at 1MHz.
3.	Nakamura et.al.,	[23]	The frequency dispersion of permeability in ferrite composite materials and observed that permeability spectra of Ni-Zn ferrite composite materials.	The sintered ferrite (volume fraction 1.0), the spin resonance was around 9 MHz and the static permeability was about 1400. As the ferrite content decreases in composite materials, the static susceptibility of the spin component decreases and the spin resonance frequency shifts to higher frequencies. The real part of the permeability in the ferrite composite materials becomes larger than that of the sintered ferrite in the RF frequency region.
4.	Su et.al.,	[25]	The effect of composition and sintering temperature on grain size, porosity and magnetic properties of the NiZn and NiCuZn ferrites.	The lowest power loss could be obtained with the equimolar composition for both NiZn and NiCuZn ferrites, which could be attributed to the lowest porosity.

S.No	Author Name	Ref. No.	Studies Carried Out	Observations/Conclusions
			Initial Permeability Studies	
5.	Bhosale et.al.,	[28]	A novel route for the preparation of high density, high permeability Cu-Mg-Zn ferrites using oxalate precursor method,	The initial permeability of all the samples increased with increase in sintering temperature and Mg^{2+} content for samples with $x \leq 0.20$.
6.	Nam et.al.,	[39]	The effect of Cu substitution on the properties of NiZn ferrites sintered at low temperature in various compositions,	The densification of NiCuZn ferrite is dependent on Cu content in these ferrites,
7.	Hua Su et.al.,	[47]	Effects of composition and sintering temperature on grain size, porosity and magnetic properties of the NiZn and NiCuZn ferrites	The lowest power loss could be obtained with the equimolar composition for both NiZn and NiCuZn ferrites, which could be attributed to the lowest porosity and a slight deficiency or excess of Fe_2O_3 content which no pronounced influence on saturation magnetic flux density (B_s).
8.	Bhosale et.al.,	[48]	The initial permeability studies on high density Cu-Mg-Zn ferrite	Variation of initial permeability (μ_i) with composition was mainly affected by variation of magnetization (M_s) and average particle size. They showed that the initial permeability increased with increase of magnesium content.
9.	Yue et.al.,	[49]	Effect of copper on the electromagnetic properties of Mg-Zn-Cu ferrites prepared by sol-gel auto combustion method	Low temperature sintered Mg-Zn-Cu ferrites possess good electro magnetic properties as well as fine grained microstructure.
10.	Yue et.al.,	[50]	the Mn-doped NiCuZn ferrites prepared by a novel sol-gel auto combustion process,	With increase of Mn content , the initial permeability is significantly increased.
11.	Ramana Muthy	[51]	the low temperature sintered of NiCuZn ferrites	For higher Cu content initial permeability increases continuously with temperature.
12.	Roy and Bera	[55]	The effect of Mg substitution on electromagnetic properties of $(Ni_{0.25-x}Mg_xCu_{0.2}Zn_{0.55})Fe_2O_4$	The permeability were found to increase and the magnetic loss decreased with Mg substitution for Ni,
13.	Zhang et.al.,	[59]	The investigation on structure and properties of low temperature sintered composite ferrites: viz., $(Ni_{0.60}Zn_{0.20}Cu_{0.20})Fe_2O_4$ spinel ferrite powder and Co_2Z were synthesized by self propagating method.	The composite ferrites have excellent electromagnetic properties in hyper frequencies more than 1.5 GHz.

S.No	Author Name	Ref. No.	Studies Carried Out	Observations/Conclusions
			Initial Permeability Studies	
14.	Qi et.al.,	[60]	the magnetic properties of composite materials with a ferrimagnetic $Ni_{0.2}Cu_{0.2}Zn_{0.6}Fe_{1.96}O_4$ phase and a ferro electric $BaTiO_3$ phase	The saturation magnetization and initial permeability decreases with increasing ferroelectric content $BaTiO_3$ phase.
15.	Wu et.al.,	[70]	the preparation of NiZn ferrites/SiO_2 nano composite powders by sol-gel auto combustion method.	
16.	Su et.al.,	[79]	Effects of composition and sintering temperature on grain size, porosity and magnetic properties of the NiZn and NiCuZn ferrites	The lowest power loss could be obtained with the equimolar composition for both NiZn and NiCuZn ferrites, which was attributed to the lowest porosity in those samples.
			Stress Insensitivity Studies	
17.	Ramamanohar Reddy et.al.,	[19]	Stress sensitivity of inductance in NiCuZn ferrites	Stress sensitivity is more in the case of iron rich and stoichiometric iron samples while in the case of iron deficient samples the stress sensitivity was found to be less.
18.	Kanada et.al.,	[27]	The effects of chemical composition and micro structure on stress sensitivity of magnetic properties to applied stress of Ni-Mg-Cu-Zn ferrite materials	They found that the application of external stress reduces the permeability.
19.	Kanada et.al.,	[36]	The effects of chemical composition and micro structure on stress sensitivity of magnetic properties with applied stress in Ni-Mg-Cu-Zn ferrite materials were investigated,	On application of stress sensitivity was reduced.
20.	Ikeda and Kumagi	[48]	Developments of stress-insensitive ferrite	The sensitivity of the magnetic property to the external stress in NiCuZn ferrite was examined from the relationship between stress and permeability.
21.	Nakano et.al.,	[49]	Development of low temperature fired NiCuZn ferrites and studied the high performance multilayer chip inductors	Controlling stress by the internal Ag-conductor and CuO_{1-x}/Ag on ferrites grain boundary is most important key point for high performance multilayer chip ferrites as well as the chemical composition of ferrite
22.	Nakano et.al.,	[51]	The magnetic properties of Mg-Cu-Zn ferrites under stress were investigated and compared with Ni-Cu-Zn ferrite under a compressive stress.	The change in permeability with stress of low temperature sintered MgCuZn ferrite was lower than that of low temperature sintered NiCuZn ferrite.

S.No	Author Name	Ref. No.	Studies Carried Out	Observations/Conclusions
			Initial Permeability Studies	
23.	Aoki et.al.,	[52]	The effect of SiO_2 on the bending strength of NiCuZn ferrite	The bending strength of NiCuZn ferrite was increased by the addition of SiO_2.
24.	Aoki et.al.,	[57]	The studies on the bending strength of sintered MgCuZn, NiCuZn, and MnZn ferrites.	The much residual carbon content affects the bending strength of these sintered ferrites.
25.	Nakano et.al.,	[59]	The development of low temperature NiCuZn ferrites and study of high performance for multilayer chip ferrites	The controlling stress by the internal Ag-conductor and CuO_{1-x}/Ag on ferrites grain boundary is most important for high performance multilayer chip ferrites as well as the chemical composition of ferrite.
26	Our Result	-----------	High permeability and Stress insensitivity of MgCuZn -NiCuZn ferrite composites for microinductor applications	Equimolar mixture of the ferrite composites showed high initial permeability (μ_i) of the order of 9619. It may be pointed out that this composition is fairly stress independent also. Not only equimolar composition, even the $Mg_{0.25}Cu_{0.25}Zn_{0.5}Fe_2O_4$ also showed stress insensitivity.

Acknowledgements

This work was financially supported by Defence Research and Development Organization (DRDO), under the grants ERIP/ER/0103301/M/01, New Delhi, India. The authors are thankful to the authorities of Sri Krishnadevaraya University, Anantapur for providing the facilities.

Author details

N. Varalaxmi[1,2] and K. Sivakumar[2]

*Address all correspondence to: narlasharma55@gmail.com

1 Department of Physics, Kakatiya University,Warangal, India

2 Ceramic Composite Materials Laboratory, Department of Physics, Sri Krishnadevaraya University, S.V. Puram, Anantapur, India

References

[1] T. Nomura and M. Takaya. Hybrids., 3 (1987) pp.15.

[2] A. Ono, T. Muruno and N. Kaihara. Jpn. Elec. Engg., 28 (1991) pp.5.

[3] T. Nomura and A. Namano."*Proc of 6th International Conference on Ferrites*" Kyoto: Japan Soc.of Powd and Powd .Metall., (1992) pp.1198.

[4] A. Nakano, H. Momoi and T.Nomura."*Proc of6thInternational Conference on Ferrites*" Kyoto: Japan Soc.of Powd and Powd. Metall., (1992) pp.1225.

[5] T. Nakamura.J. Magn. Magn. Mater., 168 (1997) pp.285.

[6] J. G. Koh and C. I. Yu.New Phys., (Korean Phys, Soc.) 24 (1984) pp.359.

[7] D. N. Bhosale, N. D. Choudhari, S. R. Swant and P. P. Bakare. J. Magn. Magn. Mater., 173 (1997). pp.51.

[8] J. Park, J. Kim and S. Cho J. Phys., (France) C1-193 (1997) pp.7.

[9] J. G. Koh and K. U. Kim. New Phys., (Korean. Phys. Soc.,) 26 (1986) pp.540.

[10] X. Qi, J. Zhou, Z.Yue, et.al., Key. Eng. Mater., 224-226 (2002) pp.593.

[11] H. Zhang, Z. Ma, J. Zhou, Z. Yue, L. Li and Z. Gui. J. Magn. Magn. Mater., 213 (2000) pp.304.

[12] X. Qi, J. Zhou, Z.Yue, Z. Gui and L. Li. J. Magn. Magn. Mater., 251 (2002) pp.316.

[13] A. Nakano, I. Nakahata and T. Murase. J. Jpn. Soc. Powd and Powd Metall., 48 (2001) pp.131.

[14] N. Rezlescu, E. Rezlescu, P. D. Popa, L. Craus and L. Rezlescu. J. Magn. Magn. Mater., 182 (1998) pp.199.

[15] N. Rezlescu, L. Sachlarie, E. Rezlescu and P. D. Popa Cryst Res. Technl., 36 (2001) pp. 157.

[16] L. Sachlarie, E. Rezlescu and N. Rezlescu.Phys. Stat. Solidi., (a) R1 (2000) pp.179

[17] R. D. Waldran. Phys.Rev., 99 (1955) pp.1727.

[18] S. S. Bellad, R. B. Pujar and B .K. Chougle. Ind. J. Pure Appl. Phys., [In press]

[19] A. Nakano, T. Suzuki and H. Momoi .J. Japan. Soc. Powd and Powd. Metall., 49 (2002) pp.77.

[20] I. Kanada, T. Aoki, T. Murase and T. Nomura. J. Japan. Soc. Powder and Powd. Metall., 46 (1999) pp.636.

[21] M. Yan, J. Hu, W. Lu and W. Zhang J. Magn. Magn. Mater., 303 (2006) pp.249.

[22] O. F. Caultn, L. Spinu and A. Stancu IEEE. Trans. Magn., 37 (2001) pp.2353.

[23] T. Nakamura, T. Sutaoka and K. Hatakeyama. J. Magn. Magn. Mater., 13 (1994) pp. 319.

[24] J. H. Oh. IEEE Trans. Magn., 35 (1999) pp.3412.

[25] H. Su, H. Zhang, X. Tang, Y. Jing, and Y. Liu. Mater. Letts., 61 (2007) pp.256.

[26] S. Murayama, S. Kumagi and Y. Ikeda. "*Proc of 6th International Conference on Ferrites*" Kyoto, Japan (1992) pp.366.

[27] I. Kanada, T. Murase and T. Nomura. J. Japan. Soc. Powd and Powd Metall.,48 (2001) pp.136.

[28] D. N. Bhosale,, S. R. Sawant Mater. S. A. Gangal, R. R. Mahajan and P. P. Bakare. Letts., 41 (1999) pp.247.

[29] M. Yan and J. Hu. J. Magn. Magn. Mater., 305 (2006) pp.171.

[30] S. A. Ghodake U. R. Ghodake, S. R. Sawant, S. S. Suryavanshi and P. P. Bakare. J. Solid. State. Chem., 179 (2006) pp.1892.

[31] X. Y. Gao, Y. Cao, Y. Zhou, W. Ding, C. Lei and J.A. Chen J. Magn. Magn. Mater., 305 (2006) pp.207.

[32] D. N. Bhosale, S. R. Sawant S. A. Gangal, R. R. Mahajan and P. P. Bakare. Mater. Sci and Engg., B65 (1999) pp.79.

[33] S. F. Wang, Y. R. Wang, C. K. Thomas Yang, C. F. Chen, C. A. Lu and C. Y. Huang. J. Magn. Magn. Mater., 220 (2000) pp.129.

[34] H. Su, H. Zhang and X. Tang. J. Magn. Mater. Devices., 3 (2004).pp.13.

[35] D. Mukesh, A. Verma, Subash C. Kashyap, D. C. Dube, Subash C. Kashyap, D. C. Dube, O. P. Thakur and Chandraprakash. Mater. Sci and Engg., B13 (2006) pp.42.

[36] A. Oi, A. Nakano and T. Nomura.J. Japan. Soc. Powd and Powd Metall., 47 (2000), pp.779.

[37] I. Nakahata, T. Murase. and A. Nakano. J. Japan. Soc. Powd and Powd and Metall., 48 (2001) pp.126.

[38] N. Ramamanohar Reddy, M. VenkataRamana, G. Rajitha, E. Rajagopal, K. V. SivaKumar, and V. R. K. Murthy. J. Magn. Magn. Mater., 292 (2005) pp.159.

[39] J. H. Nam, H. H. Jung, J. Y. Shin and J. H. Oh. IEEE Trans. Magn., 31 (1995) pp.3985.

[40] O. F. Caultn, L. Spinu, Al. Stancu, L. D. Thung and W. Zhou.J. Magn. Magn. Mater., 242-245 (2002) pp.160.

[41] S .F. Wang, Y. R. Wang, C. K. Thomas Yang, P. J. Wang and C. A. Lu. J. Magn. Magn. Mater., 217 (2000) pp.35.

[42] S. F. Wang, Y. R. Wang, C. K. Thomas Yang, C.F Chen C. A. Lu and C.Y.Huang.J. Magn. Magn. Mater., 22 (2000) pp.129.

[43] S. Ramana Murthy. Bull. Mater. Sci., 24 (2001) pp.379.

[44] S .F. Wang, Y. R. Wang, C. K. Thomas Yang, C.F Chen and C.A. Lu. Scripta Mater., 43 (2000) pp.269.

[45] D. N. Bhosale, N. D. Choudhari, S. R. Sawant, D.K. Rajendra,and P.P.Bakare. IEEE.Trans. Magn., 34 (1998) pp.535.

[46] W. C. Kim, S .L. Park, S. J. Kim S. W. Lee and C. S. Kim. J. Appl. Phys., 87 (2000) pp. 6241.

[47] H. Su, H. Zhang, X. Tang, I. Y. Jing and Y. Liu. J. Magn. Magn. Mater., 310 (2007) pp. 17.

[48] D. N. Bhosale, N. D. Choudhari, S. R. Sawant and P. P. Bakare J. Magn. Magn. Mater., 173 (1997) pp.1.

[49] Z. Yue, J. Zhou, L. Li, X. Hui and Z. Gui. Mater. Sci and Engg., B86 (2001) pp.64.

[50] Z. Yue, J. Zhou, L. Li, and Z. Gui. J. Magn. Magn. Mater., 233 (2001) pp.224.

[51] S. RamanaMurthy. J. Mater. Sci.Letts., 21 (2002) pp.657.

[52] X. Qi, J. Zhou, Z. Yue, Z. Gui and L. Li. J. Magn. Magn. Mater., 251 (2002) pp.316.

[53] A. Bhaskar, B. Rajini Kanth, and S. Ramana Murthy, J. Mater. Sci., 39 (2004) pp.3787.

[54] L. M. Hyie , I.H.S.C. Metselaar and I. I. Yaacob. Key. Engg. Mater., 306- 308 (2006) pp. 875.

[55] P. K. Roy and J. Bera J. Magn. Magn. Mater., 298 Issue 1, (2006) pp.38.

[56] Kin O. Low and Frank R. Sale. J. Magn. Magn. Mater., 256 (2003) pp.221.

[57] E. Rezlescu, N. Rezlescu, P. D. Popa, L. Rezlescu, C. Pasnicu and M. L. Craus. Bull .Mater. Res., 33 (1998) pp.915.

[58] K. S. Park, J. H. Nam and H. Lee. J. Magn. Soc.of .Japan., 22 and J. Suppl, No.S1 (1998).

[59] H. Zhang, L. Li, P. Wu, J. Zhou, Z. Ma and Z. Gui. Bull. Mater. Res., 35 (2000) pp. 2207.

[60] X. Qi, J. Zhou, Z. Yue, Z. Gui and L. Li. J. Magn. Magn. Mater., 269 (2004) pp.352.

[61] T. Aoki. T. Murase and T. Nomura. J. Japan. Soc. Powd and Powd Metall., 48 (2001) pp.140.

[62] A. Nakano and T. Nomura. "*Proc of 8th International Conference on Ferrites*"Kyoto, Japan. (2000) pp.1117.

[63] K. Amarendra Singh, T.C. Goel and G. Mendiratta. Jpn. J. Appl. Phys., 42 (2003) pp. 2690.

[64] T. T. Srinivasan , P. Ravindranathan, L. E. Cross, R. Roy , R. E. Newnham, S. G. Sankar, and K. C. Patil. J. Appl. Phys., 63 15 (1988) pp.3789.

[65] D. Y. Ju and P. Bian. "*Transactions of Nonferrous Metals Society of China*" (English Edition) 16 (Suppl.) (2006) pp. s67.

[66] H. Zhang, Z. Ma, J. Zhou, Z. Yue, L. Li and Z. Gui.J. Magn. Magn. Mater., 213 (2000) pp.304.

[67] C. S. Kim, W.C. Kim, S.Y.An, and S.W.Lee.J. Magn. Magn. Mater., 215 - 216 (2000) pp. 213.

[68] C. Y. Tsay, K. S. Liu, T. F. Lin, and I. N. Lin. J. Magn. Magn. Mater., 209 (2000) pp.189.

[69] S. F. Wang, Y. R. Wang, C. K.Thomas, C. F. Chen and C. A. Lu. Scripta Mater., 43 (2000) pp.269.

[70] K. H. Wu, Y.C. Chang and G. P. Wang. J. Magn. Magn . Mater., 269 (2004) pp.150.

[71] B. O. Ladgaonkar "*National DAE-BRNS Symposiumon Recent Trends in Electro and Magnetoceramics*" Shivaji University, Kolhapur, India. (1999).

[72] D. N. Bhosale, N. D. Choudhari, S. R. Sawant and P. P. Bakare. J. Magn. Magn. Mater., 173 (1997) pp.51.

[73] U. R. Ghodake, S. A. Ghodake, J. S. Ghodake, S. L. Darshane, P. P. Bakare, S. R. Sawant and S. S. Suryavanshi. "*Materials for Advance Technologies.: Proc. NASMAT*- Shivaji University, Kolhapur, India (2006).

[74] S. S. Suryavanshi, S .R. Sawant, R. S. Patil and S.A. Patil. Ind. J. Pure and Appl. Phys., 29 (1991) pp.644.

[75] H. Umeda, I. Kanada, T. Murase and T. Nomura. J. Japan. Soc. Powd and Powd. Metall., 47 (2000) pp.774.

[76] T. Krishnaveni, S. Komarneni, and S. R. Murthy."*Synthesis and Reactivity in Inorganic-Metal-Organic and Nano-Metal Chemistry*"., 36 (2006) pp.143.

[77] B. P. Ladgaonkar, P. P. Bakare, S. R. Sainkar and A. S. Vainganka r. Mater. Chem and Phys., 69 (2001) pp.19.

[78] Yomen Attasi and Mohamed Tally. Physics Abstract, 0602129, (2006).

[79] H. Su, H. Zhang, X. Tang ,Y. Jing and Y. Liu. Mater. Letts., 6 (2007) pp.256.

[80] D. N.Bhosale, S. R. Sawant, S. A. Gangal, R. R. Mahajan, and P. P. Bakare. Mater. Sci. Engg., B6 (1999) pp.79

[81] Souliah Zahi, Mansor Hashim and A.R.Daud. J. Magn. Magn. Mater., 30 (2007) pp.177.

[82] S. A. Ghodake, U. R. Ghodake, S. R. Sawant, S. S. Suryavanshi, J. Magn. Magn. Mater., 305 (2006) pp.110.

[83] J. Hu, M. Yan and W.Y. Zhang. Mater. Chem and Phys., 98 (2006) pp.459.

[84] T. Y. Byun, S.C. Bycon and K. S. Hong. IEEE Trans. Magn., 35 (1999) pp.3445.

[85] Z.Yue, J.Zhou, Z.Gui and L.Li. J. Magn. Magn. Mater., 264 (2003) pp.258.

[86] J. Hu and M.Yan. J. Zhejiang Univ Sci 6B (6) (2005) pp.580.

[87] T. Krishnaveni, S. R. Murthy, F. Gao, Q. Lu and S. Komarneni. J. Mater. Sci., 41 (2006) pp.1471.

[88] A. Bhaskar, B. R. Kanth, and S. R. Murthy. J. Magn. Magn. Mater., 28 (2004) pp.109.

[89] T. Murase, T. Akoi and H. Umeda. J. Japn. Soc. Powd and Powd Metall., 52 (2005) pp.199.

[90] T. Krishnaveni and S. R. Murthy. *Proc. of SPIE.*, 5650 (2005) pp.285.

[91] C. D. Mukesh, C. Subhash, D. Kashyap, C. Dube and S. K. Mohanta. J. Electro. Ceram 16 (2006) pp.331.

[92] H. Su, H. Zhang, X.Tang L. Jia and Q. Wen. Mater. Sci. Engg., B 129 (2006) pp.172.

[93] H. Su, H. Zhang, X.Tang and X. Xiang. J. Magn. Magn. Mater., 383 (2004) pp.157.

[94] X. Wang, W. Qu, L. Li and Z. Gui. Ceram. Intern., 30 (2004) pp.1615.

[95] B. Li , Z. Yue, X. Qi, J. Zhou, Z.Gui and L.Li. Mater. Sci. Engg., B99 (2003) pp.252.

[96] X. Y. Gao, Y. Cao, Y. Zhou, W. Ding, C. Li and J. A. Chen. J. Magn. Magn. Mater., 305 (2006) pp.207.

[97] Su-il Pyun and Jong-Tae Baek. J. Amer. Ceram. Soc. Bull., 6 (1985) pp.602.

[98] Zbigniew Kaczkowski. J. Magn. Magn. Mater., 41 (1984) pp.338.

[99] M. I. Abd.El-Ati and A. Tawafik. J. Ther. Anal and Calor., 37 (1991) pp.2465.

[100] Y. Enokido, H. Saitah, N. Sato,v *"Advances in Science Technology, World Ceramics Congress"*, Florence,ITALIE (14-19 June 1998).

[101] T. Yamaguchi and M. Shingawa. J. Mater. Sci., 30 (1995) pp.504.

[102] Y. Ikeda and M. Kumagi. J. Japan. Soc. Powd and Powd Metall., 48 (2001) pp.2465.

[103] M. Kumagi and Y. Ikeda. *"Proc of 5th International Conference on Ferrites"*., Bombay, India (1989) pp.625.

[104] A. Nakano. I. Nakahata J. Japan. Soc. Powd and Powd Metall., 48 (2001) pp.131.

[105] T. Akoi, H. Umeda, K. Akasura and T. Musare.J. Japan. Soc. Powd and Powd Metall., 53 (2006) pp.273.

[106] T. Akoi, T. Musare and T. Nomura J. Japan. Soc. Powd and Powd Metall., 53 (2006) pp.273.

[107] M. Fujimoto J. Am. Ceram. Soc., 77 (1994) pp.2873.

[108] J. Smit and H .P. J. Wijn. *"Ferrites"*., Philips Technical Library Eindhoven, Netherlands, (1959).

Chapter 11

Sintering and Heat Treatment of Titanium Alloys by Pulsed Electric Current Sintering

F. Zhang, Y. Quan, M. Reich, O. Kessler and E. Burkel

Additional information is available at the end of the chapter

http://dx.doi.org/10.5772/53705

1. Introduction

Pulsed electric current sintering (PECS), commonly also defined as spark plasma sintering (SPS), field assisted sintering (FAST) is a novel pressure assisted pulsed electric current sintering process utilizing ON-OFF DC pulse energizing. Due to the repeated application of the ON-OFF DC pulsed voltage and current in powder materials, the spark discharge point and the Joule heating point (local high temperature-state) are transferred and dispersed to the overall specimen [1]. The PECS process is based on the electrical spark discharge phenomenon: a high energetic, low voltage spark pulse current momentarily generating high localized temperatures, from several to ten thousand degrees between the particles resulting in high thermal and electrolytic diffusion [1, 2].

Because of the time- and cost- effective manufacturing which it permits, sinter hardening has come into being a process of high importance in powder metallurgy (PM) in the past years [3]. It is a hardening method applicable only for PM parts in which the parts are sintered and quenched directly after the sintering step saving energy and costs connected to conventional hardening where the parts have to be reheated to hardening temperature [3, 4]. Furthermore, sinter hardening is performed by gas quenching instead of oil quenching being beneficial in terms of dimensional stability and cleanliness of the specimen. Due to poor thermal transfer characteristics (lower cooling rate) of gases under normal conditions, they have to be optimized by proper adjustment of gas pressure and flow speed. In principle, gas quenching can be performed in two ways [5-8], namely: at low or atmosphere pressure with high gas velocity or at high pressure with limited gas velocities. The first part of this chapter will focus on new modifications of the PECS technique. The PECS has been modified and integrated with a novel gas quenching system. The PECS in vacuum status was quenched with high-velocity nitrogen gas. The Al-2024 alloy and ALSI-303 stainless steel have been

INTECH
open science | open minds

© 2013 The Author(s). Licensee InTech. This chapter is distributed under the terms of the Creative Commons Attribution License http://creativecommons.org/licenses/by/3.0), which permits unrestricted use, distribution, and reproduction in any medium, provided the original work is properly cited.

used for cooling power measurement and numerical simulation of the quenching process. The $\alpha+\beta$ Ti6Al4V alloy has been used as a model material to study the effects of the cooling rate on the microstructure and mechanical properties of the sintered alloy.

The second part of this chapter will focus on the preparation of Ti6Al4V foams by the PECS with a post heat treatment. Since, the mismatches of the Young's moduli of the human bones and the Ti6Al4V implants lead to stress shielding effects responsible for bone resorption and eventual implant loosening. Porous structures are expected to provide a better interaction with the bones. The fabrication of porous Ti alloy with standard techniques is difficult due to the high melting point and the extreme chemical affinity to atmospheric gases like oxygen, hydrogen, and nitrogen, especially at elevated temperatures. Currently, the fabrication of porous Ti alloy is mainly focused on powder metallurgy approaches [9]. The use of space holder materials allows a simple and accurate control of pore fraction, shape and connectivity in titanium alloys. Open-celled pure titanium foams were fabricated by vacuum hot pressing of a blend of Ti and NaCl spacer followed by NaCl removal in water [10]. However, the Ti6Al4V/NaCl mixture is difficult to be densified by the hot pressing due to the higher creep resistance of this alloy at 790 °C [10]. Promising approach to produce Ti alloys is the pulsed electric current sintering process. The PECS can easily sinter Ti and Ti alloy powders applying pressure and high pulsed DC currents. Thus the local surfaces of the particles melt, allowing junctions to be formed between particles in contact. The combination of the PECS technique with a NaCl space holder was used in porous pure Ti to fabricate macroporous Ti foams with controlled architectures recently [11]. This indicated that the PECS may be able to fabricate Ti6Al4V foams from the Ti6Al4V/NaCl powder mixture. The microstructures, mechanical properties and cell proliferation properties of the fabricated foams were investigated. Some other methods for preparation of Ti alloy foams by the PECS were also introduced.

2. Materials and methods

- The Ti6Al4V powder with a particle size about 20 μm was obtained from TLS Technik GmbH & Co, Germany. The element contents have aluminum 5.9%, vanadium 3.9%, carbon 0.01%, oxygen 0.12%, iron 0.19%, nitrogen 0.01% and hydrogen 0.004% (wt.%). The space holder materials (NaCl, Cu, Mg) for preparation of Ti6Al4V foams with 99.0% purity were obtained from Alfa Aesar and sieved in the range of 50 to 250 μm.
- The PECS experiments were conducted using a Model HP D-125 FCT spark plasma sintering system (FCT systeme GmbH, Rauenstein, Germany) installed at the Tycho Sinter Lab in the University of Rostock. The temperatures in different sites of the samples were measured by Pyrometer and thermal couples.
- The numerical simulations used a rotation-symmetric finite-element-model considering the axial symmetry by software MSC.MARC 2010.1.0.
- The hardness of the polished samples was measured with a SHIMADAZU Micro Hardness tester HMV. The compressive strength and ductility under compressive load was

performed according to DIN 50106 on an Instron 8502 testing machine at room temperature with a compression speed of 0.5 mm/min.

- The polished metallographic samples were etched by Weck's reagent (2 g ammonium acid, 100 ml distilled water, 50 ml alcohol) for 20 seconds and subjected to optical microscopy to observe microstructure morphologies. The fracture surface of the compressed samples was analyzed using scanning electron microscope (SEM, Zeiss Supra 25, Germany) at 20 keV. The Ti6Al4V foam architecture was examined by using X-ray microcomputed tomography (Micro-CT, GE, USA). The phase composition was analyzed using high energy synchrotron radiation (42.58 keV) at Beamline P02.2 (DESY/ PETRA III).
- The human osteoblast-like cell line MG-63 (ATCC, CRL-1427, LGC Promochem, Wesel, Germany) was used for the evaluation of the cellular acceptance of the surfaces

3. Ti6Al4V alloys prepared by PECS with a gas quenching system

3.1. Gas quenching system and cooling power

Figure 1 shows the image of the gas quenching system in the PECS. In the system, six nozzles in one group and eight groups up to 48 nozzles are arranged rotationally symmetric around the sintered component. The gas is distributed evenly on all nozzles. The gas nozzle field is positioned inside the PECS chamber and quenches the sintered component together with the graphite tool directly after sintering without any movement of the component [12]. The temperature is measured by a central pyrometer with a focus point at the bottom of the central borehole of the graphite up-punch and thermal couples. The Al alloy and steel cylinders (ф60 mm) were heated and quenched in the PECS as dense materials without sintering, to study the quenching power without sintering influence. The Ti6Al4V powders were pressed into a ф40 mm graphite die, and sintered at 850 °C and 50 MPa in vacuum. A heating rate of 100 K/min was adopted, and the sintering process lasted typically 6 min. The applied direct current for PECS was 1000-2000 A with pulse duration of 10 ms and an interval of 5 ms. Before cooling, the PECS furnace was flooded with argon gas keeping the sintering temperature. Afterwards, it was quenched from 850 °C using room temperature nitrogen gas at very high flowing rate of 8000 l/min out of the nozzles.

In Figure 2 the cooling curve in the core of the Al 2024-sample surrounded by the graphite form during gas quenching from 480 °C is compared to the curves that were achieved by natural cooling with or without flooding the sinter chamber. The gas quenching exerts a considerable influence, so that the Al alloy cools down to 200 °C after flooding the vacuum chamber more than twice as fast as without being quenched. A maximum cooling rate of 2.0 K/s is reached during quenching in contrast to 0.5 K/s during natural cooling. Figure 3 shows that the cooling process of the steel sample from 850 °C is affected by the gas nozzle field quenching, too. With surrounding graphite form the sample cooled down to 200 °C thrice as fast as naturally cooled after flooding the vacuum chamber. By gas quenching maximum cooling rates of 2.6 K/s with graphite form were reached, whereas a natural cooling affects only maximum cooling rates of 0.9 K/s with graphite form.

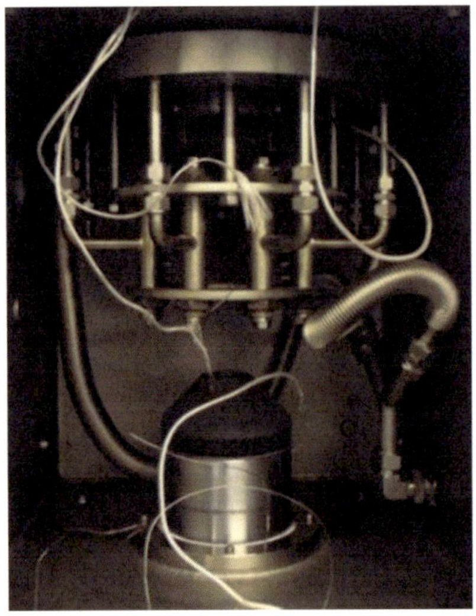

Figure 1. The image of the gas quenching system in the pulsed electric current sintering.

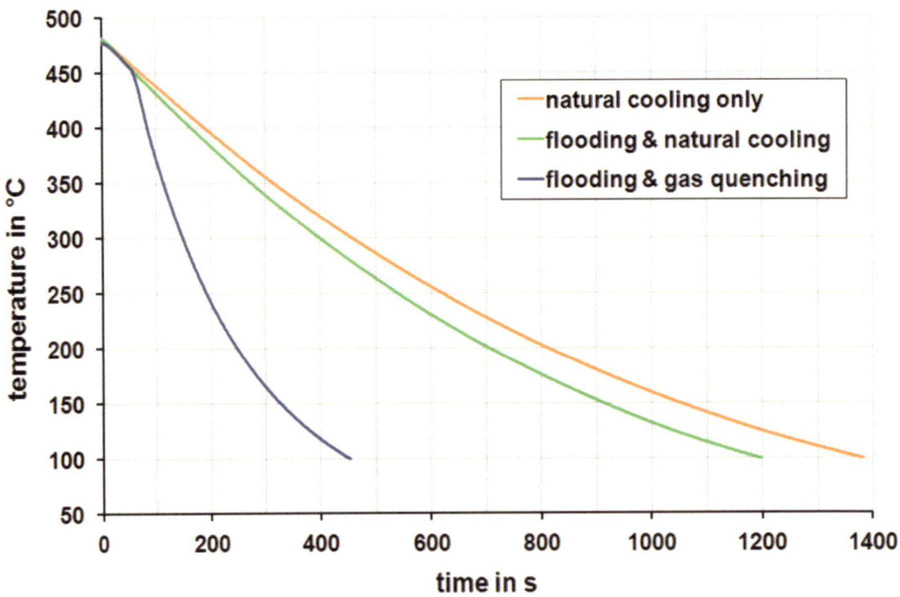

Figure 2. Measured cooling curve in the core of an Al-2024 sample surrounded by the graphite form during gas quenching with 8000 l/min (nitrogen) compared to natural cooling

http://dx.doi.org/10.5772/53705

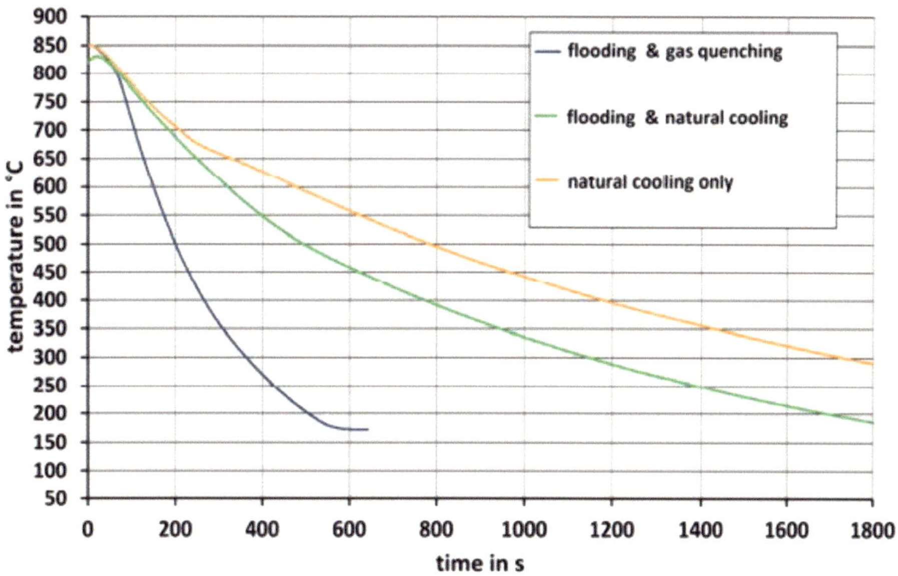

Figure 3. Measured cooling curves in the core of an AISI 303 stainless steel sample during gas quenching with 8000 (nitrogen) compared to natural cooling without the graphite form

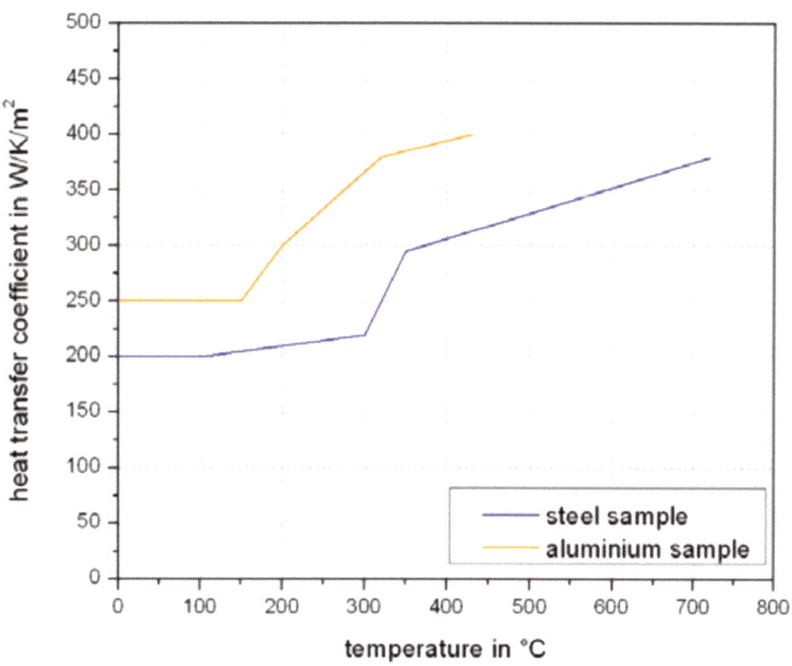

Figure 4. Calculated heat transfer coefficient of aluminium and steel samples.

From these cooling curves, heat transfer coefficients have been determined by inverse thermal simulation. Thereby the thermophysical properties of the surrounding graphite die have been considered. The determined coefficient for the aluminium and the steel sample are set out in Figure 4. Over the process time, the gas nozzle filed loses quench intensity. The heat transfer coefficient falls from about 400 W/(m^2K) to the half of initial value.

Figure 5 contains the results of the FEM-simulation for the quenching of the steel sample after 100 s. The temperature distribution reflects the different thermal behavior of steel and graphite. The heat of the sample is discharged considerably through the surface of the sinter form being cooled by gas nozzle field. However, only a small proportion of the sample heat dissipates in the massive stamps.

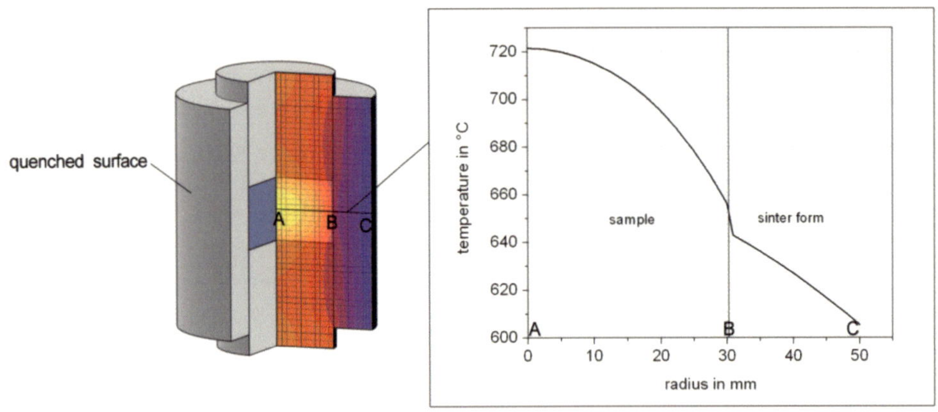

Figure 5. Quenching simulation results of steel sample, temperature distribution after 100 s.

Various cooling rates can be achieved by changing the thickness of the graphite die or the flowing rate of the nitrogen gas (0-8000 l/min). In this study, various cooling rates of the Ti6Al4V alloy are obtained with altering the thickness of the dies. The sample dimensions are 40 mm in diameter and 10-12 mm in height. Figure 6 shows the temperature-time cooling curve of the sintered Ti6Al4V alloys with various cooling rates. Below 400 °C, the pyrometer cannot measure the temperature of the sample. From the cooling curves sample mean cooling rates of samples 1.6 K/s, 4.8 K/s, 5.6 K/s and 6.9 K/s with an uncertainty of 0.1 K/s were derived. The cooling rate 1.6 K/s was achieved with a 20 mm thick graphite die with a natural cooling, 4.8 K/s with the 20 mm die with gas quenching, and 5.6 K/s with the 10 mm die, also with gas quenching. The rate of 6.9 K/s was achieved in two steps. Hereby, the powder sample was firstly sintered in the PECS at 500 °C for 5 min at 50 MPa in die to get a 70-80% relative density, and then heated to a sintering temperature of 850 °C without die to get full density and directly gas quenched to the specimen, later on.

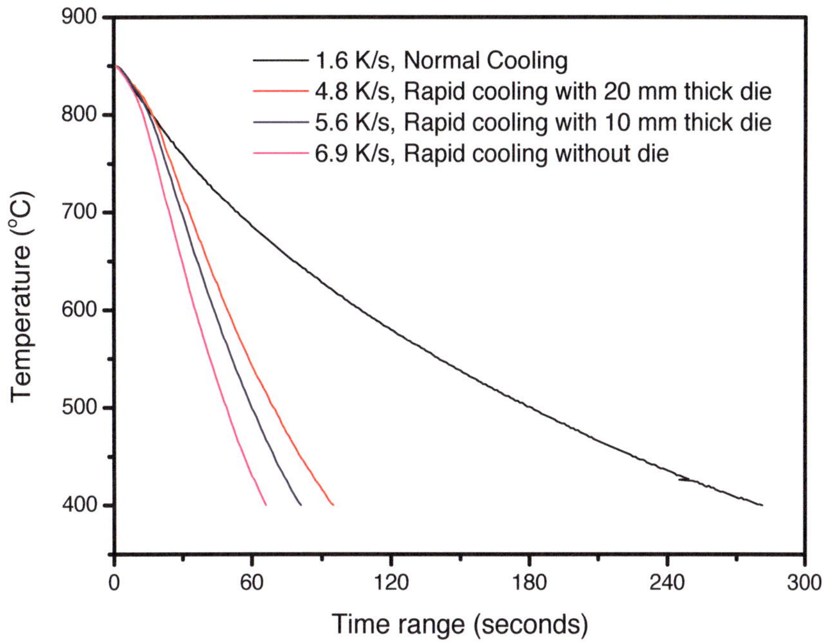

Figure 6. The cooling curves of pyrometer temperature as a function of time showing the temperatures in the center of the Ti6Al4V samples with various quenching rates.

3.2. Mechanical properties of the gas quenched Ti6Al4V alloys

Figure 7 shows the hardness (HV1) of the sinter hardened Ti6Al4V alloy with various cooling rates. The naturally cooled sample with the cooling rate of 1.6 K/s has hardness of 327±8. The gas quenched samples show hardness of 337±7, 342±4.4 and 353±3.6 for cooling rate of 4.8, 5.6 and 6.9 K/s, respectively. The hardness increases with higher cooling rate. Thus, sinter hardening of the Ti6Al4V alloys has been realized by the PECS with gas quenching.

Figure 8 shows the effects of cooling rate on the ultimate compressive strength and ductility of the Ti6Al4V alloys. The ultimate compressive strength goes up with the increase of the cooling rate. The samples of natural cooling rate 1.6 K/s show the ultimate compressive strength of 1578±80 MPa. At a cooling rate of 4.8 K/s, it reached to 1723±63 MPa. At 5.6 K/s cooling rate, it is increased to 1775±70 MPa and at 6.9 K/s, it rises to 1832±43 MPa. The rapid cooling has increased the ultimate compressive strength of the Ti6Al4V alloy. The ductility increased with higher cooling rate. The samples with natural cooling rate 1.6 K/s show a ductility of 25.7±2.3 %. At a cooling rate of 4.8 K/s, it reaches to 27.9±2.4 %, at 5.63 K/s to 29.5±2 %, and at 6.9 K/s to 34.0±2.7 %. The rapid cooling in the PECS has enhanced the ductility of the Ti6Al4V alloy.

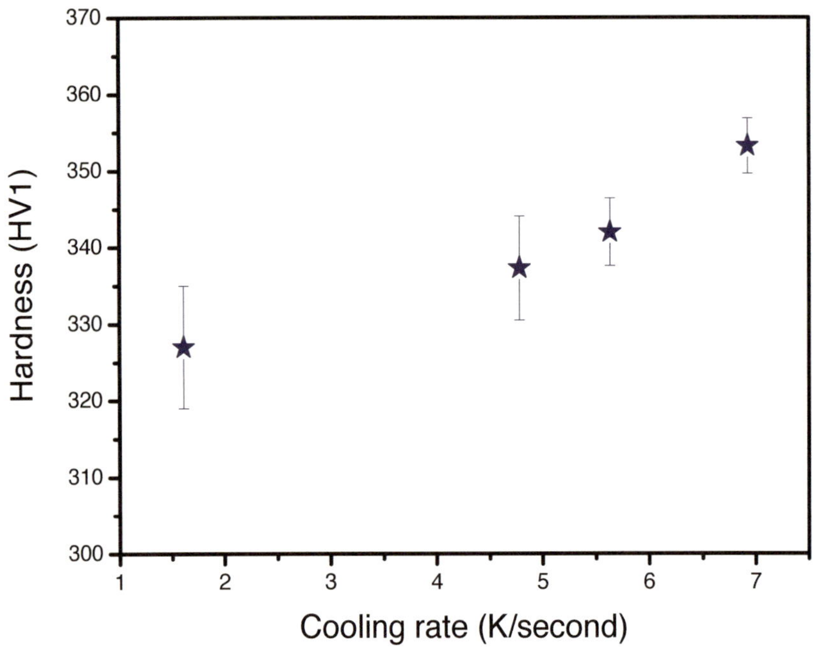

Figure 7. Hardness (HV1) of the Ti6Al4V alloys with various cooling rates.

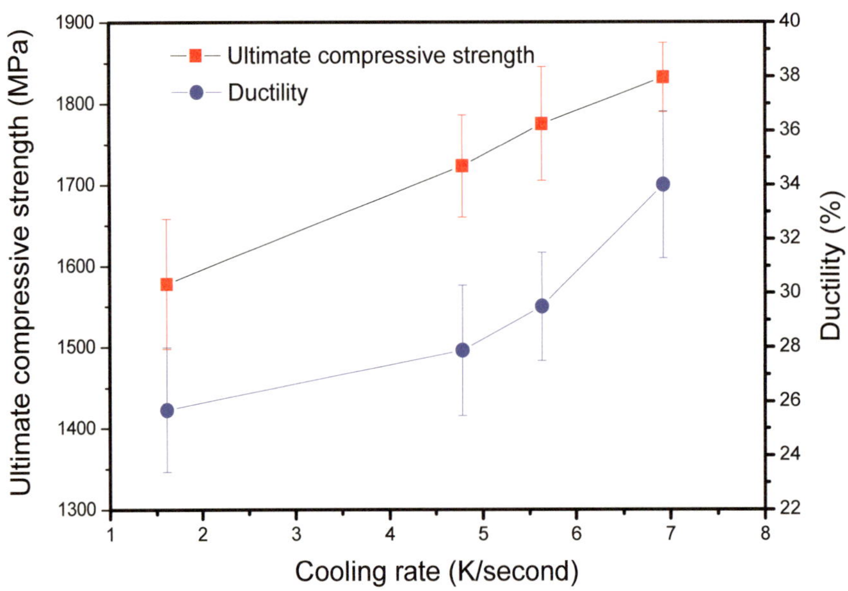

Figure 8. Ultimate compressive strength and ductility of the Ti6Al4V alloys with various cooling rates.

3.3. Microstructure of the gas quenched Ti6Al4V alloy

Figure 9 shows the SEM micrographs of the fracture surface of the Ti6Al4V alloys after PECS with various cooling rates. The samples all present a ductile fracture mode with a large amount of plastic deformation undergoing transgranular fracture. However, the fracture surfaces present more and more dimples with increase of the cooling rate (Figure 9a-d). In Figure 9 (d), the sample with cooling rate of 6.9 K/s has much more big dimples and shows more ductile than the sample with cooling rate of 1.6 K/s in Figure 9 (a). It indicates the sample becomes much ductile after the rapid cooling in the PECS. The SEM microstructures are consistent with the mechanical results in Figure 8.

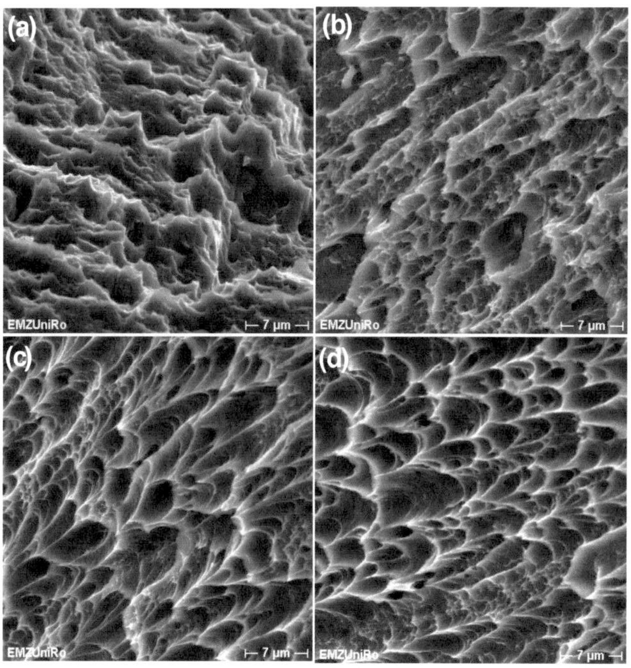

Figure 9. SEM micrographs of the fracture surfaces of the Ti6Al4V alloys with various cooling rates: (a) 1.6 K/s, (b) 4.8 K/s, (c) 5.6 K/s and (d) 6.9 K/s.

Figure 10 provides the metallographic morphology of the Ti6Al4V alloys with various cooling rates. In the etched metallographic images, the β-Ti phase (bcc) appears in white color and α-Ti phase (hdp) in dark color. The samples all show both phases in the microstructures. The microstructures form during sintering at 850°C and subsequent cooling with different rates. On sintering temperature the microstructure consists of a lower amount of α and a higher amount of β. During cooling β partially transforms to α. After cooling the microstructure consists of former α, retained β and newly formed α. The cooling rate mainly influences on the morphology of the newly formed α. The naturally cooled sample with 1.6 K/s cooling rate shows a relative homogeneous globular α+β microstructure (Figure 10a). The rapidly cooled sample shows a different microstructure with a higher fraction of lamellar α phase (Figure 10b). As the cooling rate increases, the 5.6 K/s cooled sample in Figure 10(c) has even more lamellar α phase. In Figure 10(d), the α phase is getting much more lamellar. In general, gas quenching in the SPS resulted in a higher fraction of lamellar α-Ti phase. The XRD results show the gas quenching in the SPS further resulted in the formation of intermetallic phases of $Al_3Ti_{0.8}V_{0.2}$ and Al_2Ti in the alloys [12].

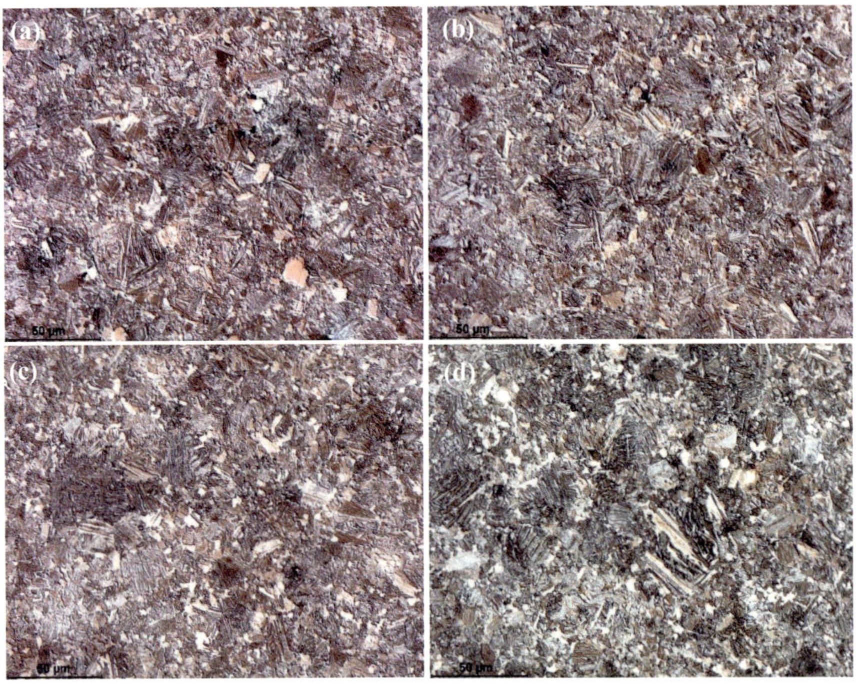

Figure 10. Metallographic micrographs of the surfaces of the Ti6Al4V alloys with various cooling rates: (a) 1.61 K/s, (b) 4.78 K/s, (c) 5.63 K/s and (d) 6.92 K/s.

3.4. Discussions

The cooling power measurments of the Al alloy and steel samples revealed that by the flooding with Ar gas and quenching with N_2 gas can effectively cool the samples. Without graphite dies can cool faster than with graphite dies. The quenching simulation results indicates that the temperature reduced from outside surface of the die gradually to the centre of the sample. At cooling time 100 s, it has about 120 °C temperature gradients from the outside surface of the die to the sample centre.

http://dx.doi.org/10.5772/53705

The hardness, ultimate compressive strength and ductility of the Ti6Al4V alloys have been increased due to the gas quenching. Sinter hardening is applicable for the Ti6Al4V alloy in which the samples are PECS sintered and gas quenched directly after the sintering. It saved energy and costs when compared with a conventional hardening in which the parts must be reheated to the hardening temperature. The microstructures consisted of former α-Ti, retained β-Ti, new α-Ti formed during quenching and intermetallic precipitates. The precipitated nanocrystalline Al_2Ti and $Al_3Ti_{0.8}V_{0.2}$ can play a role of hardening and dispersion reinforcing of the Ti6Al4V alloys. Therefore, the hardness and ultimate compressive strength of the Ti6Al4V alloy are increased. With increasing cooling rate, the newly formed α-Ti becomes more lamellar. The higher fraction of lamellar structured α-Ti phase can lead to the increase in ductility of the Ti6Al4V alloy [13].

The use of industrial gases for quenching high added value parts and offers significant environmental and performance advantages over liquid quenching (water, oil, etc.). Gas quenched parts are clean thus eliminating the need for post-cleaning operations. Due to poor thermal transfer characteristics (lower cooling rate) of gases under normal conditions, they have to be optimized by proper adjustment of gas pressure and flow speed. Gas quenching can be performed in two methods, namely: at low atmosphere pressure with high gas velocity or at high pressure with limited gas velocities [5-8]. In this study, we used the first technique in the PECS that at low or atmosphere pressure with high gas velocity. The flowing rate of the nitrogen gas from nozzles is 8000 l/min. The pressure of gas is only at normal atmosphere pressure. The thermophysical properties of the gas are also an important issue. Based on availability, density and specific heat capacity and thermal conductivity, hydrogen would appear to be good choice as a quench agent [12]. However, because of the explosive risks associated with hydrogen, it is seldom used as a quench gas in commercial heat treating. Nitrogen is the most popular choice, primarily because it is readily available and inexpensive. Argon is used in some special applications but does not quench as effectively as nitrogen and is considerably more expensive. In this study, the PECS vacuum furnace was flooded with argon gas to 40 mbar with the sintering temperature 850 °C being kept, and then quenched with nitrogen gas at high velocity.

The Ti6Al4V alloys have been used in clinic for several decades already. For uncemented knee and hip arthroplasties, Ti6Al4V alloys are the favourable implant materials. Additionally, the Ti6Al4V alloys are preferred for intramedullary rods, spinal clamps, self-drilling bone screws and other implants. One of the main problems of this alloy in biomedical applications is the insufficient ductility which lead to the difficulty in their contouring as required for pelvic and mandibular plates [14]. This investigation indicates that the traditional Ti6Al4V alloy can be modified by PECS with gas quenching leading to various microstructures and mechanical properties. The high ductility of the Ti6Al4V achieved by PECS with gas quenching leads to the easy fabrication of complex shape and plates for the medical applications. The rapid cooling-PECS combining sintering and gas quenching provides a novel method to tailor the microstructure and mechanical properties of the Ti alloys.

4. Pulsed electric current sintering of porous Ti alloys with post heat treatment

4.1. Characterization of raw powders

The Ti6Al4V and NaCl powders are shown in Figure 11. The Ti6Al4V powders have spherical morphologies and size distributions of 5-35 μm. The cuboidal sodium chloride powders have round angles and a size distribution of 125-250 μm. In the preparation process, the size of the metal powder should be smaller than the average powder size of the space holder. By choosing the size, shape and quantity of the space holder material, the mechanical properties of the metal foams can be adjusted [15]. The two kinds of powders were mixed uniformly in a mixer with a small amount of PVA (Polyvinyl alcohol) solution as binder. Then, they were subjected to the PECS machine and sintered at 700 °C under 50 MPa for 8 min in vacuum. Porous Ti6Al4V foams were obtained through dissolution of the NaCl phase in renewed deionized water for 72 h in room temperature. The alloy foams were cleaned in an ultrasonic water bath for 30 minutes and furnace dried at 120 °C for 12 h.

Figure 11. SEM images of the raw Ti6Al4V (a) and NaCl powders (b).

4.2. Post heat treatment by the pressureless PECS

The PECS post heat treatment was used to increase the density and strength of the foams. It was carried out by a pressureless PECS. The set-up of the pressureless PECS is shown in Figure 12. A small gap *d* in the range of 0–5 mm was designed to avoid damaging the porous materials. The foams after the NaCl being removed were sintered at 1100 °C for 5 min in this pressureless PECS.

4.3. Densities and microstructures of the foams

Table 1 shows a summary of the Ti6Al4V foam parameters after PECS. Different weight ratios of NaCl powders were mixed with the Ti6Al4V powders to get various porosities. After the foams were sintered at 700 °C, they show porosities of 47.6%, 57.6%, 63.9% and 72.5% (±1.0), respectively. When they were post heat treated at 1100 °C, they display porosities of 44.7%, 54.4%, 60.7% and 70.0% (±1.0)%. The porosities decreased and the densities increased after the heat treatment. The densities of the foams have been increased 5.6%, 7.4%, 8.8% and 9.0% at above specific porosities after the heat treatments.

Parameters	3 g Ti6Al4V +1.02g NaCl	3 g Ti6Al4V +1.6 g NaCl	3 g Ti6Al4V +2.2 g NaCl	3 g Ti6Al4V +3.4 g NaCl
Density after PECS at 700 °C (g/cm^3)	2.32	1.88	1.60	1.22
Density after heat treatment at 1100 °C(g/cm^3)	2.45	2.02	1.74	1.33
Relative density after PECS at 700 °C (%)	52.4	42.4	36.1	27.5
Porosity after PECS at 700 °C (%)	47.6	57.6	63.9	72.5
Relative density after heat treatment at 1100 °C (%)	55.3	45.6	39.3	30.0
Porosity after PECS at 1100 °C (%)	44.7	54.4	60.7	70.0
Increments in density	5.6 %	7.4%	8.8%	9.0%

Table 1. Summary of parameters of the Ti6Al4V foams after PECS.

Detailed microstructural views of the Ti6Al4V foam with 70.0% porosity are shown in Figure 13. Figure 13a and b display the foams fabricated at 700 °C and 50 MPa. They show the porous structure and pore walls, revealing relatively uniformly shaped macropores having rough surfaces, with many micropores (<10 μm) and the shapes and the sizes of the original alloy powders being visible. This indicates that the Ti6Al4V/NaCl mixture is difficult to be densified and, as expected, the diffusion at 700 °C was not sufficient to smooth the micropores due to the very high melting point of Ti6Al4V. Figure 13c and d display the Ti6Al4V foams post heat treated at 1100 °C at a pressureless mode. After the heat treatment, the junctions between the Ti6Al4V powder grains are formed with only few micropores left. As seen in Figure 13d, most of the micropores disappeared due to the heat treatment. This indicates that the pressureless heat treatment contributes to reduce the microporosity of the foams, since no pressure was applied and only diffusion dominated the PECS process. Thus, Ti6Al4V foams with higher density of the walls have been fabricated by the PECS at 700 °C and the post heat treatment at 1100 °C.

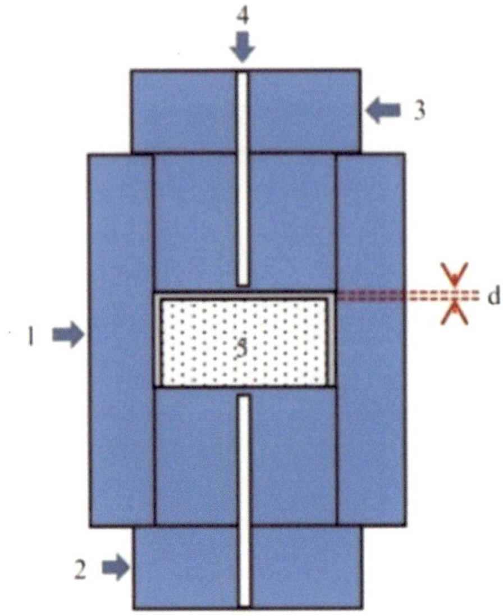

Figure 12. Schematic diagram of the graphite die set-up for the pressureless (1, graphite form; 2, graphite bottom punch; 3, graphite up punch; 4, Pyrometer measurement hole; 5, sample).

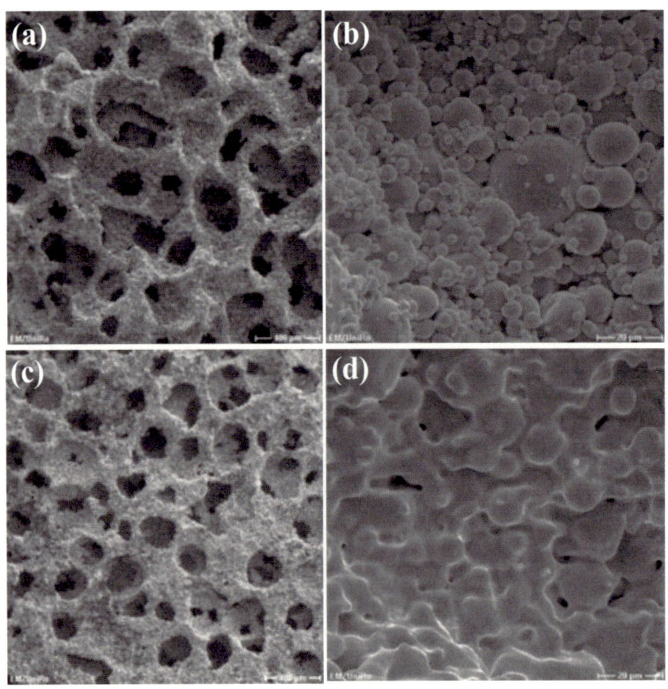

Figure 13. SEM images of the porous structures and pore walls of the Ti6Al4V foams with 70% porosity of the spark plasma sintered at 700 °C and 50 MPa (a, b), and PECS post heat treatment at 1100 °C (c, d).

http://dx.doi.org/10.5772/53705

The outer shape of the Ti6Al4V foam with 54.4% porosity fabricated by the PECS with post heat treatment is shown in Figure 14a. Figure 14b-d shows the micro-CT 3D reconstructions of the foam. The 3D cropped isometric view of cross sections in the Ti6Al4V foam shows the relative uniform pore distribution and interconnected 3D porous structures. The Micro-CT 2D top view and side views show that the macropore shapes are in square cross sections with a relative uniform distribution of pore sizes of 210±40 μm with an average cell wall thickness of 22.1±5 μm.

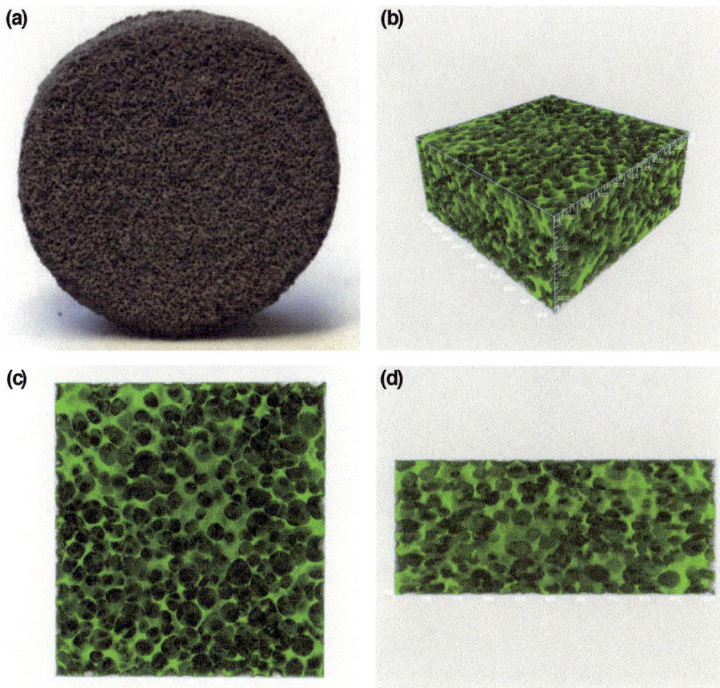

Figure 14. Micro-CT three-dimensional 3D reconstructions of the Ti6Al4V foam (54.4% of porosity) with outer shape of the alloy foam (a), an isometric view (b), top (c) and side view (d).

Figure 15 shows the SEM micrographs of the porous Ti6Al4V foams with different porosities of 44.7%, 54.4%, 60.7% and 70.0% fabricated by the PECS with post heat treatment. SEM images obtained from the porous Ti6Al4V in Figure 15 proved that the space holder NaCl phase was removed completely. The macrostructure of the foam is composed of homogeneously dispersed porous cavities and continuously connected Ti6Al4V struts. It shows mean pores with a size of 150 μm to 250 μm achieved with a NaCl spacer material in the range of 125 μm to 250 μm. The thickness of the pore walls decrease with increasing porosity. The higher porosity samples showed good interconnectivity. The primary pores replicate the size and shape of the angular NaCl particles with rounded corners. Thus, pore shapes can be controlled by using NaCl powders with various shapes. Finally, interconnected pores are visible either as black pores or as necks between adjacent pores in cross sections. These interconnected pores usually ranged 20-150 μm in sizes, indicating that osteoblasts may be able to penetrate into the porous structure.

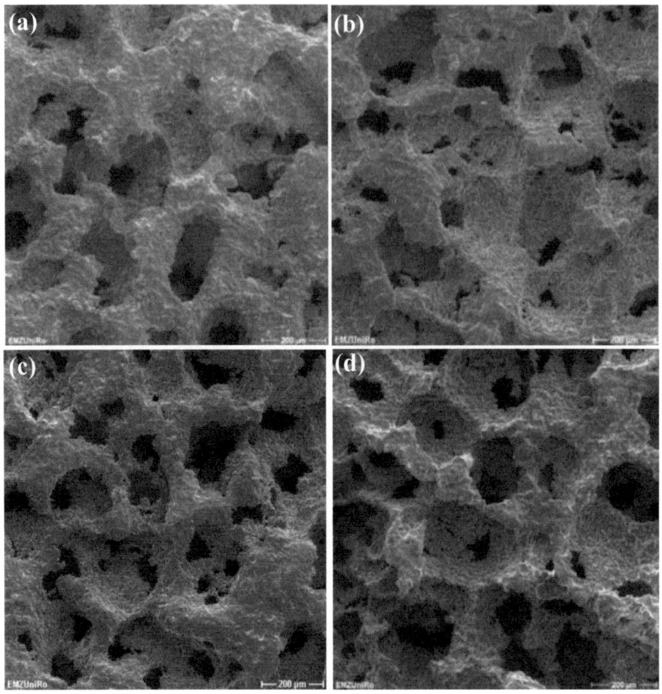

Figure 15. SEM images of the Ti6Al4V foams prepared by PECS with post heat treatment with different porosities of (a) 44.7%, (b) 54.4%, (c) 60.7%, (d) 70.0 %.

4.4. Mechanical properties of the foams

For the evaluation of the mechanical properties of the porous samples, compression tests were performed. Figure 16a shows the Young's modulus and compressive yield strength as a function of relative density for the Ti6Al4V foams and the predicted theoretical values. The Young's modulus of the porous metals is compared with a prediction according to the Gibson and Ashby model [16]:

$$\frac{E}{E_0} = C\left(\frac{\rho}{\rho_0}\right)^2 \tag{1}$$

where E and E_0 are Young's modulus of foam and bulk materials with densities ρandρ_0, respectively, using E_0 = 117 GPa and ρ_0=4.43 g/cm^3 for Ti6Al4V [17]. The proportionality constant C including data of metals, rigid polymers, elasomers, and glasses has to be considered as 1 for titanium alloys [16]. The estimated Young's moduli ranging from 34.11GPa to

9.84GPa obey exponential decays with the increase of porosity, and the actual measured Young's moduli of the foams are comparable with the calculated Young's modulus based on the density of the Ti6Al4V foam. Besides the Young's modulus, the strength is an important property of orthopedic implants, in particular in load bearing applications. According to the Gibson-Ashby model, the relationship between the yield strength and relative density is given by [16]:

$$\frac{\sigma}{\sigma_0} = C\left(\frac{\rho}{\rho_0}\right)^n \tag{2}$$

where σ_0 and ϱ_0 are the yield strength and the density of the solid material, σ and ϱ of the foam, respectively. The experimental data indicate that C from the data of cellular metals and polymers is around 0.3 and the strength of the foam is proportional to the strength of the solid (n = 3/2) [18]. The density of dense solid Ti6Al4V is 4.43 g/cm^3 with yield strength 902 MPa [19, 20]. The yield strength of Ti6Al4V foams and predicted theoretical values for open porous material are given in Figure 16b. It can be seen that the measured values for porous Ti6Al4V are slightly higher than the analytical predictions by Gibson and Ashby for open structures, also much higher than the compressive strength of cancellous bone (10-50MPa). The experimental values are all comparable with the calculated data obeying the Gibson and Ashby model.

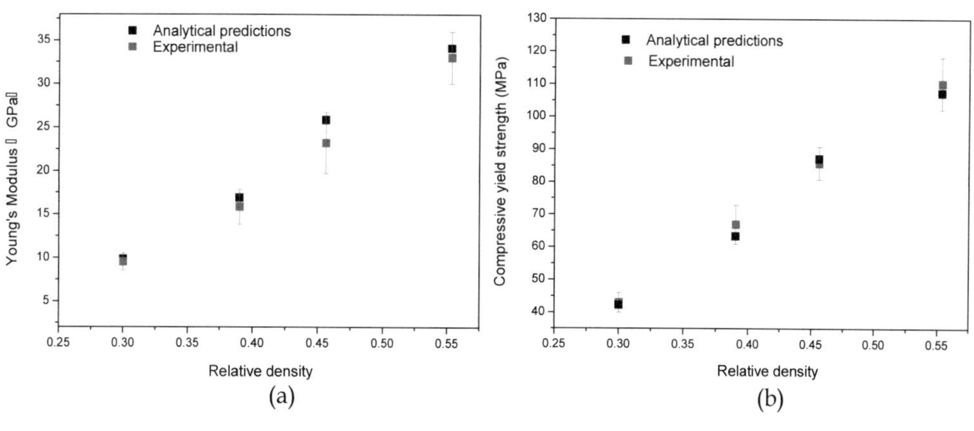

Figure 16. Young's modulus (a) and yield strength (b) of the Ti6Al4V foams and theoretically predicted values.

Figure 17 shows the SEM images of the Ti6Al4V foams (44.7%porosity) after the compressive tests. The macropores have been compressed to a crushed shape. However there were no bulk cracks been oberserved. By this way, it can adsorbe a large amount of energy. The strength of the foams depends on the thickness of the struts. The 44.7%porosity foam has the largest yield strength and Young's modulus (Figure 16) due to the the thickest struts in this group of the prepared samples.

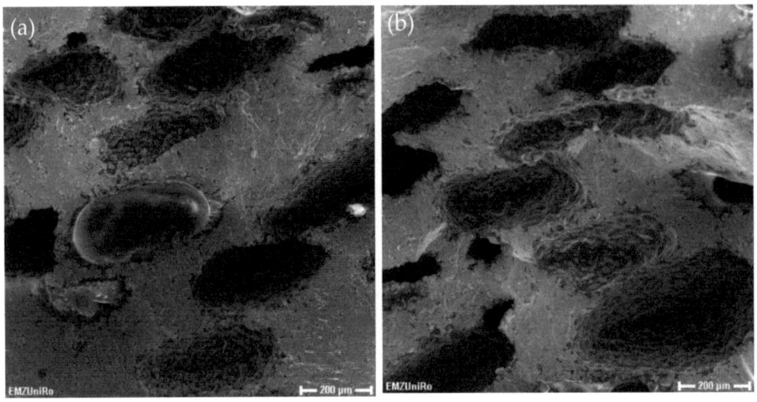

Figure 17. SEM images of the Ti6Al4V foams (44.7%porosity) after the compressive tests showing the crushed shape pores of different places (a) and (b).

4.5. Cell proliferation in the foams

Human osteoblast like cells (MG-63) were cultured on the porous Ti6Al4V samples. Figure 18 shows the SEM images of human bone-like MG-63 cells on porous Ti6Al4V foams with 44.7%, 54.4%, 60.7%, and 70.0 %. After 24 h they exhibit a well spread morphology and excellent bonding to the surface. The cells form filopodia to reach the adjacent grains of the Ti6Al4V structure. The cells grow inside the pores to a large extend implying a good integration when implanted into the bone. For a better interconnectivity, higher porosity Ti6Al4V foams (above 60%) are recommended for the biomedical applications. This cell tests display that the pore sizes and porous surface of the Ti6Al4V foams fabricated by the PECS are suitable for the osteoblasts in-growth.

4.6. Other methods for preparation of Ti6Al4V foams by PECS

Besides NaCl particles as spacer material, some other ways to fabricate Ti foams by the SPS were also tried and studied. Fig.19 shows the SEM images of the Ti6Al4V foams prepared by SPS using Cu spacer. The Ti6Al4V powders and Cu powders were mixed and sintered by PECS at 800 °C and 50 MPa for 5 min holding time. Then the Cu spacers were removed by high concentration HNO_3 acid washing. Finally, they were washed in deioned water to get a neutral PH value. The porous foams were obtained. The porous structures at various magnifications show that the pores are interconnected with pore sizes from 10 to 200 micrometers. The pore walls show that there are many TiO_2 nanostructures formed on the surfaces. Its formation is due to the strong acid oxidation. The TiO_2 nanostructures could increase the bioactivity and cell adhesion property of the Ti6Al4V foams.

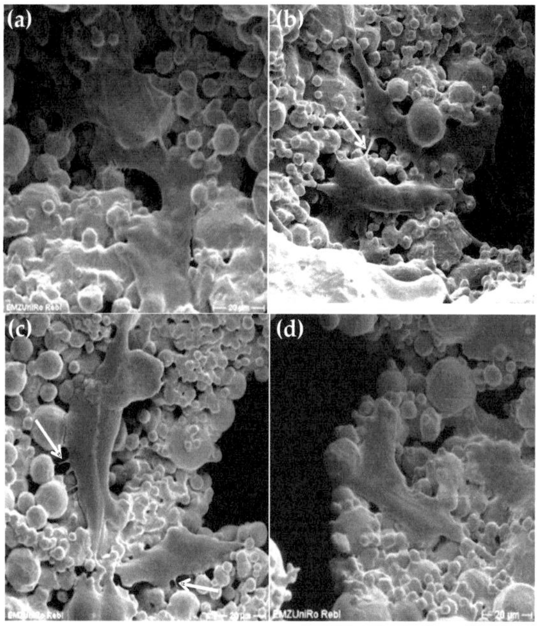

Figure 18. SEM of human bone-like MG-63 cells on porous Ti6Al4V foams (a) 44.7%, (b)54.4%, (c) 60.7%, (d) 70.0 %. The cells display a well spread morphology and moreover the cells extend various filopodia to the adjacent grains (arrow).

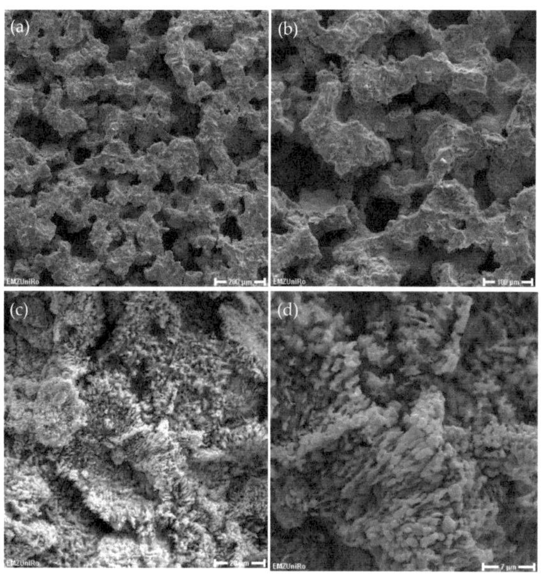

Figure 19. SEM images of the Ti6Al4V foams prepared by the PECS using Cu spacer: (a,b) porous structure at various magnifications, and (c,d) pore walls showing the TiO_2 nanostructures.

Another method to prepare the Ti foams is by the simulated body fluid soaking of the sintered TiMg solids. It is an in-situ pore formation method. The TiMg solids were prepared by ball milling of Ti and Mg powders and PECS at 600 °C and 50 MPa for 5 min. Figure 20(a) shows the XRD diffraction results of the TiMg solids with 10-40 wt% fraction of Mg. It exhibits pure Ti and Mg phases, no intermetallic phases. Afterwards, they were immersed in the simulated body fluid for several weeks. Figure 20 (b) exhibits the surface of the Ti-40 wt %Mg solids before the body fluid soaking. Figure 20 (c) shows the in-situ formed macropores after 4 weeks soaking in the simulated body fluid. The pore sizes are about 500-700 μm in dimension. The pore wall in Figure 20(d) shows the microstructures after body fluid soaking and some hints of corrosions can be found. The dense TiMg solids can load higher forces than the porous Ti foams. After they were implanted *in vivo*, the Mg composition in the TiMg solids will be degraded by body fluid immersion to form porous foams with time going. The bone tissue would grow into the foams simultaneously with the pore forming progress. The degradation rate can be tailored by the fraction of Mg and the alloying elements in the Mg phase.

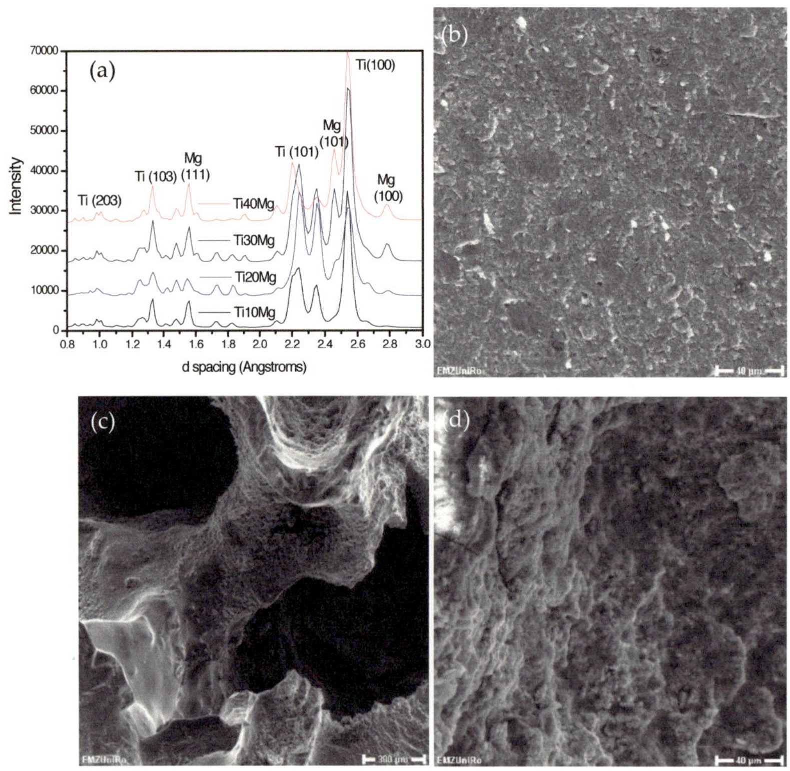

Figure 20. XRD results of the SPS prepared TiMg system (a) and SEM images of the Ti40Mg solid surface (b), in-situ formed macropores in the simulated body fluid (c) and pore wall structure (d).

4.7. Discussions

In this study, the Ti6Al4V foams were fabricated by using PECS at controlled porosity followed by dissolution of NaCl phase and post heat treatment at 1100 °C. Previous investigations showed that high-density pure Ti foams can be fabricated with NaCl as spacer by the PECS at 700 °C for 8 min under 50 MPa [11]. However, the Ti6Al4V/NaCl mixture cannot be densified at the same condition. It indicates that the Ti6Al4V/NaCl mixture is difficult to be consolidated either in the hot pressing [10] or in the PECS. The pressureless PECS method was firstly proposed by Zhang et al [21]. It can be used for the preparation of ceramic nanoscaffolds and for other inorganic porous materials [21, 22]. The Ti6Al4V foams were post heat treated by the pressureless PECS method at 1100 °C for 5 min. The densities of the foams have been increased from 5.6 to 9.0% at various porosities (Table 1). The relative density values calculated by dividing the mass of the samples by the volume can provide the information of macroporosity and microporosity in the foams. The macroporosity has been determined by the fraction of NaCl spacer after the PECS at 700 °C and dissolution in water. The post heat treatment by pressureless PECS at 1100 °C will contribute to reduce the microporosity of the foams. During the heat treatment by the pressureless PECS, there was no pressure being applied, therefore the surface diffusion, volume diffusion and grain boundary diffusion dominated the PECS process. These diffusions mainly play roles of reducing the microporosities in the foams. The SEM results in Figure 13a-d validated this scientific assumption. This post heat treatment by the pressureless PECS is similar to the technique of hot isostatic pressing (HIP) used to reduce the microporosity of metals and to increase the density of many ceramic materials, thus improving the material's mechanical properties. The post heat treatment by the pressureless PECS can be done in vacuum and argon gas up to 40 hPa. The results in this study indicate that this pressureless PECS method can be used for post heat treatment of Ti6Al4V foams, and possibly for all the metallic foams, dense metals and ceramics.

This method using PECS by dissolution of NaCl spacer and post heat treatment provides a foamed structure with a close to homogenous pore structure, high levels of porosity and high mechanical properties. The mechanical properties of the metal foam can be adjusted by choosing the size, shape and quantity of the space holder material used. Besides the NaCl as spacer material, Cu also can be used as a spacer material to get Ti6Al4V foams. It can modify the surface of the foams simultaneously with TiO_2 nanostructures. The key point of this method is how to get rid of the Cu completely. The in-situ pore forming method is very interesting. The Mg spacer degraded in the body fluid and generated some pores in the sintered TiMg solids. The key issue of this method is how to match the growth rate of the bone tissue. It can be realized by tailoring the fraction of Mg in the Ti alloy and the alloying elements in the Mg phase (e.g. Zr, Ag etc.). The highly porous nature of the alloys combining good mechanical properties with osteoconductivity makes these materials ideal for bone scaffolds. The future highlights will be net-shape processing of foams with complex forms by the PECS method. Prospectively, this kind of Ti6Al4V foams is potential to alleviate the problem of mechanical mismatch between the bone and the Ti alloy implant and may provide a new candidate as a long-term bone substitute for biomedical applications.

5. Conclusions and outlook

The pulsed electric current sintering was modified and equipped with a novel nitrogen gas rapid cooling system. The cooling power of the system was measured and numerical simulated using the Al-2024 and ALSI303 steel as model materials. The argon gas flooding and nitrogen gas quenching can effectively cooling the samples and increase their heat transfer coefficient. The Ti6Al4V alloys were prepared with various cooling rates ranged from 1.6 to 6.9 K/s by changing the thickness of the graphite dies. The gas quenching has increased the hardness, ultimate compressive strength and ductility of the Ti6Al4V alloy from 327 HV1, 1578 MPa, and 27.4% up to 353 HV1, 1832 MPa and 34%, respectively. The rapid cooling leads to much more pronounced dimples in fracture surfaces, precipitation of intermetallic phases ($Al_3Ti_{0.8}V_{0.2}$ and Al_2Ti), and more lamellar α-Ti phase in metallography. The rapid cooling-PECS combining sintering and gas quenching provides a novel method to tailor the microstructure and mechanical properties of the Ti alloys.

The Ti6Al4V foams were fabricated by the PECS with post heat treatment using a blend of Ti6Al4V and sodium chloride powders. The Ti6Al4V foams fabricated at 700 °C and 50 MPa in PECS cannot achieve high relative densities. The sintered foams were post heat treated in a pressureless mode of the PECS at 1100 °C for 5 min. This heat treatment is very effective to reduce microporosity and to fully densify the foam walls. Young's moduli of the foams were in the range of 33.0 GPa to 9.5 GPa and the yield strengths ranged from 110.2 MPa to 43.0 MPa with porosity values from 44.7% to 70.0% obeying the Gibson-Ashby models. The macropores have been crushed to flat shapes without cracks. The human osteoblast cell line MG-63 validated the cellular acceptance of the foam surfaces. This pressureless PECS method can be used for post heat treatment of Ti6Al4V foams. The Ti6Al4V foams with Cu spacer were prepared by the PECS. There pore walls were modified with TiO_2 nanostructures. The in-situ pore forming method by soaking of the TiMg solids in the simulated body fluid was also introduced.

In this century, worldwide energy shortage is a serious problem for humankind. Therefore, how to save energy is a hot issue in industry and our everyday life. The PECS is a relative new sintering technique with rapid energy transfer and less energy consumption to produce advanced materials. It is considered as a novel field assisted sintering technique for fast preparation of diverse bulk materials (metals, ceramics, polymers and their composites) with a near net shape. The sinter hardening by the PECS, in which the parts are sintered and quenched directly after the sintering step, has a perspective future in the PM industry due to energy saving and cleanness. The post heat treatment of materials in the pressureless PECS possibly should be suitable for all the porous materials and dense solids. The in-situ pore forming method represents the tendency of the future in the field of cellular solids. In the end, the future highlights will focus on the preparation of nanostructured materials and composites by the PECS with gas quenching and the computer simulation of the gas quenching process in the PECS, as well as the processing of porous nanofoams by the pressureless PECS with complex shapes.

http://dx.doi.org/10.5772/53705

Acknowledgements

This research was supported by the DFG (German Research Foundation) with grant No. GRK1505/1 (Welisa). Mr. Quan is supported by China Scholarship Council (CSC) with file No. 2010603007. The authors acknowledge Mr. C. Schweigel for his help in the mechanical tests and Ms. C. Ladewig for her help in metallography, and Ms. H. Rebl in the group of Prof. B. Nebe in Department of Cell Biology of Rostock University for the help in the cell experiments.

Author details

F. Zhang[1], Y. Quan[1], M. Reich[2], O. Kessler[2] and E. Burkel[1]

1 Chair of Physics of New Materials, University of Rostock, Rostock, Germany

2 Chair of Materials Science, University of Rostock, Rostock, Germany

References

[1] Zhang F., Burkel E. Novel titanium manganese alloys and their macroporous foams for biomedical applications prepared by field assisted sintering. In: Anthony N. (ed.) Biomedical Engineering, Trends, Researches and Technologies. Rejeka: InTech; 2011. p 203-224.

[2] Zhang F., Burkel E. Synthesis of diamond using spark plasma sintering. In: Lakshmana A. (ed.) Sintering of Ceramics - New Emerging Techniques. Rejeka: InTech; 2012. p 27-58.

[3] Dlapka M., Danniger H., Gierl C., Lindqvist B.. Sinter hardening-a special heat treatment for powder metallurgy precisions parts. Journal of Heat Treatment and Materials 2012; 67(3) 223-231.

[4] Thakuc S. N., Newkirk J. W, Fillari G.B., Murphy I.F. and Narasimhan K.S. Mechanical Properties of Sinter-Hardened Steels. International Journal of Powder Metallurgy. 2004; 40(3) 45-54.

[5] Reich M., Schöne S., Keßler O., Nowak M., Grydin O., Nürnberger F., Schaper M. Simulation of gas and spray quenching during extrusion of aluminium alloys, Key Engineering Materials. 2010; 424: 57-64.

[6] Schöne S., Reich M., Keßler O. Integration des Gasabschreckens in das Strangpressen von Aluminiumlegierungen - eine Machbarkeitsstudie, HTM Journal of Heat Treatment and Materials. 2011; 66 (3):175-181.

[7] Schöne S., Reich M., Keßler O., Gas nozzle field quenching of L-shaped aluminium extrusion profiles, Proc. 3rd International Conference on Distortion Engineering 2011, Sept 14-16, 2011, Bremen, Germany, 75-82

[8] Reich M., Kessler O. Numerical and experimental analysis of residual stresses and distortion in different quenching processes of aluminum alloy profiles, Proc. 6th Int. Quenching and Control of Distortion Conf., 9.-13.9.2012, Chicago, IL, USA, Ed. D. Scott MacKenzie, ASM International, 2012, 563-574.

[9] Dunand D.C. Processing of titanium foams. Advanced Engineering Materials 2004; 6(6): 369-376.

[10] Ye B, Dunand D. C. Titanium foams produced by solid-state replication of NaCl powders. Materials Science and Engineering A 2012; 528 691-697.

[11] Zhang F., Otterstein E., Burkel E. Spark plasma sintering, microstructures and mechanical properties of macroporous titanium foams. Advanced Engineering Materials. 2010, 12 (9): 863-872

[12] Zhang F., Reich M., Quan Y., Kessler O., Burkel E. Preparation and characterization of Ti6Al4V alloys by spark plasma sintering with integrated gas quenching. Journal of the Mechanical Behavior of Biomedical Materials.(Accepted)

[13] Kim J.H., Semiatin S.L., Lee C.S. Constitutive analysis of the high-temperature deformation of Ti–6Al–4V with a transformed microstructure. Acta Materialia 2003,51:5613-5626.

[14] Zhou Y. L., Niinomi M., Akahori Toshikazu T. Changes in mechanical properties of Ti alloys in relation to alloying additions of Ta and Hf. Materials Science and Engineering A. 2008; 483–484: 153–156

[15] Quan Y, Zhang F., Rebl H., Nebe B. Kessler O., Burkel E.Ti6Al4V foams fabricated by spark plasma sintering with post heat treatment. Materials Science and Engineering A. 2012, doi 10.1016/j.msea.2012.12.026

[16] Gibson L. J., Ashby M. F. Cellular Solids: Structure and Properties 2nd , Cambridge; 1997. p175–281.

[17] Choe H., Abkowitz S.M., Abkowitz S. and Dunand D.C. Effect of Tungsten Additions on theMechanical Properties of Ti-6Al-4V. Materials Science and Engineering A 2005; 396: 99-106.

[18] Biener J., Hoodge A. M., Hamza A. V., Hsiung L. M. and Satcher J. H., Nanoporous Au: A high yield strength material. Journal of Applied Physics 2005; 97: 024301.

[19] Bram M., Schiefer H., Bogdanski D., Koller M., Buchkremer H. P., Stover D. Implant surgery: How bone bonds to PM titanium. Metal Powder Report 2006; 61 (2) 26-28.

[20] Lee M. H., Kim K. B., Han J. H., Eckert J., Sordelet D. J. High strength porous Ti–6Al–4V foams synthesized by solid state powder processing. Journal of Physics D: Applied Physics 2008; 41: 105404.

[21] Zhang F., Lin K., Chang J., Lu J. and Ning C. Spark plasma sintering of macroporous calcium phosphate scaffolds from nanocrystalline powders. Journal of the European Ceramic Society. 2008; 28: 539-545.

[22] Ibrahim A., Zhang F., Otterstein E., Burkel E. Processing of Porous Ti and Ti5Mn Foams by Spark Plasma sintering. Materials and Design. 2011; 32: 146-153.

Section 3

Basic Sintering

Chapter 12

Mechanisms of Momentum Transport in Viscous Flow Sintering

Shiva Salem and Amin Salem

Additional information is available at the end of the chapter

http://dx.doi.org/10.5772/53259

1. Introduction

Viscous flow sintering is one of the important processes with a great variety of applications in densification of ceramics. Sintering in the presence of liquid phase makes to fabricate products with low porosity and high technical performance. The most of commercial products such as metals, glass and ceramics with low porosity are fabricated through the viscous flow sintering. The morphology of pores widely affects the characteristics of end product. Adjusting porosity can therefore, control the technical properties of products. During the sintering, the success of process depends on the control of shrinkage and porosity consequently. Many factors affect porosity and characteristics of final product. Some of the most important parameters are the chemical and mineralogical composition of raw materials and mixing ratio of them, glassy phase composition, particle size distribution, temperature and soaking time. Each of these factors can affect shrinkage of product. During the manufacturing process, modification of some parameters may be required to achieve desired porosity. Porosity and pore size distribution of products change with variation in type and content of fluxing materials. Sintering in the presence of liquid phase occurs through the melting flux materials. The molten phase diffuses into the pores by capillary forces, creating closed pores and shrinking the body. The viscosity of molten phase is able to influence shrinkage and porosity of pieces, drastically.

Liquid phase which contains different amounts of oxides greatly affects creeping flow. The mathematically study of momentum transport mechanism can be useful in improvement of microstructure as a result physical-mechanical characteristics of fine ceramics. The term of creeping flow denotes the motion of reactive phase whose Reynolds number is very low. In this chapter the importance of viscous flow in densification and vitrification of ceramics is discussed. The effect of wetting and capillary forces on the motion of liquid phase during

INTECH
open science | open minds

© 2013 The Author(s). Licensee InTech. This chapter is distributed under the terms of the Creative Commons Attribution License http://creativecommons.org/licenses/by/3.0), which permits unrestricted use, distribution, and reproduction in any medium, provided the original work is properly cited.

sintering is studied in the other section. Along above dissuasion, the continuity and momentum equations are presented for creeping flow of liquid phase in spherical systems and boundary conditions are considered to obtain velocity distribution. A mathematical model is proposed to describe isothermal variation in porosity of ceramic body according to the observed phenomena. Finally, the kinetics of viscous flow sintering and variations in activation energy and frequency factor are studied. In addition, the computation of optimum sintering time to achieve the minimum porosity at various conditions is explained by kinetic model. The effect of several factors such as amounts of fluxing agents and particle size distribution are verified.

2. Chemical reactions in pre-sintering process

Pre-sintering reactions affect the microstructural development in pieces and consequently, are important in the properties of final products. Firing of ceramics has gained wide recognition in industrial scale, reducing production costs by efficient use of energy in this process. From energy costs point of view, reducing firing temperature or time substantially influences manufacturing costs. There is obviously a maximum heating rate for any ceramic composition that allows thermal reactions to fabricate acceptable properties. Certainly, heating at a slower than maximum rate leads to consider a safe margin for firing process. The firing schedule requires knowledge of chemical reactions and microstructural changes occurring during the process. In this sections a simplified method for determining the optimum firing schedule of ceramics is demonstrated.

The firing profile of a ceramic body may be divided into three parts, representing structural changes that occur with temperature and time [1]. (i) During heating, the green body is a rather fragile and brittle. (ii) After, the formation liquid phase, the viscosity decreases as the maturation temperature rises. In this temperature range that may be called pyroplastic range, which extends into cooling period. Liquid phase loses its viscous characteristics at glass transition temperature. Deformation of body may be occur due to applied stress. (iii) The third division is the final portion of the cooling curve below the glass transition temperature, where the body is a relatively strong and brittle. Thermal shock causing fracture can occur in any of the three divisions but it is most dangerous during heating and cooling when the body is brittle. During the heating, the ceramic body is fragile because of the relatively low inter-particle bond strengths that may be compounded by residual strains from shaping process. During cooling, the ceramic body is relatively strong but is subject to brittle fracture as a result of strain such as phase inversion [2].

In order to design the optimum firing profile, the ideal curve should be determined. The corrections of firing profile can be carried out based on physical and chemical reactions. The actual rate schedule of each of three divisions can be determined. The basic data required can be obtained by simultaneous thermogravimetry and differential scanning calorimetry, TG-DSC or differential thermal analysis, DTA [3,4]. Hot stage microscopy, HSM [5-7], and dilatometeric irreversible and reversible thermal expansion also should be used to optimize heating and cooling stages [1,2].

Figure 1 is an example of TG-DSC curves for a typical porcelain stoneware body for determining physical and chemical reactions in pre-sintering process [8]. These curves show that clay minerals dominate the body in terms of non-linear thermal behavior and the particular temperatures at which the firing profile should be modified. The first and second endothermic peaks around 40–100 ºC in DSC curve is related to the loss of absorbed water. In this temperature range, the absorbed water causes a temperature depression in the body which will increase temperature gradient between the surface and center of pieces. The third endothermic peak, 550 ºC, is due to the dehydroxylation of kaolinite. The weight loss of 2.84 wt.% confirms the elevated percentage of clay minerals. The dehydroxylation of kaolinite occurs to form metakaolin. The crystal structure of kaolinite contains hydroxyl groups and the dehydroxylation of these groups to form metakaolin occurs at 550 ºC. The chemical reaction representing this process is:

$$Al_2O_3.2SiO_2.2H_2O \rightarrow Al_2O_3.2SiO_2 + 2H_2O\uparrow \quad (1)$$

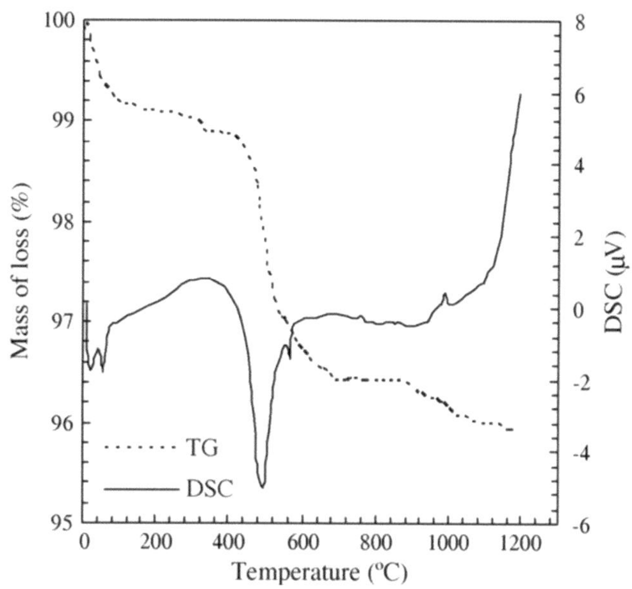

Figure 1. TG-DSC curves of a typical porcelain stoneware composition [8].

Dehydroxylation of clay minerals is observed in typical analytical investigations, such as differential scanning calorimetry and thermogravimetry [9]. These measurements also, are important in the design of fast firing profile. Dehydroxylation rate is directly proportional to the surface area of kaolin. Experiments with large pieces at very high heating rates indicated explosions caused by water vapor pressure at both the dehydration and dehydroxylation temperatures. The last endothermic peak, 571 ºC, is associated with the allotropic transformation of quartz. Because of the relatively great flexibility of the compact particles, the quartz inversion is of little consequence during the firing schedule. The only observed exothermic peak at 988 ºC is typical metakaolin decomposition, forming Al–Si spinel, amorphous silica and mullite phase as shown by following equation:

$$3\left(Al_2O_3.2SiO_2\right) \rightarrow 0.282Al_8\left(Al_{13.33}\oplus_{2.66}\right)O_{32} \tag{2}$$

where ⊕ represents a vacancy. A γ-alumina type phase recently is the predicted product. At 950-1000 ºC, the controversial exothermic peak appears where the excess silica is exolved from the metakaolin to form a precursor for primary mullite crystallization and to begin solid-state and liquid phase. The DSC-TG curves of the body confirm low weight loss, which is mainly due to the metakaolin dehydroxylation. The γ-alumina type phase, being a non-equilibrium unstable phase, certainly transform to mullite above 1000 ºC. The chemical reaction describing the mullite formation is [9]:

$$0.282Al_8\left(Al_{13.33}\oplus_{2.66}\right)O_{32} \rightarrow 3Al_2O_3.2SiO_2 + 4SiO_2 \tag{3}$$

The ceramic bodies generally contain three different mullite that are: (i) Primary mullite from decomposition of pure clay such as kaolinite. (ii) Secondary mullite from reaction of feldspar and clay, clay and quartz. (iii) Tertiary mullite may precipitate from alumina-rich liquid obtained by dissolution of alumina filler [10]. The size and shape of mullite crystals is to large extent controlled by fluidity of the local liquid matrix from which they precipitate, and in which they grow, which itself is a function of its temperature and composition. The composition of this local liquid is determined by the extent mixing of the porcelain raw materials and the role of fluxing agents is critical. The observation of polished and chemically etched surfaces points out as in the samples containing nepheline-syenite, Figure 2. The amount of secondary elongated mullite is larger than in the body prepared without nepheline-syenite. That is essentially due to the lower viscosity of the liquid phase formed in the presence of nepheline, that besides to favor the sintering, allows a better and easy growth of the elongated crystals [11].

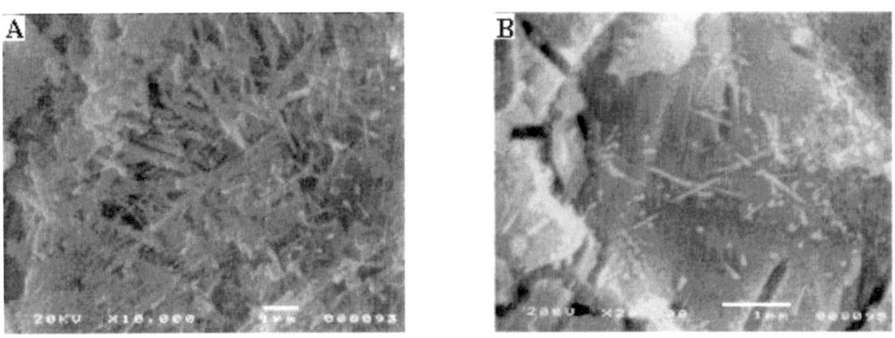

Figure 2. Comparing mullite size in porcelain stoneware body prepared (a) 10 wt.% nepheline syenite and (b) without nepheline syenite [11].

Also, the organic materials burn from 250 ºC onward. The experiments with heavy electrical porcelain result in explosions of body at 250 ºC that is a result of the ignition of volatiles from the lignites in the ball clays.

Hot stage microscopy provides an advantageous data for characterizing the firing and sintering behaviors of ceramic bodies. It allows monitoring of thermal behavior of body from room temperature to sintering steps. The application of this technique to study the sintering of different ceramic substrates was reported by many investigators [5-7]. Hot stage microscopy allows assessing the shrinkage of ceramic body. The compact sample is made from powder by the uniaxial compaction technique. Hot stage microscopy is standard, well-known devices in ceramic and glaze fields. Traditionally, the instrument has found application mainly to assess the thermal behavior, particularly about softening and melting temperatures of glazes, ceramics and other silicate materials. The cylindrical samples of the powders are obtained by pressing in a die at room temperature, usually without addition of any binder. For non-isothermal experiments the furnace of the microscope is heated to the maximum temperature usually with constant rate. The samples are placed on a small ceramic plate with the longitudinal axis coinciding with the vertical direction. By measuring the changes of length and diameter of the samples during the sintering process, either by taking photographs of the sample at pre-chosen time intervals during the sintering process, or by video recording the whole experiment, it is possible to obtain the axial and radial shrinkage. The height and diameter of the photographed or video-recorded sample images can be measured with a relative error, <1%, on suitable enlargement.

To clarify the ability of HSM technique, Figures 3 and 4 show HSM graphs for a typical porcelain stoneware body and a composition modified by 10 wt.% nepheline syenite. Listed below the images is the characteristic shape temperature. Also, the measured valves of reference temperatures for the mentioned compositions are reported in Table 1. The traditional single firing porcelain stoneware composition, reference body, is sintered about 20 ºC higher than sample containing 10 wt.% nepheline syenite. The use of fluxing agent like nepheline syenite in composition clearly shifts the sintering of ceramic body to low temperature because of its tendency to melt and decrease the viscosity of liquid phase. It is evident that the sintering speed of modified composition containing nepheline syenite is enhanced and 20 ºC is enough to obtain well sintered body. The range of dimensional stability is very high for reference composition. When 10 wt.% nepheline-syenite is added, the range of stability drastically reduces, and a negative shrinkage, expansion, is observed. The addition of nepheline causes the decrease in softening temperature.

Nepheline syenite (wt. %)	Starting shrinkage	Maximum Shrinkage	Expansion	Softening point
0.0	1120	1220	1280	1340
10.0	1100	1200	1240	1320

Table 1. The technical temperatures of porcelain stoneware compositions obtained by HSM technique.

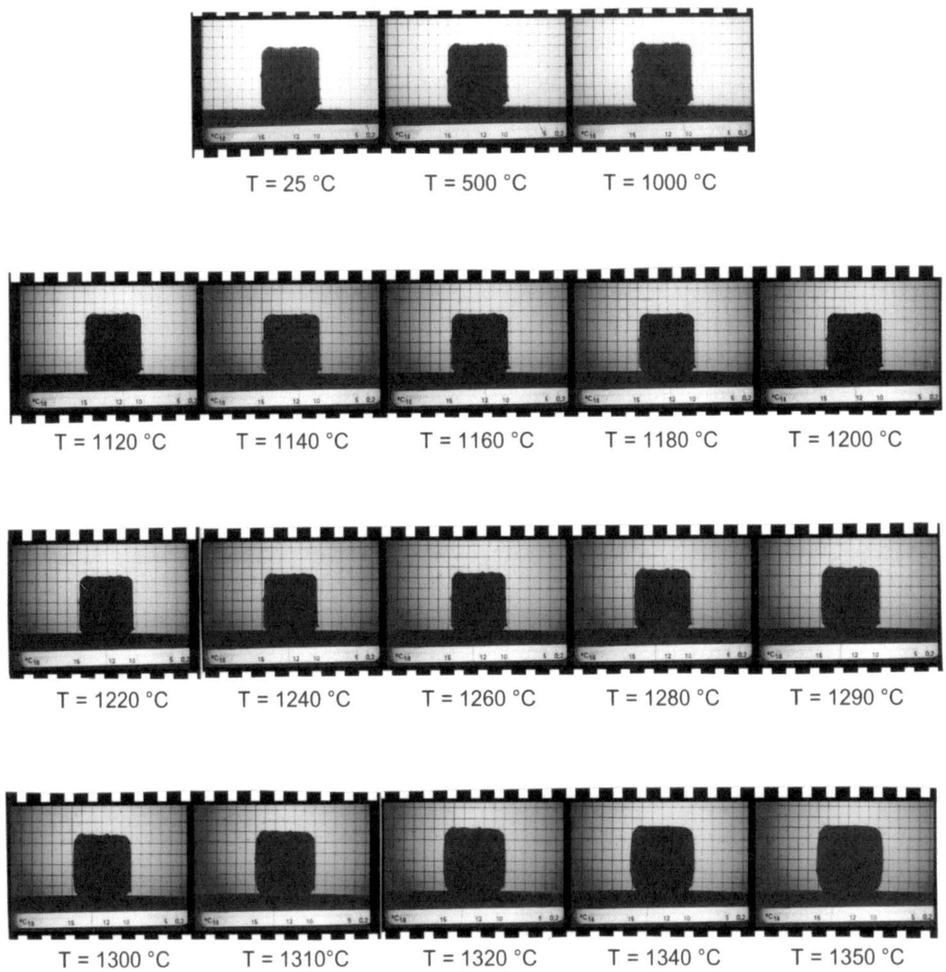

Figure 3. The HSM images of a typical porcelain stoneware composition.

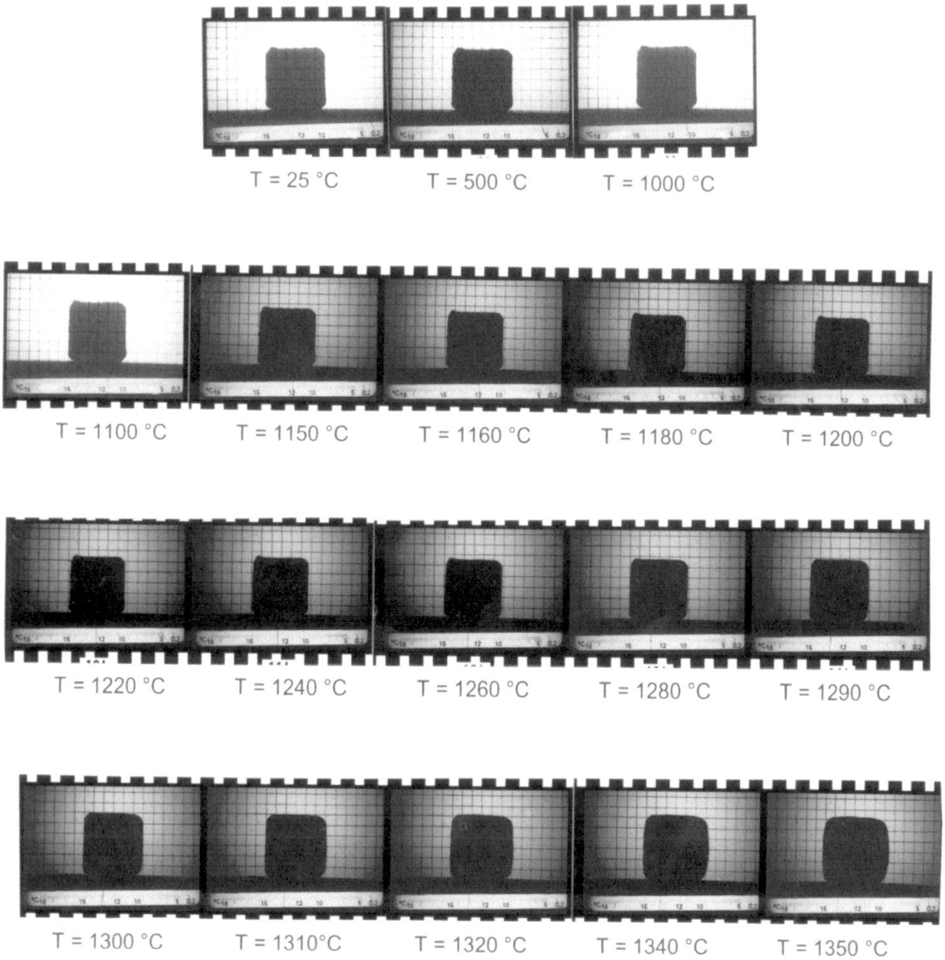

Figure 4. The HSM images of porcelain stoneware composition prepared with 10 wt.% nepheline syenite.

Irreversible thermal expansion curve really consists of behavior of all unfired materials in composition. The thermal expansions of some components are well known. For example, the α-β quartz inversion causes a sharp increase in expansion [9]. Between the 500 and 550 ºC, dehydroxylation of the clay minerals modifies the expansion as the clay is contracted. On the other hand, the fluxing agents like feldspars and nepheline syenite, which undergo no transformations at this temperature range, continue to expand almost linearly. The phenomena around 600 ºC may be quite severe for some clay based ceramics such as porcelains and porcelain stoneware bodies until the dense structure are formed. According the previously explanations, the ceramic body begins to sinter between the 850 to 900 ºC due to expulsion of excess silica released from metakaolin when γ-alumina spinel is crystallized. In this condition the body presents pyoplasticity. When the feldspars begin to melt at 1050 ºC [12,13], the contraction accelerates and pyoplasticity increases remarkably.

The dilatometric curve of a typical porcelain stoneware composition is shown in Figure 5. It is observed that expansion increases until a maximum value about 900 ºC. The sharp increase at 571 ºC is due to α-β quartz inversion. From this temperature onward, the expansion rate gradually decreases and sintering starts at 950 ºC. From 950 ºC, the shrinkage increases in an exponential form with temperature. The shrinkage of ceramic bodies varies with type and content of fluxing materials because, sintering of porcelain bodies occurs through the melting of fluxing materials. The molten phase diffuses into the pores by capillary forces, creating closed pores and shrinking the body. The viscosity of molten phase is able to influence shrinkage and porosity of porcelain stoneware bodies drastically. From dilatometric curves for compositions containing 10 wt.% nepheline syenite, Figure 6, it can be noticed that the shrinkage rate increases. When shrinkage reaches to maximum value at 1225 ºC, the open porosity of body containing nepheline syenite reaches to zero. As a result, the pores of these bodies are closed at this temperature. The use of fluxing agents like nepheline syenite favors the sintering behavior, allowing sintered materials with minimum porosity. The maximum value of shrinkage, reached by the samples containing nepheline is higher than the maximum value of composition prepared without nepheline syenite. Furthermore, it is reached in rather low temperature.

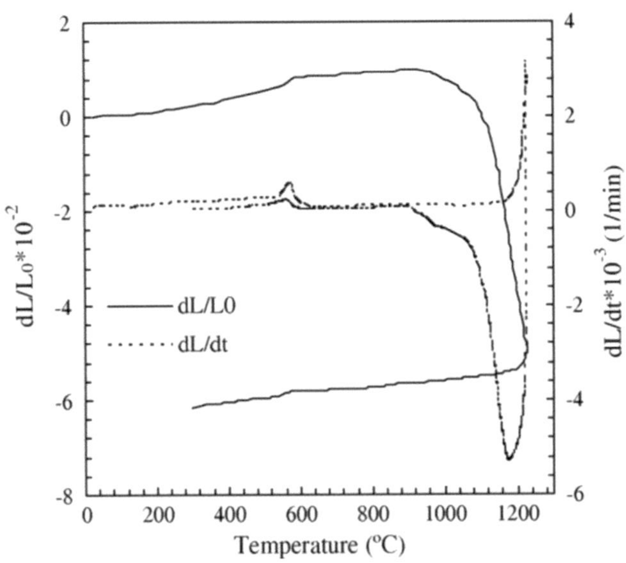

Figure 5. The dilatometeric curves of a typical porcelain stoneware composition.

http://dx.doi.org/10.5772/53259

The reversible thermal expansion occurs in fired ceramics such as electrical porcelain and porcelain stoneware due to the α-β residual quartz inversion at the glass transition temperature in which a viscoelastic material transforms to a elastic solid in cooling stage [2]. The glass content in different porcelains may be vary from 50 to 80 wt.% according the selected composition. Usually porcelain stoneware body shows the inflection at 570-600 ºC but electrical porcelains may not show this inflection clearly due to increasing glass content during firing process. It should be note that the glass phase should be annealed in the cooling to reduce the total stress when the residual quartz inverts at 570-600 ºC. The industrial experiences clearly showed that nothing can prevent the β-α quartz inversion with its consequent stress development in ceramic body. It is interesting to note that quartz particles as small as 1 μm produce crack in porcelain body. Therefore, there are two major sources for development of stress in the cooling of ceramic bodies. If there is no thermal arrest at glass transition temperature, the glass phase will be improperly annealed and strain energy will be stored. When β-α quartz inversion occurs at 570-600 ºC, the inversion stresses are added to the residual glass stresses. If stresses exceed the strength of ceramic body, fracture occurs immediately otherwise, the stresses may be large enough to cause crack growth and it may be developed hours or even months later. Fabricating an absolutely free-stress ceramic body is actually impossible but it is possible to manufacture reliable product by minimizing stress.

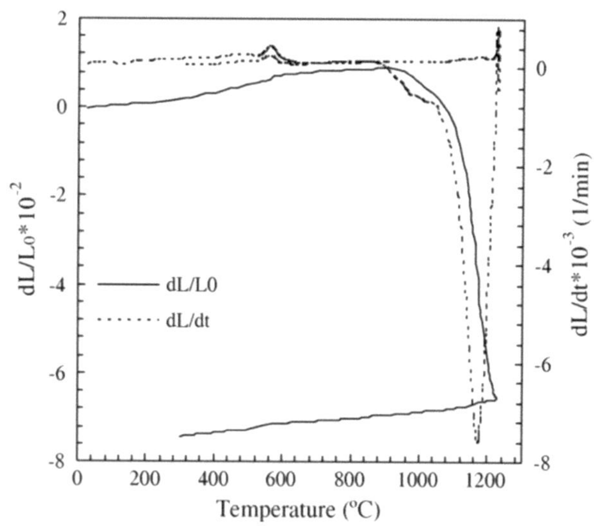

Figure 6. The dilatometeric curves of porcelain stoneware compositions prepared with 10 wt.% nepheline syenite.

In design of firing profile, the irreversible and reversible thermal expansion data should be considered in first step and then it can be modified by DSC or DTA data. This procedure establishes the shape of firing profile at three stages. The heating rate should be separately determined in each section. The irreversible thermal expansion data are used in heating step and the reversible expiation data should be used in design cooling step. Critical points on the profile are identified by DSC or DTA data. Ceramic body can be damaged by steam pressure at both adsorbed water removal and dehydroxylation of clay occurring in 100 and 550 ºC, respectively. The combustion of volatiles from lignites causes crack in very large cross section bodies. It is important that the ignition of organic materials is completed before the ceramic bodies become impervious. Because the nature of the ceramic bodies are substantially different in the three major section of firing profile, the heating and cooling rates should be determined in next step. The firing profile should be determined empirically if it is not available for a new composition. In summary, the most critical points are between 550 and 600 ºC during the heating. The pyroplastic section may be modified slightly but the maturation should be completed. The cooling profile should be adjusted separately. The ceramic body can almost be quenched from the maturing temperature to 800 ºC and then it should be annealed until 550 ºC by considering the β-α quartz inversion.

3. Wetting and liquid migration during sintering

Wetting and surface phenomena play important role in sintering in the presence liquid phase because the surface area of the compact powder reduces during the heating process. Therefore, the microstructure of product is affected remarkably by surface phenomena. The fundamental problem of sintering in the presence liquid phase is explanation of mechanisms that reduction of energy occurs which are especially important for understanding the factors affecting process. During the sintering process the particles weld together and pores between them become more nearly spherical and the porosity of compact decreases simultaneously. The driving force for both phenomena is due to excess surface free energy. An atom at a free surface is bonded to fewer neighboring atoms than an atom within the particles. Since bonding reduces the potential energy, a surface atom has extra energy, called the surface energy, which can be partially reduced by slight adjustments in the composition and bonding between the atoms in the surface. Nevertheless, surface atoms or ions are more active. Thermodynamically, the surface tension, γ, is defined as [1]:

$$\gamma = \left(\frac{\partial G}{\partial A_P} \right)_{P,T,N_i} \tag{4}$$

where, G is the Gibb's free energy of the system. During the change in particle area, A_p, the independent variables of pressure, P, temperature, T, and number of species, N_i, in the system remain constant. If the radius of pore r_1 in ceramic structure is considered as shown in Figure 7, surface tension will tend to contract the surface area and the internal volume, increasing the internal pressure by an increment ΔP. At equilibrium condition, the word of contraction $\Delta P \Delta V$ is equal to the decease in surface free energy γdA.

$$\Delta P 4\pi r_1^2 dr_1 - \gamma 8\pi r_1 dr_1 = 0 \tag{5}$$

http://dx.doi.org/10.5772/53259

Therefore,

$$\Delta P = \frac{2\gamma}{r_1} \tag{6}$$

where ΔP and ΔV are the pressure and volume differences, respectively. In sintering process ΔP are considered respect to atmospheric pressure.

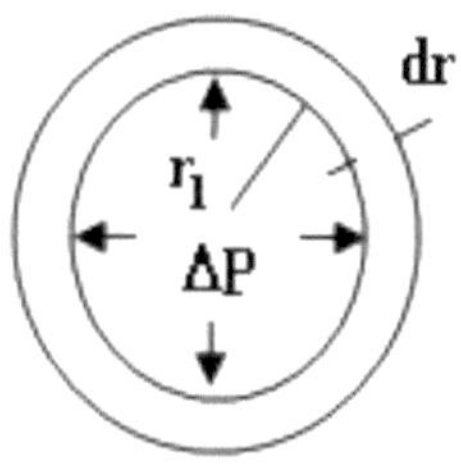

Figure 7. The pressure difference in pore produced by surface tension.

It should be note that the radius r_1 is negative when the curvature is concave. The effect of surface curvature may cause the average chemical potential of atoms in microscopic particles to be greater than that in large particles. The atoms in a microscopic region of sharp positive curvature are of a higher chemical potential than atoms in a flat surface. In this section the interfaces between a solid and liquid phase is considered to study sintering process. Consequently, the surface curvature effect is that equilibrium pressure is a function of surface energy and surface curvature. An example for processing consequence of the surface curvature effect is capillary phenomena. Spreading liquid on solid occurs when the contact angle measured through the liquid phase approaches zero. In the sintering in the presence liquid phase, wetting and spreading phenomena affect the pore morphology as a result microstructure of ceramic body. Liquid phase that wets the surface of particles will spread over the surface and concentrate in contact region, forming necks as shown in Figure 8. The pressure difference across the curved meniscus is negative and a compressive stress occurs in contact region. The pressure differences across a curved meniscus can also, cause the migration of liquid phase between the pores or migration of liquid from a saturated region to a less saturated region.

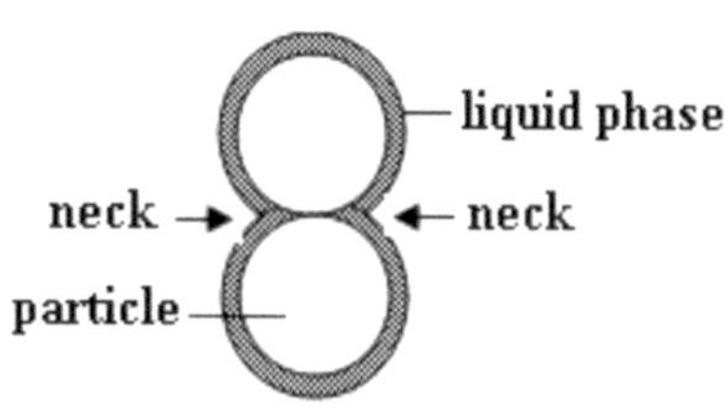

Figure 8. Liquid phase distributed on surface of spherical particles.

As illustrated in Figure 9 liquid phase will rise in capillary spaces if it wets the surface. On the other hand, it will be depressed if non-wetting of particle surface occurs. In this case, it is impossible to sinter the ceramic body. This phenomena rarely observed in sintering ceramic materials. For the fine capillary spaces, the meniscus is approximately hemispherical. At equilibrium condition, this pressure difference will offset the hydrostatic pressure which may rise above the meniscus external to the capillary space. The penetration of liquid into the porous medium will be grater for a liquid lower viscosity and higher surface tension. Finer pores produce a greater suction. An increase in temperature may reduce surface tension/viscosity ratio and improves penetration. In sintering of ceramic bodies in the presence liquid phase, the capillary force provides a mechanism for the cohesion, migration of liquid in pores and rearrangement of particles. Capillary suction produces a driving force for migration of liquid in sintering [14].

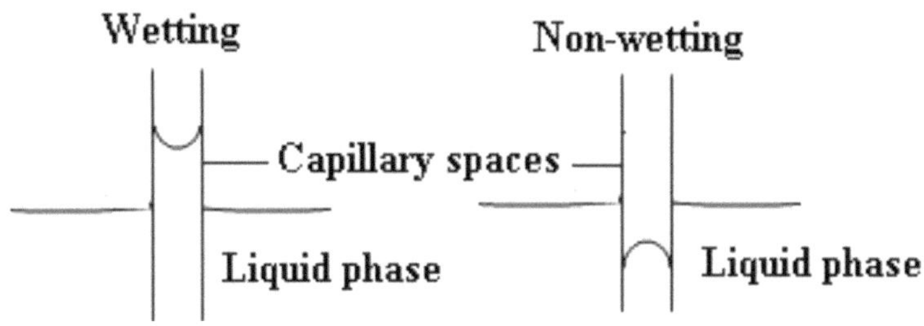

Figure 9. Wetting and non-wetting behaviors of liquid phase.

4. Vitrification and microstructural changes

The vitrification of fine ceramic products such as electrical porcelains, whitewares and porcelain stoneware bodies is complex since sintering occurs with reaction of materials and formation new crystalline and amorphous phases. The glassy phase coats quartz, new crystals such as mullite and cristobalite and remained fluxing agents such as feldspars [9]. The ceramic body shrinks when metakaolin transformed into primary or secondary mullite crystals and amorphous silica between 950 and 1000 ºC. The amorphous silica liberated during the metakaolin decomposition is highly reactive, possibly assisting eutectic melt formation at 990 ºC. Feldspars, minerals with high alkali content, are generally used as fluxing agent in the production of fine ceramics [12]. A eutectic melt of potassium feldspar with silica starts at this temperature. The eutectic temperature depends on the type of feldspar. Sodium feldspar forms eutectic melt at 1050 ºC. The lower liquid formation temperature in potassium feldspar system is beneficial for reducing the ceramic body sintering temperature. The presence of albite in potassium feldspar can reduce the liquid phase formation. Quartz in contact with the feldspar liquid dissolves slowly above 1250 ºC. The rate of mullite formation and quartz dissolving is very dependent on particle size of materials and type of impurities and secondary fluxing agents. The particular important factors that influence the rate of vitrification are viscosity, surface tension and particle size distribution.

The viscosity of the liquid phase, influenced by the kind of the fluxing agent used and the sintering cycle, is able to drastic affect the microstructure of fired products, in particular to change the amount, morphology and size of the porosity [11]. Nepheline, even if it enters in the formulation of different ceramic products, such as sanitary ware, electrical porcelain and china ware bodies, reduces the firing temperature and increases the alkali level in the glass phase [13]. Nepheline is a major component of several igneous rocks called nepheline-syenite, nepheline-monzonite and nephelinite. The basic difference among these rock types is the amount and type of feldspars present. In nepheline-syenite, feldspars are the most important phase. In the nepheline-monzonite rocks both potassium feldspars and plagioclase are present in nearly equal proportions. In nephelinites there is little feldspar present and the rock is mostly nepheline. Nepheline is a member of the feldspathoid group of silicates, minerals whose chemistry is close to that of the alkali feldspars, but they are poorer in silica. About the use of nepheline and nepheline-syenite for the production of high temperature ceramic products, such as glasses, glass-ceramics or ceramics, it is important to make some distinctions. Nepheline should not be confused with nepheline-syenite, which is actually a mixture of about 55 wt.% albite, 25 wt.% potassium-feldspar and only about 20 wt.% nepheline. The chemical and physical properties are consequently very different. The melting point of pure nepheline is very high, 1520 ºC [13], compared to that of the other feldspars: 1118 ºC for sodium-feldspar, 1150 ºC for potassium-feldspar and, 1223 ºC for nepheline-syenite [12]. Compared to pure feldspars, the advantages coming from the use of nepheline-syenite are: (i) the content of potassium and sodium is higher, K_2O + Na_2O is about 9–12 wt.% in feldspars, whereas it is larger than 14 wt.%, in nepheline syenite, and (ii) the melting temperature is generally lower than that of potassium-feldspar, which always contains other phases, such as quartz, which shift the melting point to higher temperatures. In the glass production, the use of nepheline-syenite provides the necessary additives of alumina and alkali for the glass formulation, and it is low in silica and does not contain free quartz. Furthermore, due to the lower melting point of nepheline-syenite, in comparison with potassium feldspar, the glass batches obtained have lower viscosity and easier workability. The content of Al_2O_3 is high and the content of SiO_2 is lower in nepheline-syenite with respect to feldspars (considering that in feldspar Al_2O_3/ SiO_2 is about 0.2, whereas in nepheline Al_2O_3/SiO_2 is 0.5).

When higher amounts of nepheline-syenite are added, the range of dimensional stability drastically reduces, and a negative shrinkage, expansion, is observed [11]. The fracture surface of a typical porcelain stoneware body is presented in Figure 10(a). It is characterized by the presence of round pores, essentially closed porosity, whose sizes are significantly larger than the ones present in the nepheline-syenite modified composition, Figure 10(b). Spherical pores indicate a mature microstructure, where a sort of equilibrium is reached, from the equilibrium between the pressure of gas and the viscosity of the liquid phase, spherical pore results.

The microstructures of bodies prepared with high amounts of nepheline syenite do not present substantial differences among them [11]. The presence of nepheline-syenite results in fired bodies with a larger degree of vitrification. Even 5 wt.% of nepheline-syenite produces nearly 10 wt.% more glass than the body prepared without nepheline syenite. The glass fraction linearly increases with the amount of nepheline syenite. The crystalline phases, such as quartz and mullite, which are unstable in the alkali oversaturated glass matrix, decompose more easily. The fraction of albite linearly increases with the addition of nepheline-syenite. The replacement of the potassium feldspathic sand with nepheline-syenite strongly favors the sintering behavior of material. Even only a replacement of 5 wt.% of nepheline syenite, causes a drastic decrease of the soaking time necessary to reach a minimum total porosity.

The lower viscosity of the liquid phase, that decreases with the increase of nepheline-syenite, favors the shrinkage, but also the growing of rather large closed pores, that, trapped in the glassy matrix during cooling, cause the observed expansion, for the longer soaking times. The presence of the nepheline-syenite in the body mix strongly favors the sintering behavior to obtain sintered materials with a minimum porosity. The fired modified compositions show homogeneous microstructures, characterized by smaller pores, with a narrow pore size distribution. The use of a 5 wt.% of nepheline-syenite allows reaching the best results. Higher percentages reduce, in an unacceptable way, the range of dimensional stability.

For clay based ceramics the surface tension is not changed much by composition [15,16]. However, surface tension is not a variable that normally causes difficulty during the design of composition or control of sintering process.

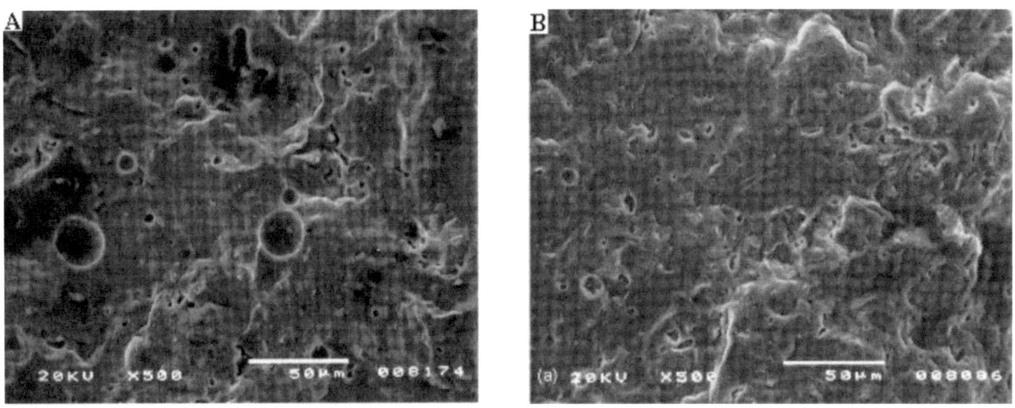

Figure 10. SEM micrograph of the fracture surface of the porcelain stoneware composition prepared (a) without nepheline syenite and (b) 5 wt.% nepheline syenite.

The particle size has a strong effect on the sintering rate and should be controlled if vitrification process is going to be controlled. In changing from 10 to 1 μm particle, the rate of sintering is increases by a factor of 10 [2]. When the porcelain stoneware composition - milled at longer time, the compact microstructure is formed with very few and narrows pores, Figure

11. The increase of the milling time, allowed obtaining a finer powder. As a consequence, more reactive particles are produced, that allows producing a very compact and homogeneous material.

5. Creeping flow

For liquid phase, the physical property that characterizes the resistance to flow is viscosity. Laminar flow is orderly type of flow that is usually observed, in contrast to turbulent flow. In order to characterize the type of fluid flow Reynolds number is used as a follows [17]:

$$N_{Re} = \frac{\varrho u l}{\eta} \tag{7}$$

where N_{Re} is Reynolds number, ϱ is fluid density, u is velocity, l is characteristic length such as diameter and η is viscosity of fluid. Experimental observations show that there are actually three flow regimes and these may be classified according to the Reynolds number for the flow. The three flow regimes are: (i) laminar flow with negligible rippling, $N_{Re}<20$, (ii) laminar flow with pronounced rippling, $20< N_{Re}<1500$, and (iii) turbulent flow with $N_{Re}>1500$. For the Reynolds number less than 20, the ripples are very long. This number is the desired method that provides the flow regime. When is less than 0.1, $N_{Re}<<0.1$, and the flow is carried out slowly, this type of flow is referred as creeping flow or Stokes flow. If the flow of incompressible fluid into the spherical pore is considered, such as viscous flow sintering, the fluid approaches the center of pore diametrically. In this case, the creeping flow means that Reynolds number is less than 0.1.

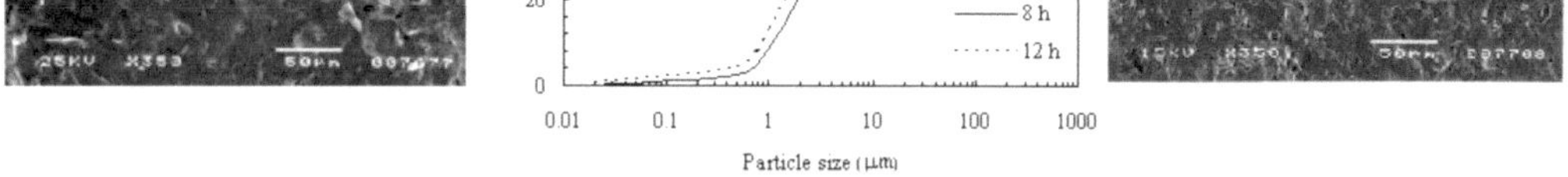

Figure 11. SEM micrograph of the fracture surface of a typical porcelain stoneware composition containing 10 wt.% nepheline syenite milled (a) 8 h (b) 12 h.

6. Continuity and motion equations

In order to calculate the flow characteristics such as average velocity and force, the velocity distribution should be determined by the shell momentum balance method. It is tedious to set up a shell balance for each flow. A general mass and momentum balances that can be applied to each flow are needed, including cases with non-rectilinear motion. These are the main points of this section. The continuity and motion equations that are related to mass and momentum balances respectively, can be used as a starting point for studying the viscous flow. The equation of continuity is developed by making a mass balance over a small element of volume through which the fluid is flowing. Then the desired partial differential equation is generated. The following equation describes the time rate of fluid density at a fixed point in space. This equation can be written more concisely by using vector notation as a follows [17,18]:

$$\frac{\partial \varrho}{\partial t} = -\left(\nabla \varrho u\right) \tag{8}$$

where t is time. This equation shows that the rate of increase of mass per unit volume is equal to net rate of mass addition per unit volume by convection. A very important special form of continuity equation is that for an incompressible fluid. For example, the following particularly simple form is considered in spherical coordinates (r, θ, φ):

$$\frac{1}{r^2}\frac{\partial}{\partial r}\left(r^2 u_r\right) + \frac{1}{r\sin\theta}\frac{\partial}{\partial \theta}\left(u_\theta \sin\theta\right) + \frac{1}{r\sin\theta}\frac{\partial}{\partial \phi}\left(u_\phi\right) = 0 \tag{9}$$

Of course, no fluid is truly incompressible, but the assumption of constant density in viscous flow sintering results in considerable simplification.

The equation of motion is developed by making a momentum balance over a small element of volume and letting the volume element become infinitesimally small. Again a partial differential equation is generated. The equation of motion can be used along with continuity equation to solve many more complicated flow problems. It is a key equation in transport phenomena. The following motion equation was obtained by momentum balance:

$$\frac{\partial}{\partial t}\varrho u = -\left[\nabla.\varrho uu\right] - \nabla.p - \left[\nabla.\tau\right] + f_g \tag{10}$$

where $\nabla.p$ is pressure gradient, $\nabla.\tau$ is divergence of shear stress and f_g is external force acting on fluid per unit volume. For incompressible fluid with Newtonian behavior the motion equation can be written as very famous Navier-Stokes equation. If the acceleration and external force terms in this equation are neglected, the following equation is obtained:

http://dx.doi.org/10.5772/53259

$$\nabla .p=\eta \nabla^2 u \tag{11}$$

which is called the creeping or Stokes flow equation.

7. Velocity distribution in spherical systems

The sintering in the presence of liquid phase is complex process, because a lot of phenomena simultaneously occur during sintering. Raw materials react and new crystalline phases are formed. The melting process of raw material produces a liquid phase whose viscosity decreases by increasing the sintering temperature so that it can enter the pores and eliminate them. Some crystals such as quartz tend to dissolve in the liquid phase. A characteristic shrinkage is observed when the metakaolin formed from the clay minerals at high temperature transforms into needle-shaped mullite crystals in the presence silica glassy phase between 950 and 1000 °C. A liquid phase forms between 950 and 1150 °C when the fluxing agents such as feldspars are presented in contact with silica, eutectic point [1]. The above considerations clearly illustrate that it is impossible to develop a theoretical kinetic model only on the basis of the chemical reactions that occur during sintering. Very few models are developed for the case in which the solid phase partially reacts with the liquid phase but theoretical models are developed by considering the pore size and the shrinkage variations. In those models [19,20] some geometrical assumptions are exaggerated respect to microstructure of compact body. Therefore, the results of these equations have an approximate validity regard to the influence of kinetic parameters such as temperature and soaking time on sintering rate. In the sintering in the presence liquid phase, the viscosity changes continuously by increasing soaking time due to the formation of new crystalline phases and melting of some crystals such as quartz. The average pore size increases progressively and kinetic models were proposed to describe the isothermal sintering based on the average pore size [21-23]. In this section a kinetic model is developed to describe the changes in porosity of compact body during the soaking time, using the Navier-Stokes equation. The sintering process is due to the liquid phase diffusion, by capillary pressure, in the interconnected pores.

The type of pores in compact materials is divided to open and closed pores. An open pore is cavity or channel that communicates with surface of body. Closed pores are located inside of the compact body and are completely isolated from the external surface. The summation of open and closed porosity gives the value of total porosity. The open, total and closed porosity of a special ceramic body versus soaking time are shown in Figure 12. With increasing soaking time both open and total porosity decrease progressively. The closed porosity, however, increases with rising soaking time. The trends of open and total porosity at lower soaking times are very similar, indicating that all pores are open. As the soaking time increases, the value of closed porosity increases, reaching a point where all pores are closed. Scanning electron microscopy observations of fracture surfaces of the bodies show a microstructure which lacks homogeneity, characterized by some closed pores. Longer soaking times favor the development of larger amounts of glassy phase, so the sample becomes more compact and interconnected pores tends to disappear, in agreement with minimum porosity. At the sintering conditions able to reach the minimum porosity, the pores are generally spherical and they are small. It is interesting to note that a too large increase in soaking time causes abnormal growth of the closed pores which in turn, influences mechanical strength. Indeed, the total porosity is affected by two factors. (i) The capillary pressure, due to the surface tension of liquid phase, tends to reduce the pore size, which in turn, reduces the open and total porosity. (ii) The pressure of gas inside the closed pores tends to expand the pores, when the minimum porosity is reached, therefore, the total and closed porosity increase simultaneously [16].

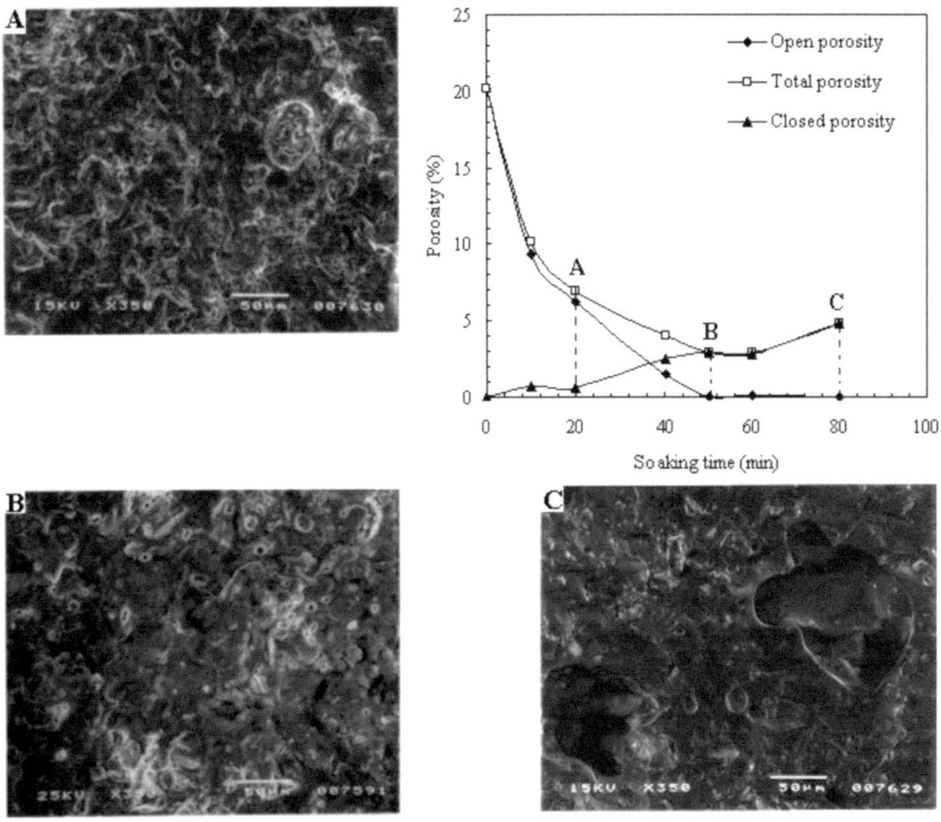

Figure 12. The variations of open, total and closed porosity as a function of soaking time for a typical porcelain stoneware composition.

In the sintering in the presence liquid phase two types of pressure act on the pores during the process. One of those is the capillary pressure, P_c, that is produced in the fine pores. The other one is the gas pressure inside the pores, P_g. The capillary pressure and gas pressure work in opposite directions. Therefore, the total pressure, ΔP, is the difference between the two pressures, $\Delta P = P_c - P_g$. The general capillary pressure for spherical pores is given by equation 6 where r_1 is the average pore radius and γ is the surface free energy. In the first and intermediate stages of the sintering process the effect of gas pressure inside the pores on the sintering rate is assumed to be negligible as compared with the capillary pressure. In the final sintering stage, where the pores are almost closed, the pressure of the gas inside the pores becomes high and influences the sintering rate. The minimum porosity is where the gas pressure becomes equal to the capillary pressure. After this point, the total and closed porosity both increase and the pieces expand with increasing soaking time. Since the sintering process is finished when minimum porosity is reached, in this section the change in total porosity is investigated until this point. In developing the model for this phenomenon we considered an average radius, r_1, for a total of pores which are surrounded by a spherical shell and an equal amount of real incompressible material with radius of r_2 [24].

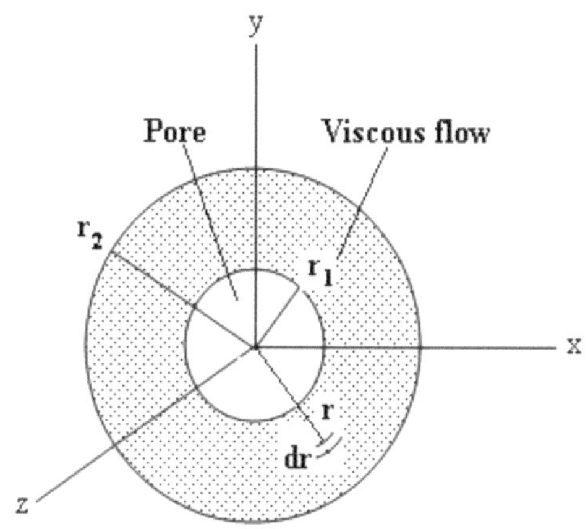

Figure 13. The diffusion of materials into spherical pore.

When an external or a negative internal pressure is applied, the flow of the material inside the shell decreases the pore volume by radial movement. Also, it is assumed that the variation in density of the liquid phase is negligible. Therefore, the porosity of the system, ε, is expressed as:

$$\varepsilon = \left(\frac{r_1}{r_2}\right)^3 \tag{12}$$

If we assume Newtonian behavior and creeping conditions for the system, the mass and momentum balance can be written in spherical coordinates according to equations 9 and 11:

Mass balance:

$$u_r r^2 = u_1 r_1^2 \tag{13}$$

Momentum balance:

$$-\frac{\partial P}{\partial r} + \eta \frac{1}{r^2} \frac{\partial}{\partial r}\left(r^2 \frac{\partial u_r}{\partial r} \right) = 0 \tag{14}$$

where u_r and u_1 are the velocity of real material at radiuses r and r_1, respectively. Substituting equation 13 into equation 14 and integrating between capillary and atmospheric pressures, we obtain the flow velocity at the boundary between the pore and real material, u_1, as a function of total porosity:

$$\int_{-\frac{2\gamma}{r_1}}^{0} dP = \int_{r_1}^{r_2} 2u_1 \eta r_1^2 \frac{dr}{r^4} \tag{15}$$

$$u_1 = -\frac{3\gamma}{\eta} \frac{1}{1-\varepsilon} \tag{16}$$

The equation 16 shows that the velocity of pore and material boundary decreases as sintering reach minimum total porosity.

8. Porosity variation in isothermal viscous flow sintering

In the sintering process the compact powder is usually held at constant temperature and the porosity is measured as a function of soaking time. The volume of real material approximately remains constant and the total number of pores does not change, if they are all equal in size but, the pores really are not equal in size. The small pores disappear more rapidly than the larger ones, so that the total number of pores decreases as soaking time increases [22]. Therefore, it is important to evaluate the number of pores per unit volume of real material, n, to find the relationship between the different parameters. In this case the relationship between pore radius and porosity is given by:

$$r_1 = \left(\frac{3}{4\pi} \right)^{\frac{1}{3}} \left(\frac{\varepsilon}{1-\varepsilon} \right)^{\frac{1}{3}} \left(\frac{1}{n} \right)^{\frac{1}{3}} \tag{17}$$

and, since $u_1 = dr_1/dt$, substituting equation 17 into equation 16 we find:

$$-\frac{d\varepsilon}{dt}=\left(\frac{9\gamma}{\eta}\right)n^{\frac{1}{3}}\left(\frac{4\pi}{3}\right)^{\frac{1}{3}}\varepsilon^{\frac{2}{3}}(1-\varepsilon)^{\frac{1}{3}} \tag{18}$$

The densification process of compact body is usually carried out at constant temperature in experimental scale and total porosity is measured as a function of soaking time. The obtained results by Orts et al. showed that the pores with small dimensions are eliminated during sintering process and the average pore size increases continuously [22]. Theoretically, it is assumed that total number of the pores per unit of real volume of material do not change if their dimensions are equal during sintering process. Also, the average volume is considered for all of the pores. If the equation 18 is integrated, it is possible to calculate the soaking time as a function of total porosity:

$$\int_{\varepsilon_0}^{\varepsilon}-\frac{d\varepsilon}{\varepsilon^{\frac{2}{3}}(1-\varepsilon)^{\frac{1}{3}}}=a\int_{0}^{t}dt \tag{19}$$

where $a=(9\gamma/\eta)n^{1/3}(4\pi/3)^{1/3}$. The left part of above equation can be easily evaluated by substituting $x^3 = \varepsilon/(1-\varepsilon)$ and the result is summarized as a follows:

$$F(x)-F(x_0)=at \tag{20}$$

$$F(x)=\frac{1}{2}\ln\frac{x^3+1}{(x+1)^3}-\sqrt{3}\tan^{-1}\left(\frac{2x-1}{\sqrt{3}}\right) \tag{21}$$

$F(x_0)$ is the value of $F(x)$ if the sintering process is started at $\varepsilon = \varepsilon_0$. The effective viscosity of the system largely depends on the amount and viscosity of liquid phase formed from the quartz and fluxing agents present in the mix. Consequently, evaluation of the effective viscosity is very complex. According to Sack and Vora [25] the effective viscosity of a system non-linearly varies with time. Also, it was shown that the average pore size of compact body changes non-linearly in isothermal conditions [22]. The surface tension is assumed to be constant [15,16]. Therefore, the generalized form of equation 20, which is applicable for all the materials that are sintered in the presence of liquid phase, can be expressed by following equation:

$$F(x)-F(x_0)=at^b \tag{22}$$

The parameters of equation 22 can be obtained by non-linear regression method. The constant of a must be related to surface tension/viscosity ratio of system and total number of the pores per unit volume of real material.

The total porosity–soaking time plots shown in Figure 14 refer the sintering of a special porcelain stoneware composition at three firing temperatures. All curves are identical in nature and are characterized by exponential behavior. This kind of plot is generally observed in sintering in the presence of liquid phase. The observed exponential behavior can be attributed to diffusing liquid phase formed at high temperatures. The total porosity decreases as soaking time rises, expect for temperature and soaking times that body undergoes over firing followed by expansion of air inside the occluded pores. The decrease in total porosity is result of diffusing liquid phase into the open pores due to capillary pressure. As the sintering progresses, the closed pores grow and open pores are transformed into the closed pores. Finally, the total porosity increases due to expansion of air into the closed pores, leading to an expansion in body dimensions. It is obvious that there is a clear increase in densification rate with increase in temperature [24].

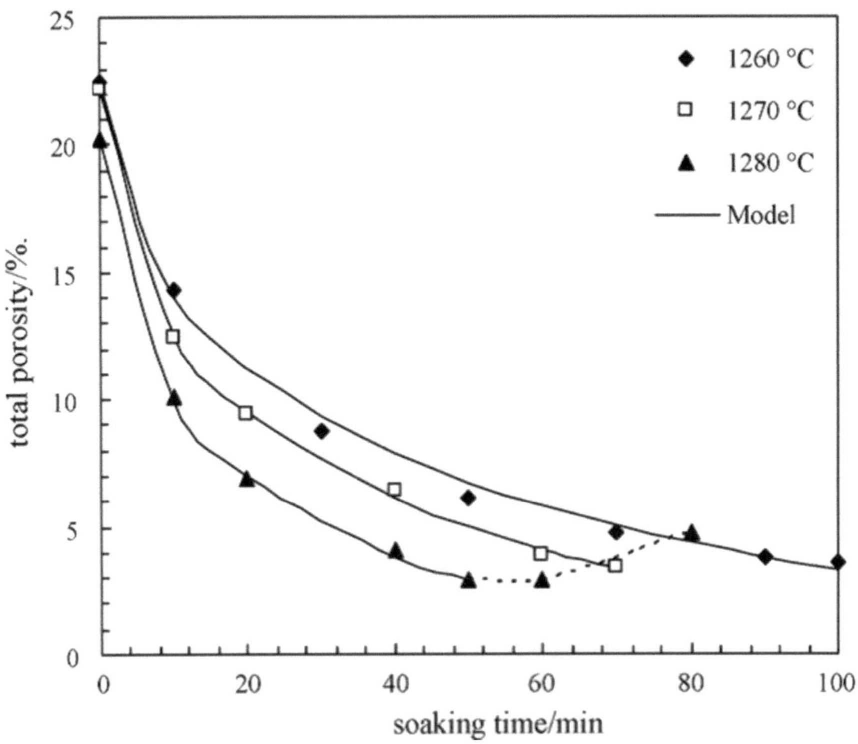

Figure 14. The variations of total porosity as a function of soaking time for a typical porcelain stoneware composition sintered at three different temperatures [26].

In order to obtain the parameters of equation 22 and thereby to calculate the total porosity for each body composition, the modified model is used to correlate total porosity-soaking time data. This equation is significantly complex model and depicts structural changes taking place in the pores of body during sintering process. Equation 22 also is used for estimating exponential behavior of total porosity-soaking time data. Since the sintering of body occurs by diffusing liquid phase and takes place by capillary forces, application of this model is justified. Model parameter, a, is related to kinetic rate constant and depends on temperature. Parameter b is related to physical changes occurring in ceramic body matrix. This parameter is very important since its value determines the total porosity characteristics. Thus for b=1, the total porosity changes follows homogenous first order kinetic. For b<1 a exponential behavior is guaranteed. The model given by equation 22 is a general one and is applicable to all temperatures and compositions. A non-linear plot of $F(x)-F(x_0)$ versus t will give the values of a and b [26]. The total porosity–soaking time data fit the model very nicely. From the parameters of the non-linear correlations, the constants of a and b are easily calculated. By use of the constant parameters the rate constant is computed in each temperature. For example, the values of a and b are tabulated in Table 2 for a typical porcelain stoneware body and compositions containing different amounts of nepheline syenite. The value of b virtually remains constant for each composition whilst the parameter a increases considerably as sintering temperature rises.

9. Kinetics of viscous flow sintering

If the values of a are plotted versus the inverse of temperature on semi-logarithm scale according to equation 23, the plots fit straight lines well, which indicate that the variation of a, with temperature may be represented by an exponential equation form as a follows:

$$a = a_0 \exp\left(-\frac{Q}{T}\right) \tag{23}$$

where a_0 and Q are the constant parameters. Table 3 details the values of these parameters. Since sintering process takes place in the presence of liquid phase, the rate of densification defined in term of rate constant, $k(\varepsilon)$, varies continuously with soaking time due to continuous structural changes in pores of ceramic body. The densification rate can be expressed as following equation:

$$-\frac{d}{dt} = k(\varepsilon) \tag{24}$$

The expression for the rate constant is given by following equation:

$$k(\varepsilon) = ba^{\frac{1}{b}} \frac{\left[F\left(\sqrt[3]{\frac{\varepsilon}{1-\varepsilon}}\right) - F\left(\sqrt[3]{\frac{\varepsilon_0}{1-\varepsilon_0}}\right)\right]^{\frac{b-1}{b}}}{\sqrt[3]{\frac{\varepsilon}{1-\varepsilon}}} \tag{25}$$

k(ε) can be calculated at each total porosity, for example in minimum total porosity. Arrhenius plots can be applied for different compositions according the following equation:

$$k(\varepsilon)=k_0\exp\left(-\frac{E_a}{RT}\right) \quad (26)$$

Nepheline syenite (wt. %)	Temperature (°C)	a	b
0.0	1260	0.0897	0.5035
	1270	0.1073	0.4940
	1280	0.1231	0.4973
5.0	1240	0.0435	0.7043
	1250	0.0578	0.7089
	1260	0.0702	0.7134
10.0	1240	0.0357	0.8182
	1250	0.0480	0.8196
	1260	0.0593	0.8127
15.5	1240	0.0235	0.9438
	1250	0.0306	0.9587
	1260	0.0442	0.9205

Table 2. The constant values of equation 22 for a typical and modified porcelain stoneware compositions [26].

Nepheline syenite (wt. %)	a_0	Q (K)
0.0	4.3×10^{9}	37689
5.0	2.2×10^{14}	54719
10.0	2.9×10^{15}	58871
15.5	2.4×10^{19}	73235

Table 3. The parameters of equation 26 for a typical and modified porcelain stoneware compositions [26].

where E_a is activation energy, k_0 is the frequency factor and R is the constant of ideal gas. The values of kinetic parameters are presented in Table 4. Also, the values of activation energy special compositions are shown in Figure 15 at three values of total porosity as a function of nepheline syenite content. This range of porosity is considered because it covers all of the minimum porosity values for special compositions. The value of activation energy approximately remains constant for each composition in the domain of porosity. The value of this parameter increases, reaching to maximum value in the presence 5.0 wt.% of nepheline syenite. This phenomenon indicates that the mechanism of pore changes in ceramic structure is different in the presence mentioned amounts of nepheline syenite.

The total porosity changes involve typically with surface tension/viscosity ratio of liquid phase that increases as nepheline content reaches to 5.0 wt.% in body composition. As the amount of nepheline syenite rises in body composition this ratio falls down because of dissolving quartz and mullite crystals in melted phase [11]. The surface tension/viscosity ratio of liquid phase grows and reaches maximum value in the presence of 5.0 wt.% nepheline syenite, Figure 16 [16]. The increase in surface tension/viscosity ratio accelerates the removal rate of porosity but the more increase in nepheline content cannot positively increase the surface tension/viscosity ratio. Therefore, the increment of nepheline syenite content in body composition has negligible role on densification rate of body. The trends of these changes in kinetic parameters show an overall positive effect on densification rate and as a result, total porosity decreases considerably in the presence 5.0 wt.% nepheline syenite. The addition of nepheline syenite in porcelain stoneware composition results an increase in activation energy and frequency factor for composition prepared with 5 wt.% nepheline syenite. Finally, the variations of kinetic parameters improve the densification rate in the presence of 5.0 wt.% nepheline syenite. The densification of a porcelain stoneware body is governed by the viscosity of the liquid phase formed at high temperature which is controlled by the Na_2O+K_2O content. The Na_2O/K_2O ratio is also a controllable factor on viscosity of liquid glassy phase. In the compositions modified by nepheline syenite, the bodies contain more Na_2O with lower amount of K_2O and SiO_2. Although, potassium oxide leads to a liquid phase with less viscosity compared to sodium oxide but, increasing Na_2O+K_2O content and low content of silica in modified compositions are compensated this negative effect. These variations bring about a lower viscosity of liquid phase in materials containing nepheline syenite. Fluxing agent like nepheline syenite should help in enhancing the densification rate of ceramic bodies if melted phase viscosity is reduced by addition fluxing agent. The rate constant is improved in the presence 5.0 wt.% nepheline syenite. The addition of nepheline syenite content more than 5.0 wt.% has negligible role in densification rate. Therefore, the viscous melted phase is not able to diffuse into the pores by capillary pressure since viscosity of liquid phase is not effectively influenced by nepheline syenite content especially at low sintering temperature.

Nepheline syenite (wt. %)	k_0 (min^{-1})	E_a (kJ/mol)
0.0	7.4×10^{19}	76.7
5.0	8.8×10^{24}	92.3
10.0	1.8×10^{19}	71.7
15.5	9.6×10^{17}	66.9

Table 4. The kinetic parameters of Arrhenius equation for a typical and modified porcelain stoneware compositions [26].

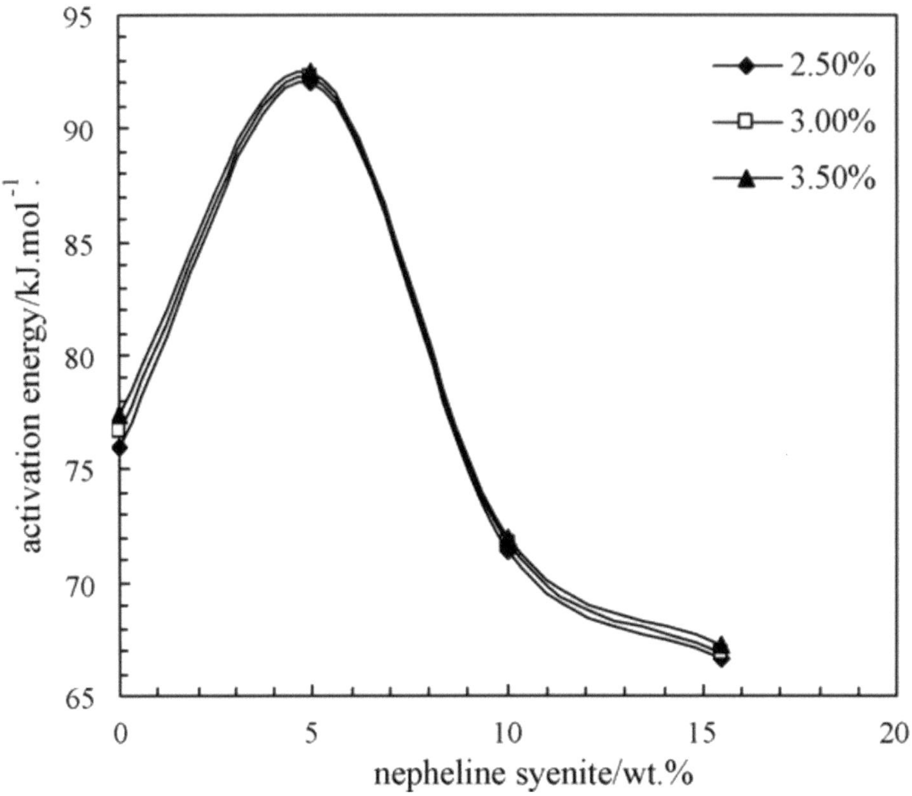

Figure 15. The variation of activation energy as a function of nepheline syenite content at three different total porosity [26].

10. Role of particle size distribution on sintering

As previously discussed in section 4, the particle size distribution of composition can effectively influence sintering process. For example - if the composition containing 10 wt.% nepheline syenite milled during 8 and 12 h is considered, Figure 11, the kinetic parameters of equations 23 and 26 are summarized in Tables 5 and 6, respectively. The values of activation energy this composition show that the increase in milling time changes the activation energy values considerably, indicating that the mechanism of sintering in these particular cases are different due to changes in surface tension/viscosity ratio. An increase in the specific surface area of particles is sufficient to increase the sintering rate. The particle size distribution of raw materials is more important factor in densification rate and plays a beneficial role on specific surface area of materials to absorb energy. This is evident from the fact that the value of frequency factor which is the function of specific surface area is more in the composition milled for 12 h. The increase in milling time ensures a higher surface area to improve the densification rate.

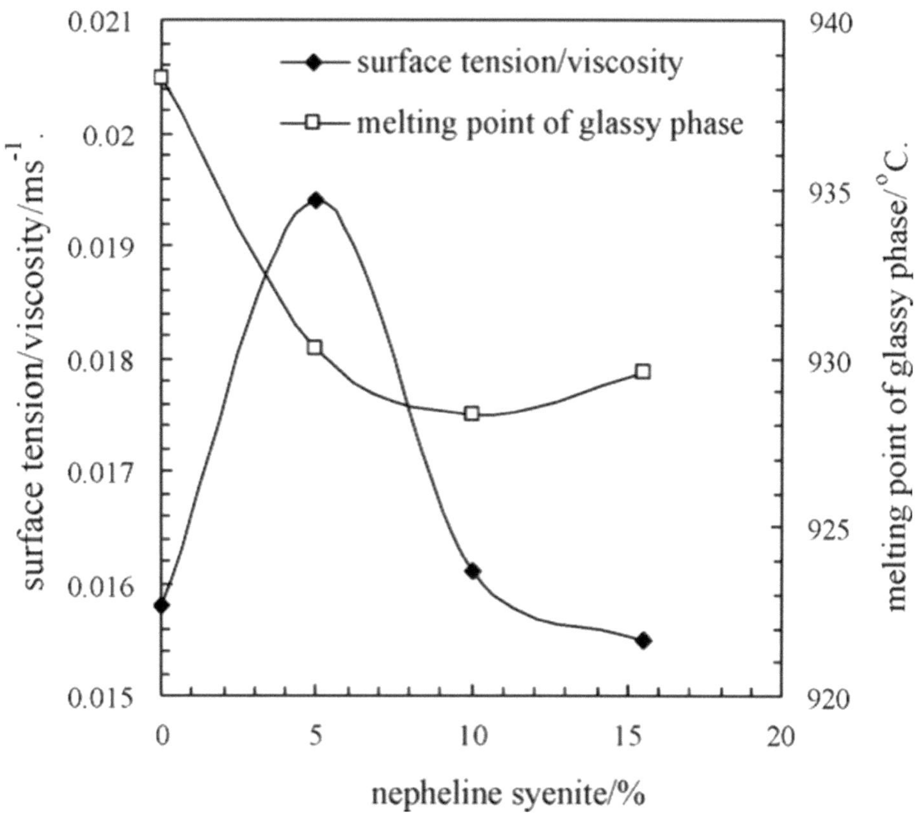

Figure 16. The variations of surface tension/viscosity ratio of liquid phase and melting point of glassy phase of porcelain stoneware body with nepheline syenite content [16].

Milling time (h)	a_0	Q (K)
8	2.9×10^{15}	58871
12	7.0×10^{17}	66623

Table 5. The parameters of equation 23 for a porcelain stoneware composition containing 10 wt.% nepheline syenite milled at different time [26].

Milling time (h)	k_0 (min^{-1})	E_a (kJ/mol)
8	1.8×10^{19}	71.7
12	1.2×10^{25}	91.9

Table 6. The kinetic parameters of Arrhenius equation for a porcelain stoneware composition containing 10 wt.% nepheline syenite milled at different time [26].

After substation equation 23 into equation 22, the following equation is developed to relate the optimum soaking time to temperature:

$$\ln t_{opt} = A + \frac{B}{T} \tag{27}$$

where t_{opt} is the optimum soaking time to achieve minimum value of total porosity and A, B are the model constants that are reported in Table 7 for mentioned compositions. The validity of this model for obtaining optimum soaking time in other temperatures for a typical porcelain stoneware composition is summarized in Table 8. The agreement between experimental data and model predictions is again excellent [26].

Nepheline syenite (wt. %)	Milling time (h)	A (h)	$B(K^{-1})$
0.0	8	-4444.6	697.7
5.0	8	-2858.8	442.6
10.0	8	-1198.0	187.7
15.5	8	-1923.1	297.9
10.0	12	-1912.6	295.6

Table 7. The parameters of equation 27 for a typical and modified porcelain stoneware compositions [26].

Soaking time (min)	1230 °C		1240 °C		1250 °C	
	$\varepsilon_{exp.}$ (%)	$\varepsilon_{cal.}$ (%)	$\varepsilon_{exp.}$ (%)	$\varepsilon_{cal.}$ (%)	$\varepsilon_{exp.}$ (%)	$\varepsilon_{cal.}$ (%)
120					3.51	3.64
180			3.22	3.51		
260	3.51	3.88				

Table 8. Experimental and calculated values of total porosity for a typical porcelain stoneware composition corresponding to minimum value of 3.51 % [26].

11. Conclusions

Because of the complex interplay between the ceramic materials and the kinetics of sintering, the viscous flow represents some of most complicated systems. Due to importance of liquid phase motion, momentum balance is required to develop fundamental equation for understanding sintering. Major challenges are the development of kinetic model and improvement of viscous flow sintering theory. The liquid phase diffusion may produce unsatisfied microstructure. The sintering behavior of ceramic materials is affected by surface tension/viscosity ratio. The surface tension is not altered by material composition in silicate systems such as porcelain and porcelain stoneware bodies. Surface wetting of liquid phase produces capillary phenomena in porous ceramic systems. In this chapter a complicated model was presented for studying viscous flow sintering. For achieve an acceptable firing profile the relation between soaking time and temperature should be determined. Since the surface tension/viscosity ratio controls total porosity during the sintering, increasing fluxing oxides enhances the removal rate of total porosity. A modified kinetic model was proposed to describe the variation of total porosity during the isothermal sintering with temperature and soaking time. Also, the value of frequency factor which is function of specific surface area, increases as the composition is prepared at high milling time. The proposed model can be used for the bodies that are sintered in the presence of melted phase to achieve minimum porosity at each temperature. It is possible to estimate the proper soaking time to obtain minimum total porosity at a given firing temperature. The data of theoretical porosity are very similar to the experimental one. In summary, the alkali oxides accelerate the sintering process by reducing viscosity of liquid phase. In effect at constant temperature the soaking time is reduced by increasing temperature.

Nomenclature

constant parameter (s)	A
particle area (m^2)	A_p
constant parameter	a
constant parameter	a_0
constant parameter (K)	B
constant parameter	b
activation energy (kJ/mol)	E_a
porosity function	$F(x)$
value of F(x) for $\varepsilon = \varepsilon_0$	$F(x_0)$
external force per unit volume (N/m^3)	f_g
Gibb's free energy (kJ/mol)	G
rate constant (s^{-1})	$k(\varepsilon)$
frequency factor (s^{-1})	k_0
characteristic length (m)	l
number of species	N_i
Reynolds number	N_{Re}
number of pores per unit volume of real material	n
pressure (Pa)	P
capillary pressure (Pa)	P_c
gas pressure (Pa)	P_g
pressure difference (Pa)	ΔP
constant parameter (K)	Q
constant of ideal gas (kJ/mol. K)	R
average radius of pore (m)	r_1
average radius of material (m)	r_2

temperature (K)	T
time (s)	t
optimum soaking time (s)	t_{opt}
velocity (m/s)	u
velocity of real material at radius of r_1 (m/s)	u_1
velocity of real material at radius of r (m/s)	u_r
volume difference (m^3)	ΔV
porosity variable	x
quartz allotrope	α
quartz allotrope	β
surface tension (N/m)	γ
total porosity	ε
initial porosity	ε_0
fluid viscosity (Pa. s)	η
fluid density (kg/m^3)	ρ
shear stress (Pa)	τ

Acknowledgements

Many peoples have made valuable information in ceramic center of Bologna, Italy. We wish to express our thanks to: Prof. G. Timellini, Dr. E. Rastelli, Dr. A. Tucci, Dr. L. Esposito, Dr. A. Albertazzi, Mr. S., Degli Esposti and Mr. D. Naldi.

Author details

Shiva Salem[1] and Amin Salem[2,3*]

*Address all correspondence to: a_salem@iust.ac.ir

1 Chemical Engineering Group, Orumiyeh Industrial University, Orumiyeh, Iran

2 Center of Excellence for Color Science and Technology, Tehran, Iran

3 Chemical Engineering Department, Sahand University of Technology, Tabriz, Iran

References

[1] Reed JS. Introduction to Principles of Ceramic Processing. 2nd ed., John Wiley and Sons, New York, USA; 1995.

[2] Funk JE. Designing the Optimum Firing Curve for Porcelains. Ceramic Bulletin 1982; 62(6) 632-635.

[3] Mackenzie RC. Differential Thermal Analysis. 2nd ed., Academic Press, London, UK; 1973.

[4] Mclaughlin RJW. Differential Thermal Analysis of Kaolinite – Illite Mixtures. British Ceramic Transactions1960; 59 178-187.

[5] Aldo R. Boccaccini AR., Trusty PA. In Situ Characterization of the Shrinkage Behavior of Ceramic Powder Compacts during Sintering by Using Heating Microscopy. Materials Characterization 1998; 41 109-121.

[6] Ahmed M., Earl DA. Characterizing Glaze Melting Behaviour via HSM. American Ceramic Society Bulletin 2002; 81(3) 47-51.

[7] Salem, Sh., Jazayeri SH., Bondioli F., Allahverdi A., Shirvani, M. Characterizing Thermal Behaviour of Ceramic Glaze Containing Nano-Sized Cobalt-Aluminate Pigment by Hot Stage Microscopy. Thermochimica Acta 2011; 521(1-2) 191-196.

[8] Salem A., Jazayeri SH., Rastelli E., Timellini G. Dilatometeric Study of Shrinkage during Sintering Process for Porcelain Stoneware Body in Presence of Nepheline syenite. Journal of Materials Processing Technology 2009; 209 1240-1246.

[9] Carty WM., Senapati U. Porcelain Raw Materials, Processing, Phase Evolution and Mechanical Behavior. Journal of American Ceramic Society 1998; 81(1) 3–20.

[10] Lee WE., Iqbal Y. Influence of Mixing on Mullite Formation in Porcelain. Journal of European Ceramic Society 2001; 21 2583-2586.

[11] Esposito L., Salem A., Tucci A., Gualtieri A., Jazayeri SH. The Use of Nepheline-Syenite in a Body Mix for Porcelain Stoneware Tiles. Ceramics International 2005; 31 233–240.

[12] Klein G. Application of Feldspar Raw Materials in the Silicate Ceramics Industry. Interceram 2001; 50 (1–2) 8–11.

[13] Rogers WZ. Feldspar and Nepheline Syenite. Ceramic Engineering Science Proceeding. 2003; 24 272–283.

[14] Kingery WD., Bowen HK., Uhlmann DR. Introduction to Ceramics. Wiley-Interscience, New York, USA; 1976.

[15] Matteucci F., Dondi M., Guarini G. Effect of Soda – Lime Glass on Sintering and Technological Properties of Porcelain Stoneware Tiles. Ceramics International 2002; 28 873-880.

[16] Salem A., Jazayeri SH., E. Rastelli E., Timellini G. Effect of Nepheline Syenite on the Colorant Behavior of Porcelain Stoneware Body. Journal of Ceramic Processing Research 2009; 10(5) 621-627.

[17] Slattery JC. Advanced Transport Phenomena. Cambridge University Press, Cambridge, UK; 1999.

[18] Bird RB., Stewart WE., Lightfoot EN. Transport Phenomena. 2nd ed., John Wiley and Sons, New York, USA; 2002.

[19] Barsoum WM. Fundamentals of Ceramics. McGraw Hill, New York, USA; 1996.

[20] Somiya S., Moriyoshi Y. Sintering Key Papers. Elsevier Applied Science, London, UK; 1990.

[21] Orts MJ., Amoros JL., Escardino A., Gozalbo A., Feliu C. Kinetic Model for Isothermal Sintering of Low Porosity Floor Tiles. Applied Clay Science 1993; 8 231-245.

[22] Orts MJ., Escardino A., Amoros JL., Negre F. Microstructural Changes during the Firing of Stoneware Floor Tiles. Applied Clay Science 1993; 8 193-205.

[23] Ducamp VC., Raj R. Shear and Densification of Glass Powder Compacts. Journal of American Ceramic Socity 1989; 72 798-804.

[24] Jazayeri SH., Salem A., Timellini G., E. Rastelli E. A Kinetic Study on the Development of Porosity in Porcelain Stoneware Tile Sintering. Boletin de la Sociedad Espanola de Cerámica y Vidrio 2007; 46(1) 1-6.

[25] Sacks MD., Vora SD. Preparation of SiO2 Glass from Model Powder Compact. Part III: Enhanced Densification by Sol Infiltration. Journal of American Ceramic Society 1988; 71, 245-249.

[26] Salem A., Jazayeri SH., E. Rastelli E., Timellini G. Kinetic Model for Isothermal Sintering of Porcelain Stoneware Body in Presence of Nepheline Syenite. Thermochimica Acta 2010; 503–504 1-7.

Chapter 13

Challenges and Opportunities for Spark Plasma Sintering: A Key Technology for a New Generation of Materials

M. Suárez, A. Fernández, J.L. Menéndez,
R. Torrecillas, H. U. Kessel, J. Hennicke,
R. Kirchner and T. Kessel

Additional information is available at the end of the chapter

http://dx.doi.org/10.5772/53706

1. Introduction

Spark plasma sintering (SPS) or pulsed electric current sintering (PECS) is a sintering technique utilizing uniaxial force and a pulsed (on-off) direct electrical current (DC) under low atmospheric pressure to perform high speed consolidation of the powder. This direct way of heating allows the application of very high heating and cooling rates, enhancing densification over grain growth promoting diffusion mechanisms (see Fig. 1), allowing maintaining the intrinsic properties of nanopowders in their fully dense products.

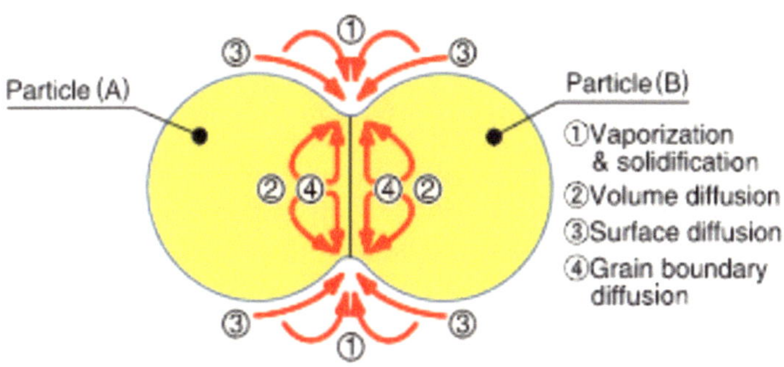

Figure 1. Material transfer path during sintering

INTECH
open science | open minds

© 2013 The Author(s). Licensee InTech. This chapter is distributed under the terms of the Creative Commons Attribution License http://creativecommons.org/licenses/by/3.0), which permits unrestricted use, distribution, and reproduction in any medium, provided the original work is properly cited.

It is regarded as a rapid sintering method in which the heating power is not only distributed over the volume of the powder compact homogeneously in a macroscopic scale, but moreover the heating power is dissipated exactly at the locations in the microscopic scale, where energy is required for the sintering process, namely at the contact points of the powder particles (see Fig. 2). This fact results in a favourable sintering behaviour with less grain growth and suppressed powder decomposition. Depending on the type of the powder, additional advantageous effects at the contact points are assumed by a couple of authors.

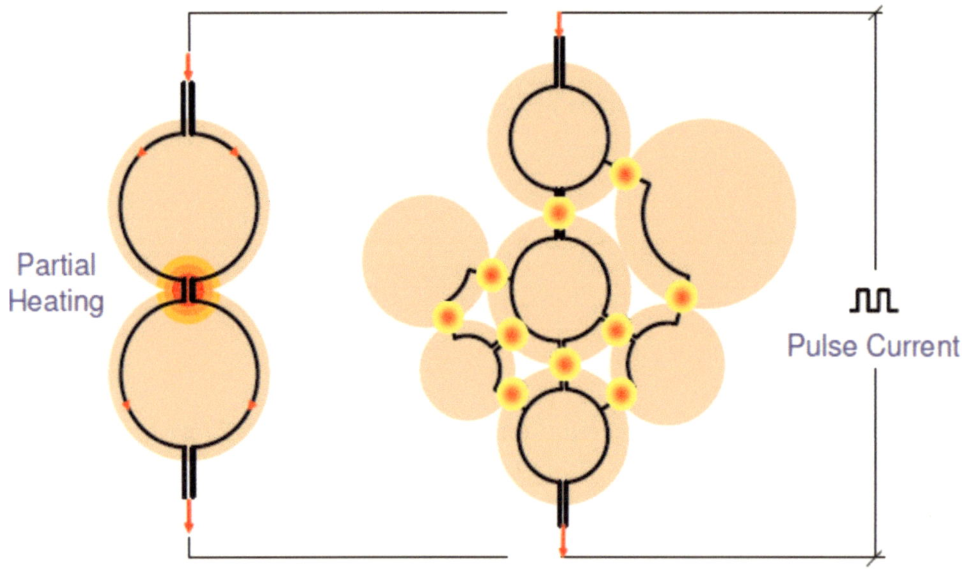

Figure 2. Energy dissipation in the microscopic scale

SPS systems offer many advantages over conventional systems using hot press (HP) sintering, hot isostatic pressing (HIP) or atmospheric furnaces, including ease of operation and accurate control of sintering energy as well as high sintering speed, high reproducibility, safety and reliability. While similar in some aspects to HP, the SPS process is characterized by the application of the electric current through a power supply, leading to very rapid and efficient heating (see Fig. 3). The heating rate during the SPS process depends on the geometry of the container/sample ensemble, its thermal and electrical properties, and on the electric power supplier. Heating rates as high as 1000 °C/min can be achieved. As a consequence, the processing time typically takes some minutes depending on the material, dimensions of the piece, configuration, and equipment capacity.

http://dx.doi.org/10.5772/53706

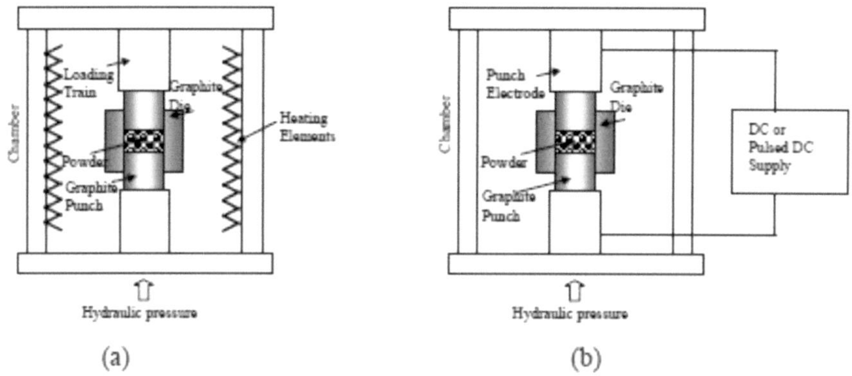

Figure 3. Schematic representation of (a) HP and (b) SPS

On the contrary, in conventional HP techniques, the powder container is typically heated by radiation from the enclosing furnace through external heating elements and convection of inert gases if applicable. Therefore, the sample is heated as a consequence of the heat transfer occurring by conduction from the external surface of the container to the powders. The resulting heating rate is then typically slow and the process can last hours. In addition, a lot of heat is wasted as the whole volume of space is heated and the compact indirectly receives heat from the hot environment. On the other hand, SPS processes are characterized by the efficient use of the heat input, particularly when electrically insulating powders is used and the pulsed electric current is applied.

It should be however mentioned that in SPS processes the problem of adequate electrical conductivity of the powders and the achievement of homogenous temperature distribution is particularly acute. In this way, the electric current delivered during SPS processes can in general assume different intensity and waveform which depend upon the power supply characteristics. In order to permit a homogeneous sintering behaviour, the temperature gradients inside the specimen should be minimized. Important parameters that are drastically determining the temperature distribution inside the sample are the sample material's electrical conductivity, the die wall thickness and the presence of graphite papers used to prevent direct contact between graphite parts and the specimen and used to guarantee electrical contacts between all parts.

The application of external electric current to assist sintering was initiated by Taylor in 1933, who incorporated the idea of resistance sintering during the hot pressing of cemented carbides [Taylor, 1933]. Later, Cramer patented a resistance sintering method to consolidate copper, brass and bronze in 1944 in a spot welding machine [Cremer, 1944]. The concept of compacting metallic materials to a relatively high density (>90% of theoretical) by an electric discharge process was originally proposed by Inoue in the 1960s [Inoue, 1965]. Inoue argued that a pulsed current was effective for densification at the initial sintering stages for low melting point metals (e.g., bismuth, cadmium, lead, tin) and at the later sintering stage for high melting metals (e.g., chromium, molybdenum, tungsten). In the United States, Lenel also used a spot welding machine for the sintering of metals [Lenel, 1955]. In addition to continuous pulses, some researchers also investigated a single discharge method, i.e., the powders were densified by a single discharge generated from a capacitor bank. In the late 1970s, Clyens et al. [Clyens et al., 1976], Raichenko et al. [Raichenko et al., 1973] and Geguzin et al. [Geguzin et al., 1975] studied the compaction of metal powders using electric discharge compaction (EDC) or electric discharge sintering (EDS). In all the methods cited, electrically conductive powders are heated by Joule heating generated by an electric current.

In 1990 Sumitomo Heavy Industries Ltd. (Japan), developed the first commercially operated plasma activated sintering (PAS) and spark plasma sintering (SPS) machines with punches and dies made from electrically conductive graphite [Yanagisawa et al., 1994]. One of the salient features of these machines was that, in addition to electrically conductive powders, high density was also achieved in insulating materials. In PAS process, a pulsed direct current is normally applied at room temperature for a short period of time followed by a constant DC applied during the remainder of the sintering process (see Fig. 4a). This procedure is often referred to in the literature as a "single pulse cycle process". In the SPS process, a pulsed DC is applied repeatedly from the beginning to the end of the sintering cycle (see Fig. 4b). In this case the procedure is referred to as a "multiple pulse cycle process".

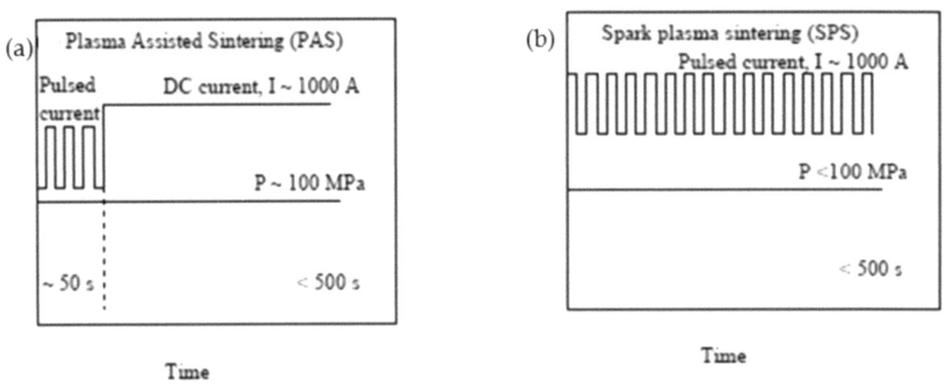

Figure 4. Sintering techniques

2. The basic SPS configuration and process

The basic configuration of a typical SPS system is shown in Figure 5. The system consists of a SPS sintering machine with vertical single-axis pressurization and built-in water-cooled special energizing mechanism, a water-cooled vacuum chamber, atmosphere controls, vacuum exhaust unit, special sintering DC pulse generator and a SPS controller. The powder materials are stacked between the die and punch on the sintering stage in the chamber and held between the electrodes. Under pressure and pulse energized, the temperature quickly rises to 1000~2500 °C above the ambient temperature, resulting in the production of a high quality sintered compact in only a few minutes.

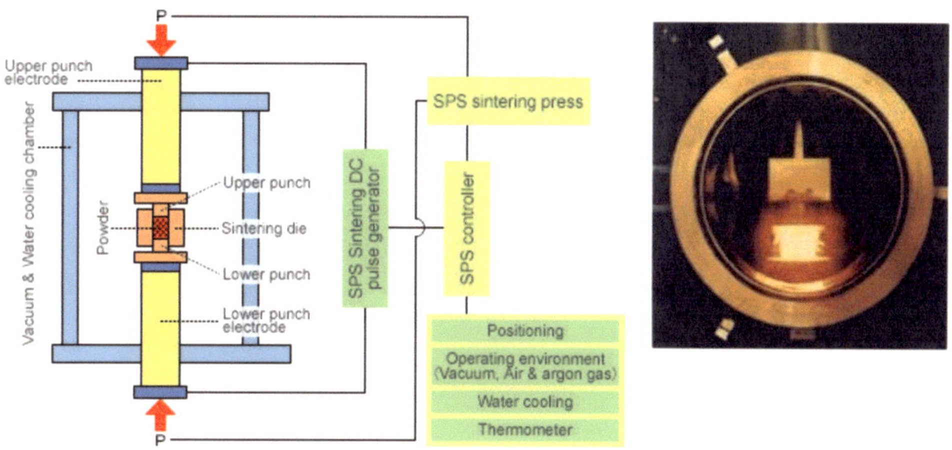

Figure 5. SPS system configuration and vaccum chamber

3. Principles and mechanism of the SPS process

The SPS process is based on the electrical spark discharge phenomenon: a high energy, low voltage spark pulse current momentarily generates spark plasma at high localized temperatures, from several to ten thousand ℃ between the particles resulting in optimum thermal and electrolytic diffusion. SPS sintering temperatures range from low to over 2000 ℃ which are 200 to 500 ℃ lower than with conventional sintering. Vaporization, melting and sintering are completed in short periods of approximately 5 to 20 minutes, including temperature rise and holding times. Several explanations have been proposed for the effect of SPS:

3.1. Plasma generation

It was originally claimed by Inoue and the SPS process inventors that the pulses generated sparks and even plasma discharges between the particle contacts, which were the reason that the processes were named, spark plasma sintering and plasma activated sintering [Inoue, 1965, Yanagisawa et al., 1994]. They claimed that ionization at the particle contact due to spark discharges developed "impulsive pressures" that facilitated diffusion of the atoms at contacts. Groza [Groza et al, 1999] suggested that a pulsed current had a cleaning effect on the particle surfaces based on the observation of a grain boundary without oxidation formed between particles. Whether plasma is generated or not has not yet been confirmed directly by experiments. Therefore, there is no conclusive evidence for the effect of a plasma generation in SPS. The occurrence of a plasma discharge is still debated, but it seems to be widely accepted that occasional electric discharges may take place on a microscopic level [Groza et al., 2000].

3.2. Electroplastic effect

Metal powders have been observed to exhibit lower yield strength under an electric field. Raichenko et al. [Raichenko et al., 1973] and Conrad [Conrad, 2002] independently studied electroplastic phenomena.

3.3. Joule heating

Joule heating due to the passage of electric current through particles assists in the welding of the particles under mechanical pressure. The intense joule heating effect at the particle conducting surface can often result in reaching the boiling point and therefore leads to localized vaporization or cleaning of powder surfaces [Tiwari et al., 2009]. Such phenomenon ensures favourable path for current flow.

3.4. Pulsed current

The ON-OFF DC pulse energizing method generates: (1) spark plasma, (2) spark impact pressure, (3) Joule heating, and (4) an electrical field diffusion effect (see Table 1).

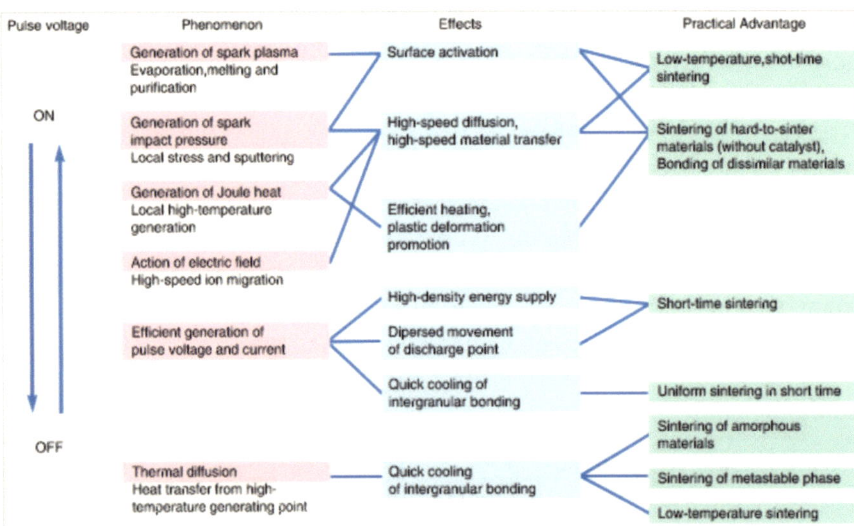

Table 1. Effects of ON-OFF DC

In the SPS process, the powder particle surfaces are more easily purified and activated than in conventional electrical sintering processes and material transfers at both the micro and macro levels are promoted, so a high-quality sintered compact is obtained at a lower temperature and in a shorter time than with conventional processes. Figure 6 illustrates how pulse current flows through powder particles inside the SPS sintering die.

http://dx.doi.org/10.5772/53706

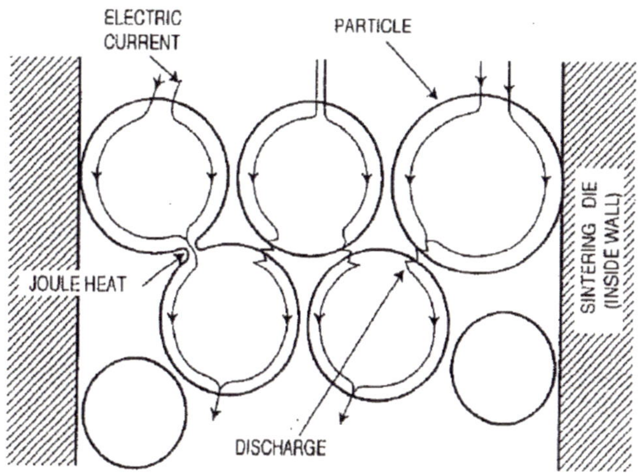

Figure 6. Pulsed current flow through powder particles

The SPS process is an electrical sintering technique which applies an ON-OFF DC pulse voltage and current from a special pulse generator to a powder of particles (see fig. 7), and in addition to the factors promoting sintering described above, also effectively discharges between particles of powder occurring at the initial stage of the pulse energizing for sintering.

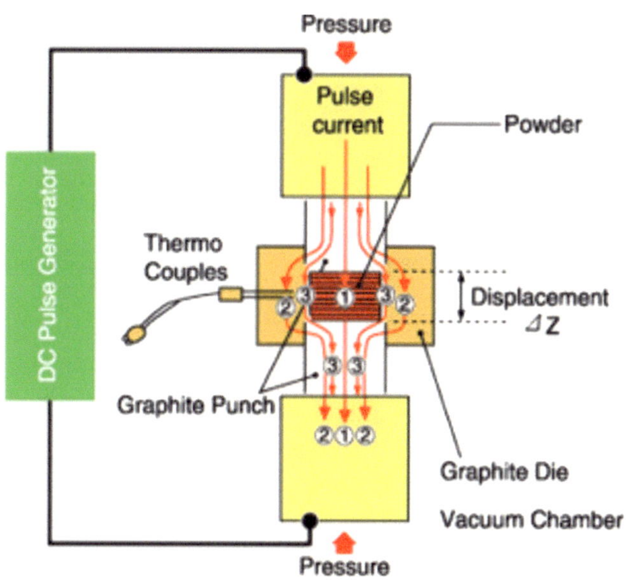

Figure 7. ON-OFF pulsed current path through the spark plasma sintering machine

High temperature sputtering phenomenon generated by spark plasma and spark impact pressure eliminates adsorptive gas and impurities existing on the surface of the powder particles. The action of the electrical field causes high-speed diffusion due to the high-speed migration of ions.

3.5. Mechanical pressure

When a spark discharge appears in a gap or at the contact point between the particles of a material, a local high temperature-state (discharge column) of several to ten thousands of degrees centigrade is generated momentarily. This causes evaporation and melting on the surface of powder particles in the SPS process, and "necks" are formed around the area of contact between particles. Figure 8 shows the basic mechanism of neck formation by spark plasma.

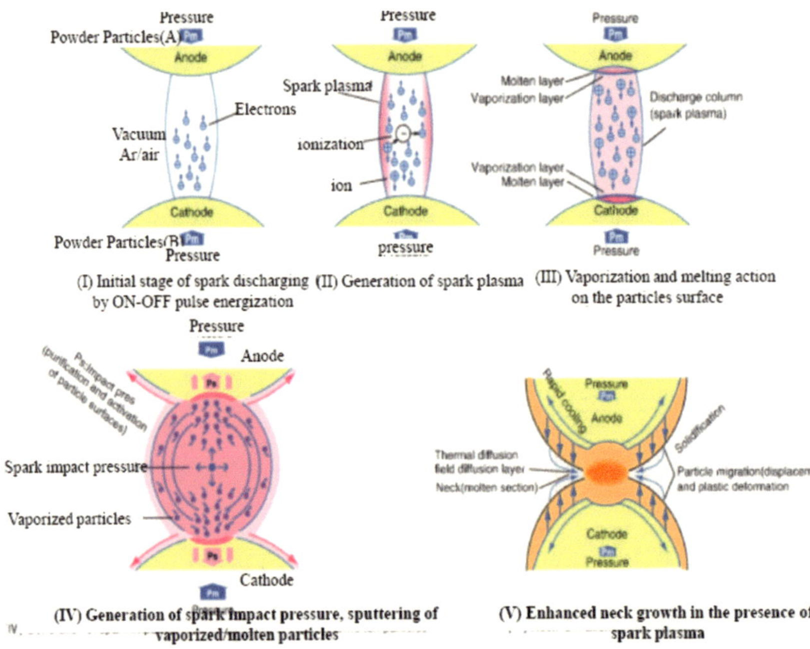

Figure 8. Basic mechanism of neck formation by spark plasma

Figure 9a shows the behavior in the initial stage of neck formation due to sparks in the plasma. The heat is transferred immediately from the center of the spark discharge column to the sphere surface and diffused so that intergranular bonding portion is quickly cooled. As seen in Figure 9b which show several necks, the pulse energizing method causes spark discharges one after another between particles. Even with a single particle, the number of positions where necks are formed between adjacent particles increases as the discharges are repeated. Figure 9c shows the condition of an SPS sintered grain boundary which is plastic deformed after the sintering has progressed further.

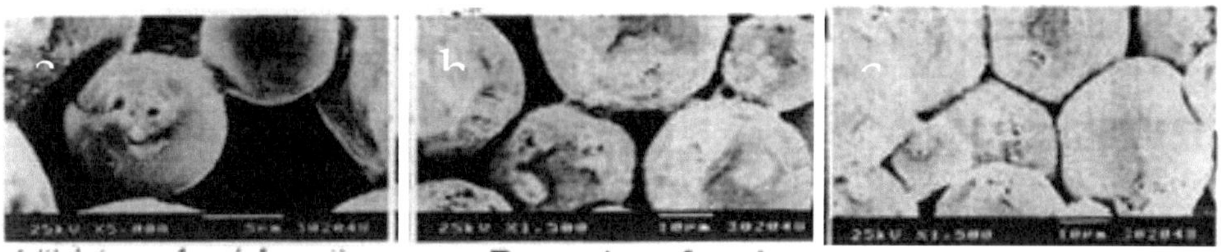

Initial state Expansion Start of neck formation

Figure 9. SPS sintering steps

4. Materials

The increasing importance of the electric current assisted sintering method as a tool for consolidation of powders is demonstrated by the large number of papers published in the recent years. It should be noted that references written in English language, at least at the abstract level, since 1922 to 2007 show an exponential growth since the late 90's [Orru et al., 2009]. While there are other reviews focused on a large list of materials sintered by SPS, we concentrate here on some relevant applications, mainly hard materials.

4.1. Cutting tools – Carbon nitrides (Materials Science and Engineering A 543 (2012) 173–179); hard materials

4.1.1. Titanium nitride

One of the most direct applications of SPS is the sintering of high melting point materials such as titanium nitride. This ceramic has received little attention in its bulk form due to its poor sinterability and inherent brittleness. Even more, all the works reported in the literature used the hot press technique to achieve densification. In particular, it has been shown [Yamada et al., 1980] that by employing extremely high pressures on the order of 1 to 5 GPa, it is possible to sinter pure titanium nitride to high density. In that work, they claimed relative densities of 98% for course powders and 95% for fine powders, respectively with a pressure of 5.0 GPa and a temperature of 1500 °C. More recent works showed [Graziani et al., 1995] that after 30 minutes at 1850 °C with a pressure of 30MPa, it was possible to obtain relative densities around 95%.

4.1.2. Transition metal carbonitrides

Similarly, transition metal carbonitrides are becoming increasingly important materials due to their excellent properties in the fields of superhardness, tribology, superconductivity and, electrical and thermal conductivities. Titanium carbonitride is the most widely employed material because it is the main component in commercial cermets. These cermets are prepared from a Ti(C, N) powder, or a combination of TiC and TiN powders, with Ni as binder and are manufactured by sintering the compacted powder mixture at a temperature at which a liquid phase is formed. Again, due to the high temperatures required to sinter this material, Ti(C, N)-based cermets, hot pressing, hot isostatic pressing, or a combination of HIP'ing and sintering, under vacuum, nitrogen or argon atmosphere are currently used. One of the most common ways to prepare titanium carbonitride is to hot press blended mixtures of TiC and TiN powders in vacuum or argon atmosphere at 1700–2400 °C [Monteverde et al., 2002]. For this reason, SPS has emerged as a useful technology and has already been used to densify titanium carbonitride powders at a relatively low temperature [Borrell et al., 2012].

Titanium–silicon–nitrogen (Ti–Si–N) composites were SPSed from a mixture of Si_3N_4, titanium nitride (TiN), and titanium thus developing wear-resistant materials with good environmental and biological compatibilities in water and sea water [Hibi et al., 2002]. SPS compaction experiments of nanostructured titanium carbonitride powders, synthesized through rapid condensation from the gas-phase (high-frequency plasma), have been performed at 1600 and 1800 °C (sintering time = 1 min) [Angerer et al., 2005]. The sintering results were compared with data obtained by various conventional sintering techniques such as pressureless sintering, gas pressure sintering, and hot pressing. The experiments showed that the SPS method is capable of obtaining high densities (about 94% of theoretical density) combined with small grain size quotient d/d_0 of 5.4–6.5.

4.1.3. *Al_2O_3-SiC*

Al_2O_3–SiC composites are usually sintered by hot pressing of powder mixtures [Sun et al., 2005]. Although the simultaneous heating and uniaxial pressure applied facilitates the densification of these composites, high temperatures up to 1800 °C are usually required. Due to the high sintering temperatures needed, alumina grain growth and the subsequent final microstructure control is limited. Two temperatures were used for spark plasma sintering tests (1400 °C and 1550 °C) [Borrell et al., 2012]. In both cases, nearly fully dense samples (>99.0% relative density) were obtained for all Al_2O_3–17 vol.% SiC composites prepared through the different raw materials combinations tested. It was then shown that when similar composite powders (Al_2O_3–SiC) are sintered by hot press, at least 1650 °C and 1 h soaking time are required for total densification, even if the total SiC content is lower (5 vol.%) [Wang et al., 2000].

4.1.4. *Transparent ceramics*

4.1.4.1. *Alumina*

As in the previous case, grain growth is an important issue determining the mechanical performance of transparent ceramics. Even more, in the case of non cubic ceramics such as alumina or zirconia, the birefringence is a critical factor leading to a dispersion that is directly related to the grain size of the ceramics. For this reason, fine-grained transparent Al_2O_3 ceramics have recently attracted much attention due to their superior mechanical and optical properties [Apetz et al., 2003]. Conventional sintering requires very high temperatures and long holding time, which leads to a deficient mechanical performance [Wei et al., 2001]. Hot isostatic pressing reduces the temperature required, leading to high strength and translucent alumina [Mizuta et al., 1992]. It has been recently demonstrated [Jin et al., 2010] that it is possible to sinter to transparency common-grade commercial powders by spark plasma sintering. The powders, previously treated in HF, were sintered between 1300 and 1400 °C for 3 minutes under an applied pressure of 80 MPa and a heating rate of 100 °C min^{-1}. Another alternative reported in the literature is the use of the so-called self-doping process [Suarez et al., 2009] in which alumina powders are doped with aluminium ethoxide which will transform to alumina after calcination. These powders were sintered at 1200 °C, with maximum shrinkage velocity at 1000 °C, under an applied pressure of 80 MPa with a holding time of 20 min. In another doping process, the precipitation of CeO_2 nanoparticles (< 5 nm) on the surface of the starting alumina nanopowder using cerium(III) acetate as precursor has been studied [Alvarez et al., 2010]. It was then shown that the ceria nanoparticles strongly enhance the transparency of the spark plasma sintered compacts due to both the ceria nanoparticles acting as powder lubricant, increasing by around 15% the initial density of the powder in the SPS die and the CeO_2 nanoparticles, locating at grain boundaries, hindering alumina grain growth by pinning during SPS sintering at 1430 °C, under an applied pressure of 80 MPa for 2 min. This effect was found to be effective only under SPS vacuum conditions.

4.1.5. Yttrium aluminium garnet

Yttrium aluminium garnet (YAG) is a material showing a large transparency range from 250 nm (UV range) up to 5 μm. It is often produced in the single crystal form, but this limits the size and shape of the components produced. It was shown [Suarez et al., 2009b] that lyophilized YAG powders could be sintered to transparency (transmittance values of 82% in the infrared region and 56% at 680 nm) by SPS at 1500 °C for 3 minutes, keeping a constant pressure during the sintering cycle of 50 MPa and a heating rate of 100 °C min^{-1}.

4.1.6. Zirconia

As a hard-to-sinter ceramic, tetragonal zirconia polycrystal (TZP) samples are usually opaque even after high-temperature sintering because the existence of residual pores considerably scatters incident light and deteriorates the transparency. However, although it is possible to control the porosity through the sintering conditions (for example, using HP [Duran et al., 1989] SPS [Casolco et al., 2008] and HIP [Klimke et al., 2011]), it is still very important to keep the grain size as small as possible. Analogously to alumina, the non-cubic crystal structure leads to an optical anisotropy, birefringence in TZP is one order of magnitude higher than that of alumina, that causes diffuse scattering at grain boundaries. It has been shown [Zhang et al., 2011] that, compared to the method of pre-sintering plus HIP, high-pressure spark plasma sintering (HP-SPS) can be an extremely simple and effective route to obtain highly IR transparent tetragonal ZrO_2.

4.1.7. Magnesium aluminate spinel

Magnesium aluminate spinel is a ceramic material showing a large transparency range, covering from 0.2-5 μm. The densification to transparency (50% at 550 nm) by SPS of a fine-grained high-purity spinel was successfully demonstrated [Morita et al., 2008]. The temperature used in that study was 1300 °C with a holding time of 20 minutes and a low heating rate: 10 °C min^{-1}. Consequently, the versatility of SPS concerning heating rates can also be considered as a key factor to obtain transparent ceramics as it may cover a wide range of heating rates. Other works [Frage et al., 2007] showed very high optical transmissions in the visible range of spark plasma sintered powders of magnesium aluminate spinel powders doped with LiF. Undoped powders sintered at the same temperature of 1220 °C showed a smaller light transmission.

4.2. Biomaterials

Bulk hydroxyapatite (HA) compacts and composites with HA matrix may find a wide variety of applications in implantology [Martz et al., 1997]. However, HA shows a limited stability at high temperatures and will dissociate into tricalcium phosphate and tetracalcium phosphate at 1300 °C in air or 1000 °C in vacuum [Chaki et al., 1994]. On the other hand, high temperatures and long sintering duration required for consolidation of HA powders by conventional techniques often result in extreme grain coarsening or surface contamination, which can degrade the desired mechanical properties. These problems can be alleviated by using spark plasma sintering to avoid exposing the compacts to high temperatures for a long duration. It has been shown [Gu et al., 2002] that it is possible to sinter HA powders by SPS at temperatures as low as 950 °C while still keeping a good performance in terms of fracture toughness, microhardness and Young's modulus. Densities obtained at 950 °C after 5 minutes of holding time reached 99.6%. Also, these authors reproted that no phase change was detected at 950 °C, whereas HA started to decompose at 1000 °C. Also, no noticeable grain growth was observed at sintering temperatures between 850 and 1100 °C.

However, the intrinsic mechanical properties of HA, particularly its low strength and high brittleness, may still limit the applications of HA ceramics in loadbearing areas of the human body that HA should be strengthened. One attractive way to overcome these mechanical limitations is to use bioactive HA's ceramic/metal composites so as to achieve the necessary mechanical strength and bioactive properties at the same time [Lynn et al., 2002], which include that the incorporation of bioinert ceramics and the addition of biocompatible glass into HA matrix [Gautier et al., 1999]. However, due to the addition of a secondary phase, the phase changes of the composites at the usual sintering temperatures may still take place. For this reason, the sintering by SPS of TiO_2/HA composites was studied [Que et al., 2008]. These authors successfully prepared TiO_2/HA composites by combining high-energy ball milling with SPS. Their results indicate that the addition of TiO_2 had a positive effect on improving both the hardness and the Young's modulus of the HA, while still keeping a bioactivity as confirmed by in vitro tests.

4.3. Materials for nuclear energy applications

The encapsulation of radioisotope materials such as $^{238}PuO_2$ and $^{241}AmO_2$ within tungsten cermets is of particular interest for the production of radioisotope heat sources for thermal management and radioisotope power systems [O'Brien et al., 2009]. The production of nuclear fuels for fission reactor systems based upon W–UO_2 or W–UN cermets is also of particular interest for increased operational safety and security. Conventional sintering techniques (such as hot isostatic pressing) required processing of materials at temperatures that were high enough to exceed the dissociation temperatures of the radioisotope oxides, which would lead to the formation of complete or partial non-cermet regions such as those described in the fuel element development summary in the General Electric report on the development of the 710 High-Temperature Gas Reactor [General Electric, 1967]. However, it has been shown in the literature that SPS is also able to densify materials at lower temperatures and shorter times than those required in conventional processes. Also, the presence of large grains observed in sintering processes such as hot isostatic pressing lead to a loss of mechanical properties. Spark plasma sintering minimizes grain growth and it is therefore ideal to be used in materials requiring a top mechanical performance. Other works also reported on the encapsulation of plutonium dioxide or americium dioxide within a tungsten-based cermet using the SPS technique [O'Brien et al., 2008].

4.4. Materials with low coefficient of thermal expansion

Materials with a very low coefficient of thermal expansion (CTE) are of great interest because of their many different applications, from cookware to aerospace applications. An interesting route to obtain this kind of materials is based on the combination of phases with positive and negative CTE [Garcia-Moreno et al., 2010]. The lithium aluminosilicate family (LAS) is the most studied system to prepare materials with very low CTE properties. Eucryptite and spodumene are the most used and studied phases with these characteristics [Moya et al., 1974] and it is well known that sintering of these materials to obtain dense ceramic bodies is quite complicated [Mandal et al., 2004], due to their narrow range of sintering temperatures and the easy formation of a vitreous phase. Therefore, by conventional methods of pressureless natural sintering, the ceramic materials thus obtained have usually low mechanical properties and Young's modulus. The fabrication of submicron LAS-alumina composites by spark plasma sintering has been suggested as a solution to this problem [Garcia-Moreno et al., 2011]. They found that it was possible to sinter the composites to the theoretical density, reaching higher strength values at lower temperatures than required by conventional sintering. In particular, composites comprising 15.65 wt.% alumina and 84.35 wt.% β-eucryptite gave the closest to zero CTE value in a wide temperature range.

5. Past, present and future of SPS process

5.1. Origin of spark plasma sintering

As previously mentioned, the technology related to current assisted sintering processes started in the late 1930s when a sintering process using electrical heating was introduced in the United States. In Japan, a similar process based on the pulsed current applied sintering method was researched and patented in the 60s and is known as spark sintering [Inoue, 1963] [Inoue, 1961]. Nevertheless, the lack of application technology at that time limited fields where it could be applied. Unsolved problems associated with industrial production, equipment cost and sintering efficiency were key points. There was little literature on research into this process until the latter half of the 70s. The second generation was developed from the middle of the 80s to the early 90s. These units were small experimental systems with maximum sintering pressure of around 5 tons and pulse generators of up to 800 A, used primarily for materials research. The emerging of SPS process took place along the 90s when the third generation of this advanced technology was developed. These systems had large DC pulse generators of 10 to 100 tons and 2,000 to 20,000 A and more. They gained a reputation as new industrial processes for synthetic processing of gradient and composite materials [Omori et al., 1994] [Omori et al, 1994].

The evolution of interest in SPS process can be followed by analysis of fair indicators such as patents or scientific publications. The explosion of scientific research on this technology started less than two decades ago and it continues nowadays. In Figure 10, it can be observed the number of published items and citations in scientific papers with "Spark Plasma Sintering" as topic [Wok, 2012].

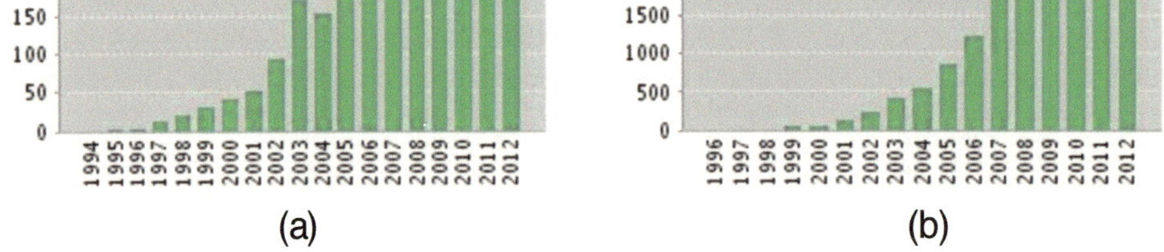

Figure 10. Published items (a) and citations (b) of Spark Plasma Sintering topic papers

The excellent properties that can be obtained when this technique is used for consolidating materials leads to an exponential growing of research publications on this topic. It has passed from an exotic technique at the beginning of 90s to a fundamental tool for advanced material preparation, especially in the case of nanostructured materials. But, the interest in this technology is not limited to research area. Thus, 30% of patents related with SPS technology have priority dates of less than 10 years. Moreover, SPS has been progressively incorporated as one of the key steps in patents for obtaining advanced materials, especially for ceramic and metallic materials. The use of SPS for solving very diverse technological problems is a reality and first commercial products based on this technology are already in the market.

5.2. Industrial production requirements

As it was previously mentioned an enormous amount of reported FAST/SPS applications are still in the area of material development and more than enough opportunities for an industrial implementation were generated. Taking the next step forward to an industrial production of novel materials by FAST/SPS is currently highly dependent on the availability of suitable equipment. The industrial application of the FAST/SPS sintering method for the rapid consolidation of novel materials require special features, which have to be fulfilled by the equipment and are different from the requirements of scientific work to some extent. The main issues are discussed below:

5.2.1. Electrical output power

In order to assure a cost effective production, equivalent with high throughput (amongst other things), a sufficient electrical output power must be provided by the system. It is important, that the electric losses in the system are low in order to generate high heating power at the location, where it is needed. The actual value of the required power depends on the size and material of the powder compact and the pressing tool as well as on the intended heating rates and maximum temperatures.

5.2.2. Flexible power supply

Depending on the type of the powder, several different sintering mechanisms are possible. Some of them can be influenced by the type of the heating current. Therefore a power supply with high flexibility is important in order to achieve optimum sintering results in terms of throughput and material quality. FAST/SPS systems are capable of generating a wide range of pulsed DC current with computer controlled, arbitrary pulse parameters to the point of pure DC current (examples see Fig. 11).

5.2.3. Precise temperature measurement & control

The correct sintering temperature is the most important process parameter besides time and heating rate. Due to a special design FAST/SPS systems are measuring the temperature in the vicinity of the powder compact centre, which gives a much more significant value than the measurement of the die temperature [Vanmeensel et al., 2005].

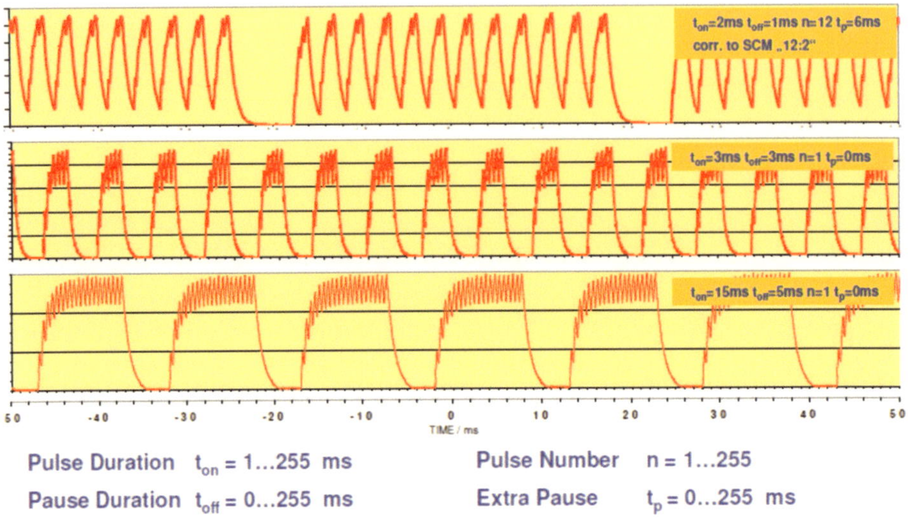

Figure 11. Flexible output of pulsed DC current

5.2.4. Optimized pressing tool systems

Due to the special construction of FAST/SPS systems, the pressing tool system, consisting of the two pressing punches, the die and other auxiliary components, is the "heart" of the system, because it not only contains the powder compact but also acts as the "heater" (in interaction with the compact). Even though the temperature gradients in the system are significantly lower than for conventional sintering methods, e.g. hot pressing (see Fig. 12), a design optimization is advantageous anyhow, especially if higher heating rates, minimized dwell time and optimum material quality are desired. A helpful tool for design optimization is the numerical simulation (finite element method "FEM") of the heating behaviour, taking into account the temperature dependent thermal and electrical properties of the applied tool materials as well as the powder compact [Vanmeensel, et al., 2006]. As an example Figure 12 shows the temperature distribution in a pressing tool system containing two powder compact circular disks of 200 mm diameter after heating to 1500°C within 12 min and 5 min dwell time. With the standard tool design (left) the remaining temperature difference in the compact amounts to 160 K, which can be reduced to 60 K by design optimization (right).

http://dx.doi.org/10.5772/53706

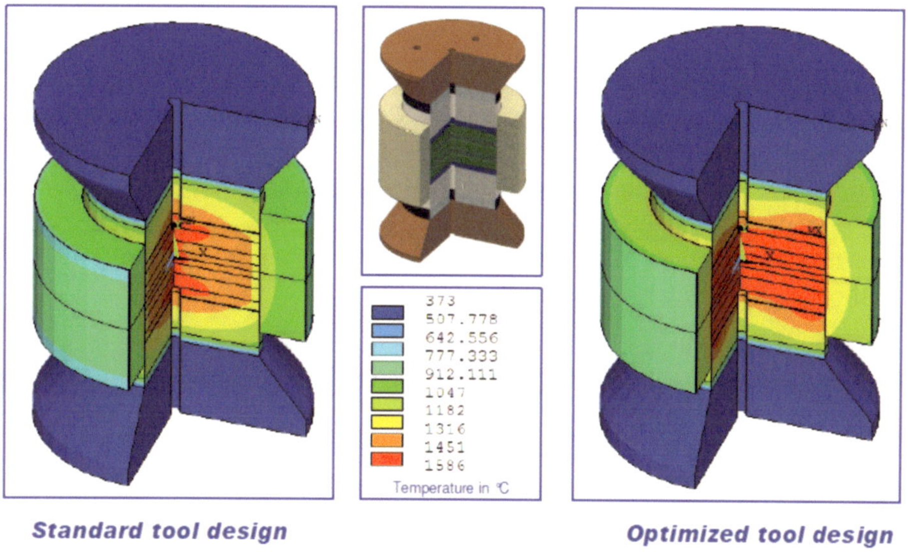

Figure 12. Temperature distribution in standard and optimized pressing tool system

The benefit of optimized pressing tool systems is a superior material quality and homogeneity, e.g. reflected by an even distribution of high hardness values across the diameter of a 200 mm circular disk compared with the standard pressing tool situation (Fig. 13). Furthermore the highest heating rates made possible that way are an essential condition for the realization of nano-structured materials, which are often impossible to sinter by conventional methods due to significantly longer sintering cycles.

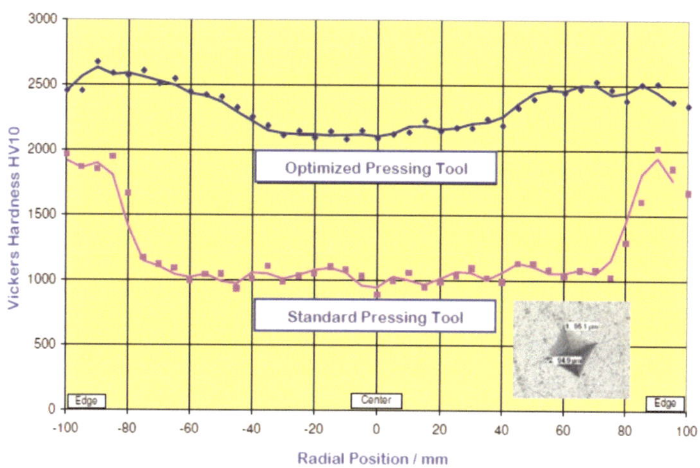

Figure 13. Hardness distributions generated by standard and optimized pressing tool systems

5.2.5. Hybrid heating

The so called "Hybrid Heating" is a combination of the FAST/SPS method with one or several additional heating systems, which act usually from the outside of the pressing tool systems, as illustrated in Figure 14. Thus the thermal gradients of FAST/SPS, which are directed from from the inside outwards typically, can be compensated by the inversely directed gradients of the additional heating system. As shown in Figure 15 the superposition of the gradients (left side) results in an extensively minimization of these gradients (right side). This allows further enhancement of the heating rates at simultaneously optimized homogeneity with all the advantages pointed out before.

A practical example showing the positive effect of hybrid heating can be found in Figure 16, which compares the sintering behaviour of rectangular plates made of binderless tungsten carbide (size 150 x 175 mm). The light grey curves show the densification byuse of FAST/SPS, whereas the dark grey curves show the enhanced sintering behaviour by use of hybrid heating.

5.2.6. Fast cooling system

The production capacity of an industrial FAST/SPS system is not only governed by the maximum possible heating rate and a minimized dwell time, but also by a fast cooling facility, which allows early discharge of the completed pressing tool. This is realized by an additional cooling chamber, separated from the actual sintering chamber by a gas/vacuum-proof, gate and equipped with special fast cooling rams. An automatically working handling system shifts the hot pressing tool system from the sintering chamber to the cooling chamber. After automatic closing of the gate the sintering chamber is ready for charging the next sintering cycle during cooling of the previous pressing tool.

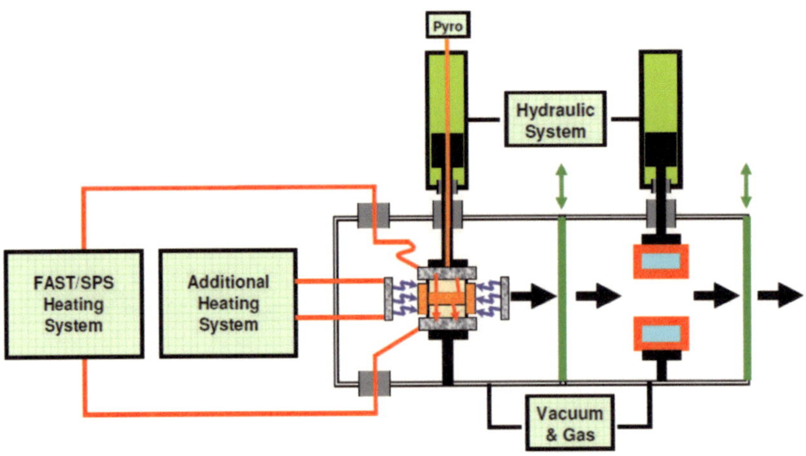

Figure 14. Schema of an industrial high throughput sintering system with hybrid heating, separate cooling chamber and semi-continuous operation facility

http://dx.doi.org/10.5772/53706

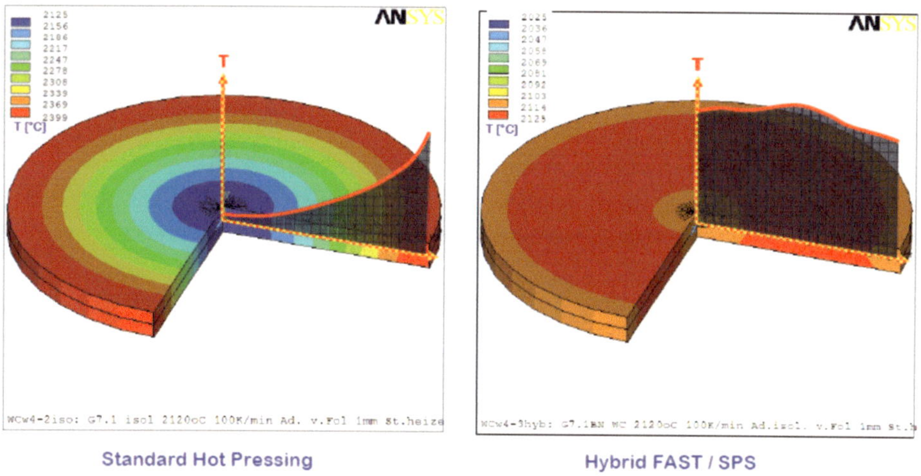

Figure 15. Compensation of residual temperature gradients by hybrid heating

5.2.7. Automatic operation

In order to realize a cost efficient industrial application of FAST/SPS sintering systems, the automation is an essential prerequisite. An important step is the semi-continuous operation mode mentioned above in conjunction with the fast cooling system. Due to a combination with robots and manipulators a fully automatic operation can be realized.

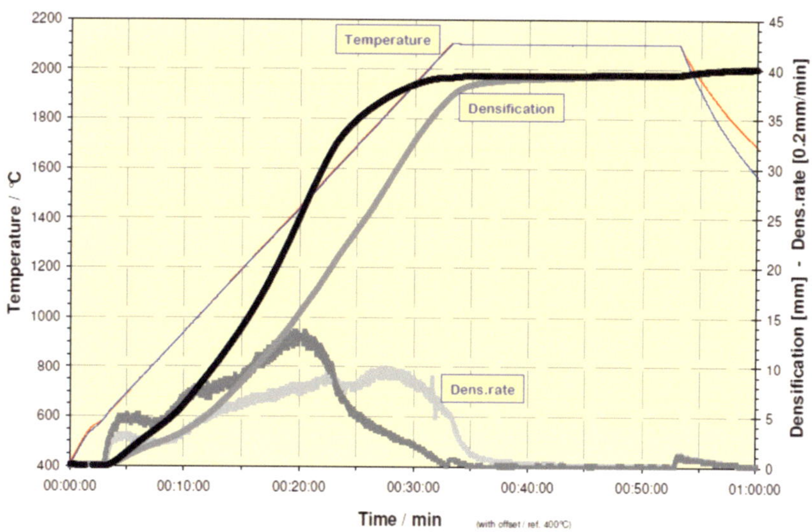

Figure 16. Comparison of sintering behaviour using FAST/SPS (light grey) and hybrid heating (dark grey)

Since 2010, the Nanomaterials and Nanotechnology Research Center (CINN-CSIC), in collaboration with FCT Systeme, the SPS technology leader in Europe, are working in the development of new advanced multifunctional materials obtained by using hybrid heating equipment. The potential applications of the components developed are very diverse covering fields such as space and aeronautics, automation, mechanical engineering or biomedicine. As examples of technical prototypes already developed, they can be mentioned ultrastable components for satellite mirrors and high precision and accuracy optical devices, ultrahard protection components, nanostructured cutting tools or biomaterials.

Author details

M. Suárez[1], A. Fernández[1], J.L. Menéndez[1], R. Torrecillas[1], H. U. Kessel[2], J. Hennicke[2], R. Kirchner[2] and T. Kessel[2]

1 Centro de Investigación en Nanomateriales y Nanotecnología (CINN). Consejo Superior de Investigaciones Científicas (CSIC) – Universidad de Oviedo (UO) – Principado de Asturias, Spain

2 FCT Systeme GmbH, Rauenstein, Germany

References

[1] Alvarez-Clemares I., Mata-Osoro G., Fernández A., Lopez-Esteban S., Pecharromán C., Palomares J., Torrecillas R. & Serafín Moya J. (2010). Transparent Alumina/Ceria Nanocomposites By Spark Plasma Sintering. Advanced Engineering Materials, 12, 11, 1154–1160

[2] Angerer P., Yu L. G., Khor K. A., Korb G. & Zalite I. (2005). Spark-plasma-sintering (SPS) of nanostructured titanium carbonitride powders. Journal of The European Ceramic Society, 25, 11, 1919-1927.

[3] Apetz R. van Bruggen M. P. B. (2003). Transparent Alumina: A Light-Scattering Model. Journal of The American Ceramic Society, 86, 3, 480–6

[4] Borrell A., Salvador M.D., García-Rocha V., Fernández A., Chicardi E., Gotor F.J. (2012). Spark plasma sintering of $Ti_yNb_{1-y}C_xN_{1-x}$ monolithic ceramics obtained by mechanically induced self-sustaining reaction. Materials Science and Engineering A, 543, 173– 179.

[5] Casolco S. R., Xu J. & Garay J. E. (2008). Transparent/Translucent Polycrystalline Nanostructured Yttria Stabilized Zirconia With Varying Colors. Scripta Materialia, 58, 6, 516–519

[6] Chaki T. K. & Wang P.E. (1994). Densification and strengthening of silver reinforced hydroxyapatite-matrix composite prepared by sintering. Journal of Materials Science: Materials in Medicine, 5, 8, 533–42.

[7] Clyens S., Al-Hassani S.T.S., Johnson W. (1976). The compaction of powder metallurgy bars using high voltage electrical discharges. International Journal of Mechanical Sciences, 18, 1, 37-40

[8] Conrad H. (2002). Thermally activated plastic flow of metals and ceramics with an electric field or current. Materials Science and Engineering A, 322, 1-2, 100-107

[9] Cremer G.D. (1944). Sintering Together Powders Metals Such as Bronze, Brass or Aluminum. US Patent No. 2,355,954, August

[10] Duran P., Reico P., Jurado J. R., Pascual C., & Moure C. (1989). Preparation, Sintering, and Properties of Translucent Er2O3-Doped Tetragonal Zirconia. Journal of The American Ceramic Society, 72, 11, 2088–2093

[11] García-Moreno O., Borrell A., Bittmann B., Fernández A., Torrecillas R. (2011). Alumina reinforced eucryptite ceramics: Very low thermal expansion material with improved mechanical properties. Journal of the European Ceramic Society, 31, 9, 1641–1648

[12] García-Moreno O, Fernández A. & Torrecillas R. (2010). Conventional sintering of LAS–SiC nanocomposites with very low thermal expansion coefficient. Journal of The European Ceramic Society, 30, 15, 3219–25

[13] Gautier, S., Champion, E. & Bernache-Assollant, D. (1999). Toughening characterization in alumina platelet-hydroxyapatite matrix composites. , Journal of Materials Science: Materials in Medicine10, 9, 533-540.

[14] General Electric. 710 High-Temperature Gas Reactor Program Summary Report, Volume III, Fuel Element Development, General Electric, Pages 146, GEMP-600 (Vol. 3).

[15] Geguzin Y.A.E., Kaganovsky Y.U.S., Onoprienko A.A., Zung F.N. (1975). Behaviour of metallic granules at the surface of an ionic crystal in an external electric field. Soviet Physics Solid State, 17, 3, 457-458

[16] Graziani T., Bellosi A., (1995). Densification and characteristics of TiN ceramics. Journal of Materials Science Letters, 14, 15, 1078-1081

[17] Groza J., Anderson K.R., Fendorf M. & Echer C.J. (1999). Surface oxide debonding in field assisted powder sintering. Materials Science and Engineering A, 270, 2, 278-282

[18] Groza J. & Zavaliangos A. (2000). Sintering activation by external electrical field. Materials Science and Engineering A, 287, 2, 171-177

[19] Hibi Y., Enomoto Y., Sato H. & Sasaki S. (2002). Titanium–Silicon–Nitrogen Composites with High Wear Resistance in Water and in Artificial Sea WaterJournal of The American Ceramic Society, 85, 9, 2373-2375.

[20] Mizuta H., Oda K., Shibasaki Y., Maeda M., Machida M. & Ohshima K. (1992). Preparation of High-Strength and Translucent Alumina by Hot Isostatic Pressing, Journal of the American Ceramic Society, 75, 2, 469–473

[21] Inoue K. (1965). Electric Discharge Heat Treatment of Metals in Electrolytes. US Patent No. 3,188,245, June

[22] Inoue K. (1963). Electric discharge sintering, Japan, US3241956, Filled Oct. 29 1963, claims priority application Japan, May 30 1963

[23] Inoue K. (1961). Method of and apparatus for controlling the porosity of electrically sintered bodies, Japan, US3317705, Filled Nov 29 1963, claims priority application Japan, Dec. 26 1961 5J.

[24] Jin X., Gao L., & Sun J. (2010). Highly Transparent Alumina Spark Plasma Sintered from Common-Grade Commercial Powder: The Effect of Powder Treatment. Journal of The American Ceramic Society, 93, 5, 1232–1236

[25] Klimke J., Trunec M. & Krell A. (2011). Transparent Tetragonal Yttria-stabilized Zirconia Ceramics: Influence of Scattering Caused by Birefringence. Journal of The American Ceramic Society, 94, 6, 1850-1858

[26] Lenel F.V. (1955). Resistance sintering under pressure. Transactions of the American Institute of Mining and Metallurgical Engineers, 203, 158-167

[27] Lynn, A. K. & Duquesnay, D. L. (2002). Hydroxyapatite-coated Ti–6Al–4V:Part 1: the effect of coating thickness on mechanical fatigue behaviour. Biomaterials, 23, 9, 1937-1946.

[28] Mandal S., Chakrabarti S., Das S. & Ghatak S. (2004). Sintering characteristics of in situ formed low expansion ceramics from a powder precursor in the form of hydroxy hydrogel. Ceramic International, 30, 8, 2147–55.

[29] Martz E.O., Goel V.K., Pope M.H. & Park J.B. (1997). Materials and design of spinal implants-a review. Journal of Biomedical Materials Research, 38, 3, 267–88.

[30] Monteverde F., Medri V. & Bellosi A. (2002). Microstructure of hot-pressed Ti(C,N)-based cermets. Journal of The European Ceramic Society, 22, 14, 2587–2593.

[31] Moya J.S., García Verduch A. & Hortal M. (1974). Thermal expansion of betaeucriptite solid solutions. Transactions and Journal of the British Ceramic Society, 73, 6, 177–8.

[32] O'Brien R. C., Ambrosi R. M., Bannister N.P., Howe S.D. & Atkinson H.V. (2008). Journal of Nuclear Materials, 377, 3, 506-521

[33] O'Brien R. C., Ambrosi R. M., Bannister N. P., Howe S. D. & Atkinson H.V. (2009). Spark Plasma Sintering of simulated radioisotope materials within tungsten cermets Journal of Nuclear Materials, 393, 108–113

[34] Omori, M., Sakai, H., Okubo, A,, Tokita, M., Kawahara, M. & Hirai, T. (1994). Preparation of Functional Gradient Materials by Spark Plasma Sintering. Symposium of Materials Research Society of Japan

[35] Omori M., Sakai H., Okubo A., Kawahara M., Tokita M. & Hirai, T. (1994). Preparation and Properties of ZrOz (3Y)lNi FGM. Proceedings of the 3rd International Symposium on Structural and Functionally Gradient Materials, Lausanne, Switzerland, pp. 99- 104

[36] Orrù R., Richeri, R., Locci A.M., Cincotti A., Cao G. (2009). Consolidation/synthesis of materials by electric current activated/assisted sintering. Materials Science and Engineering R 63, 127–287.

[37] Que W., Khor K.A., Xu J.L. & Yu L.G. (2008). Hydroxyapatite/titania nanocomposites derived by combining high-energy ball milling with spark plasma sintering processes. Journal of the European Ceramic Society, 28, 16, 3083–3090

[38] Raichenko A.I., Burenkov G.L. & Leshchinsky V.I. (1976). Theoretical analysis of the elmentary act of electric discharge sintering. Phisics sinter, 5, 2, 215-225

[39] Suarez M., Fernandez A., Menendez J. L. & Torrecillas R. (2009). Grain growth control and transparency in spark plasma sintered self-doped alumina materials. Scripta Materialia, 61, 10, 931–934.

[40] Suarez M., Fernandez A., Menendez J. L. & Torrecillas R. (2009). Transparent Yttrium Aluminium Garnet Obtained by Spark Plasma Sintering of Lyophilized Gels. Journal of Nanomaterials, 2009, 138490.

[41] Sun X., Li J.G., Guo S., Xiu Z., Duan K. & Hu X.Z. (2005). Intragranular particle residual stress strengthening of Al_2O_3-SiC nanocomposites. Journal of The American Ceramic Society, 88, 6, 1536–1543

[42] Taylor G.F. (1933). Apparatus for Making Hard Metal Compositions. US Patent No. 1,896,854, February

[43] Tiwari D., Basu B. & Biswas K. (2009). Simulation of thermal and electrical field evolution during spark plasma sintering. Ceramics International, 35, 2, 699-708

[44] Vanmeensel K., Echeberria J., Sanchez J.M., Martinez V., Bourgeois L., Hennicke J., Kessel H.U., Harden P., Van der Biest O. & Vleugels J. (2006). Field Assisted Sintering of Cubic Boron Nitride Dispersed Cemented Carbide (CDCC) Composites, EuroPM 2006

[45] Vanmeensel K., Laptev A., Hennicke J., Vleugels J. & Van der Biest O. (2005). Modelling of the temperature distribution during field assisted sintering. Acta Materialia, 53, 16, 4379-4388

[46] Wang H.Z., Gao L. & Guo J.K. (2000). The effect of nanoscale SiC particles on the microstructure of Al_2O_3 ceramics. ics International, 26, 391–396.

[47] Wei G.C., Hecker A., Goodman D.A. (2001). Translucent Polycrystalline Alumina with Improved Resistance to Sodium AttackJournal of The American Ceramic Society, 84, 12, 2853-2862

[48] WOK (Web of Knowledge) citation report. September 2012.

[49] Yamada T., Shimada M. & Koizumi M. (1980). Fabrication and Characterization of Titanium Nitride by High Pressure Hot Pressing, Ceram. Bull, 59, 6, 611-616.

[50] Yanagisawa O., Hatayama T. & Matsugi K. (1994). Recent research on spark sintering. Mateira Japan, 33, 12, 1489-1496

[51] Zhang H., Li Z., Kim B.-N., Morita K., Yoshida H., Hiraga K. & Sakka Y. (2011). Highly Infrared Transparent Nanometric Tetragonal Zirconia Prepared by High-Pressure Spark Plasma Sintering. Journal of The American Ceramic Society, 94, 9, 2739–2741